Learning from Shenzhen

Learning from Shenzhen

China's Post-Mao Experiment from Special Zone to Model City

EDITED BY MARY ANN O'DONNELL,
WINNIE WONG, AND
JONATHAN BACH

The University of Chicago Press
Chicago and London

The University of Chicago Press, Chicago 60637
The University of Chicago Press, Ltd., London
© 2017 by The University of Chicago
All rights reserved. Published 2017.
Printed in the United States of America

25 24 23 22 21 20 19 18 17 16   1 2 3 4 5

ISBN-13: 978-0-226-40109-6 (cloth)
ISBN-13: 978-0-226-40112-6 (paper)
ISBN-13: 978-0-226-40126-3 (e-book)
DOI: 10.7208/chicago/9780226401263.001.0001

Library of Congress Cataloging-in-Publication Data

Names: O'Donnell, Mary Ann, 1964– author, editor. | Wong, Winnie Won Yin, author, editor. | Bach, Jonathan P. G., author, editor.
Title: Learning from Shenzhen : China's post-Mao experiment from special zone to model city / edited by Mary Ann O'Donnell, Winnie Wong, and Jonathan Bach.
Description: Chicago : The University of Chicago Press, 2016. | "This volume developed out of two conferences: 'Shenzhen+China, Utopias+Dystopias,' held at the Massachusetts Institute of Technology in 2011, and 'Learning from Shenzhen,' held at the Shenzhen Land Use Resources and Planning Commission in 2011 as part of the Shenzhen Urbanism Biennale"—ECIP data. | Includes index.
Identifiers: LCCN 2016019384 | ISBN 9780226401096 (cloth : alk. paper) | ISBN 9780226401126 (pbk. : alk. paper) | ISBN 9780226401263 (e-book)
Subjects: LCSH: Shenzhen Shi (China)—Congresses. | Shenzhen Jingji Tequ (Shenzhen Shi, China)—Congresses. | Urban renewal—China—Shenzhen Shi—Congresses. | Free ports and zones—China—Congresses. | Municipal government—China—Shenzhen Shi—Congresses.
Classification: LCC HT178.C62 S54 2016 | DDC 307.3/416095127—dc23 LC record available at https://lccn.loc.gov/2016019384

♾ This paper meets the requirements of ANSI/NISO z39.48-1992 (Permanence of Paper).

Contents

FOREWORD vii
Ezra F. Vogel

INTRODUCTION:
Experiments, Exceptions, and Extensions 1
Mary Ann O'Donnell, Winnie Wong, and Jonathan Bach

PART 1 Experiments (1979–92)

1 Shenzhen: From Exception to Rule   23
   Jonathan Bach

2 Heroes of the Special Zone: Modeling Reform and Its Limits   39
   Mary Ann O'Donnell

3 The Tripartite Origins of Shenzhen: Beijing, Hong Kong, and Bao'an   65
   Weiwen Huang

4 How to Be a Shenzhener: Representations of Migrant Labor in
   Shenzhen's Second Decade   86
   Eric Florence

PART 2 Exceptions (1992–2004)

5 Laying Siege to the Villages: The Vernacular Geography of Shenzhen   107
   Mary Ann O'Donnell

6 The Political Architecture of the First and Second Lines   124
   Emma Xin Ma and Adrian Blackwell

7 "They Come in Peasants and Leave Citizens":
   Urban Villages and the Making of Shenzhen   138
   Jonathan Bach

8 Sex Work, Migration, and Mental Health in Shenzhen   171
   Willa Dong and Yu Cheng

PART 3 Extensions (2004–Present)

9 **Shenzhen's Model Bohemia and the Creative China Dream** — 193
   Winnie Wong

10 **Preparedness and the Shenzhen Model of Public Health** — 213
   Katherine A. Mason

11 **Simulating Global Mobility at Shenzhen "International" Airport** — 228
   Max Hirsh

Conclusion: Learning from Shenzhen 250
Mary Ann O'Donnell, Winnie Wong, and Jonathan Bach

A Shenzhen Glossary 261
Contributors 265
Acknowledgments 269
Index 271

Foreword
EZRA F. VOGEL

No city in the world has ever grown as rapidly as Shenzhen, China's southern gateway to the outside world. In 1978, when the Reform and Opening policy was introduced in China, Shenzhen was a small town of some thirty thousand people, surrounded by paddy fields. By 2010, it had a population of more than ten million people—more than New York, America's largest city. No tall buildings were more than thirty years old. It glittered with modern stores, hotels, offices, and restaurants.

Mao, convinced that the Chinese situation was ripe for Communism, announced that a single spark can light a prairie fire. When many Chinese were tired of the failures of the Great Leap Forward and the Cultural Revolution, Shenzhen became the spark that ignited reforms in China. In 1979 it was made into a Special Economic Zone (SEZ) and became, as the authors of this book relate, a new model for all of China. The story of Shenzhen in these thirty years is an unusual mixture of urban development, of a rigid socialist system stretched to incorporate a market economy. It became the crucible where new ideas from the entire world flowed into a country that had stagnated for almost two decades. After 1979, Shenzhen became an experimental ground, a key link in the transmission of new ideas where high Chinese high officials came to visit to learn about market economies and observe the introduction of modern architecture and industry.

Shenzhen brought together officials from Beijing, Guangdong provincial headquarters in Guangzhou, and Bao'an County, as well as outsiders from Hong Kong and the entire world. In Shenzhen, people who had been worlds apart struggled to find ways to communicate with each other and groped for a way to make the Reform and Opening policy work without a rebellion of local people accustomed to older ways of getting things done. The story of Shenzhen is a remarkable story, and we are fortunate now to have a full-length book, edited by Mary Ann O'Donnell, Winnie Wong, and Jonathan Bach, that brings together many of the strands, as migrants rushed in and local planners tried to keep up with unprecedented urban growth.

Shenzhen on the Eve of Reform

When I first entered China in May 1973, five years before the Reform and Opening policy began, I entered by foot, like other travelers, stepping into China at Shenzhen, just across the bridge from Hong Kong. Because there was no through train, we took the train from the Hong Kong rail station in downtown Kowloon to the border of China. We disembarked at Lo Wu (*Luohu*) and walked across the railroad bridge separating Hong Kong and mainland China, entering China at Shenzhen. I then joined the whole trainload of people walking across a railroad bridge into Shenzhen where, after a few hours, we were allowed to board a train to Guangzhou. There was a special waiting room and the border outpost was located a couple kilometers away from the town of Shenzhen. While waiting several hours for the train, we were served a Chinese meal that only two decades later would be considered well below acceptable standards.

That night, May 21, 1973, after I first saw Shenzhen, I dictated my notes on an old-fashioned tape recorder, and when it was typed up it became my diary of my first trip to China. About Shenzhen I said:

> One can distinguish several groups of people in this border town. One group is composed of officials connected with the customs. Most of these people seem relatively young, and they include men as well as women. They live immediately near the customs house on the west side. They seem to be moderately well educated and fairly pleasant to the guests they serve. They make a point of telling guests that they do not have to fill out forms and that no one will look at their baggage.
>
> A second group are the construction workers who live nearby and are actively engaged in carrying the raw materials, putting up the big bamboo poles, and all the other work connected with the large new buildings [which were at most several stories high, in old Soviet style, nothing like the new buildings that would begin going up a few years later] which are going up to accommodate the increased traffic coming across the border.
>
> A third group is the PLA [People's Liberation Army] and one can see them marching in groups of about 15 or 20 with rifles over their shoulders. There are also dark blue soldier-like uniforms worn by the public security personnel.

Public security personnel are divided into two main groups: One are those connected with the railroad and serve under the direction of public security personnel who are in turn divided into two main groups. One group is connected with the railroad and serve under the direction of the provincial railway administration. They work out of the *paichusuo* [the local police stations]. One *paichusuo* is located in the county capitol (*xiancheng*) roughly 3 kilometers to the northwest of the customs house.

About 2 kilometers to the northeast of the customs house is the market town of Shenzhen, the county seat with a population of somewhat over 30,000. In the market town are large numbers of state owned stores, and there is a factory making cement, another making bricks, one making limited kinds of machinery which is used to repair various tractors and other machinery. There are also two small fertilizer factories which make various kinds of fertilizer. In addition to the large number of state-owned stores, there is a private market held on the 1st, 4th, 7th, 11th, 14th, etc., throughout the month, where farmers can bring the tiny amount of vegetables they grow on their private plots. Most of the adults in the village were engaged in industry, commerce and service work, serving the surrounding countryside. However, the town is a separate administrative unit and is not the headquarters for the surrounding communes. The county capitol (*xian cheng*) for Bao'an County is strictly an administrative unit and does not contain the commercial facilities located in the market town of Shenzhen. The commune headquarters are also located in the county headquarters.

Most of the buildings connected with the market town seem to be old buildings built before 1949, but many of the newer housing facilities [at most several stories high, in old Communist style] immediately next to the customs area have been built in the last several years to house the military and other personnel connected with the customs. People in the area are mostly engaged in rice growing, and one can see well-cared for ponds and rice paddies, much as one can see on the train all the way to Guangzhou. The farmers grow chickens, ducks and pigs, which are sold to the marketing co-op which are in turn sold to the state and thence shipped elsewhere, much of it to Hong Kong. Most of the chickens are owned privately, but most of the ducks and pigs are owned by the production teams under the commune. The soldiers sometimes help the farmers in the field, especially in the busy season. Harvesting of the rice is normally done in late June.

Another group one sees is composed of about twenty girls with white blouses and dark blue pants who live in Guangzhou and come to the border every day on the morning train to the border, spend two or three hours in the border area, and they go back to Guangzhou on the 1 p.m. train. One can see them marching in loose formation in rows of two or three from one locality to another. They appeared very neat and smiling and are among those honored by the Youth League in Guangzhou as model workers.

At about 12 noon one could see a couple hundred elementary school students (elementary school is now grades one to five, in line with Cultural Revolution directives). They marched to classes in groups of thirty to forty. Although they marched in a loose formation, they were singing patriotic songs, some boldly as if they enjoyed it, others not singing at all. These are children of customs area officials who go to the state operated elementary school on the nearest side of the market town. They are marching home to lunch and when they arrive at the customs area they disperse to their private families, and assemble an hour or so later to march back to their afternoon classes.

The children of people living in the market town go to a different school, and children of the rural people mostly attend the more humble "people operated schools," financed by the local production team or brigade, the subdivision of the commune. The commune also has some schools supported by the state and some supported at least partly by funds from the commune itself. The best graduates of these local elementary school can proceed to attend the single commune junior middle school, hereby the commune headquarters.

One can see large numbers of trucks, some large and small three wheelers, driven by PLA (People's Liberation Army) soldiers. One can see large numbers of bicycles owned by ordinary peasants. There are a few pedicabs (powered by bicycle peddling, with seats attached to the back wheel of the bicycle that hold two or three people). The passengers pay 10 cents (*fen*). Although one occasionally sees people walking individually, most of the people one sees are in a group.

The Beginnings of a New Era

I next visited Shenzhen in the summer of 1980, a year after the four SEZs had been formally established. I was on my way to Zhongshan University where my wife Charlotte Ikels and I would spend two months researching overall developments in the vicinity of Guangzhou, some sixty miles northwest of Shenzhen. In the next few years, I passed through Shenzhen almost every year, and in 1987 I was able to spend more time investigating Shenzhen as the one foreign researcher attached to the Guangdong Provincial Economic Commission. Having written a book on Guangdong,[1] I had been made a member of the Guangdong-Massachusetts Sister State Committee after it was formed in 1983 and served as the guide to Governor Michael Dukakis when he visited Guangdong. I served as a host to Guangdong leaders who visited Massachusetts, and I was then invited by the province to study the economic and social development of the province. I made many visits to Guangdong, one of which was seven months long, and wrote up my observations in a book titled *One Step Ahead in China: Guangdong under Reform*.[2]

When I spent some time in Shenzhen in 1987 as part of my research on provincial developments, it had begun to take shape. The Shenzhen SEZ included not only the old market town of Shenzhen but almost eight hundred square miles, much of what had been Bao'an County. During the Maoist period, when the border was tightly closed and contact with the outside was strictly controlled, Bao'an County had stagnated and remained poor because of the misguided policies of the Great Leap Forward, which tried to establish large collective units that destroyed incentives.

By 1987, the main roads inside Shenzhen had been laid out through the city from east to west. When Deng Xiaoping visited Shenzhen on the eve of his 1978 Reform and Opening policy, he had disagreed with those who said that the flow of refugees, some risking their lives to swim from Shenzhen across the border to Hong Kong, was basically a public security issue to be dealt with by clamping down and policing the border. The basic problem, he argued, was economic. The economy on the Chinese side of the border had been too far behind and it was, he said, understandable that young men should flee to Hong Kong where they found opportunities for a better way of life. The answer, he said, was in economic development on the Chinese side of the border.[3]

By 1987, a small industrial area had grown up in Shekou, the western tip of Shenzhen, under the direction of Yuan Geng, the head of the China Merchant Steam Navigation Company in Hong Kong. He realized that China's ships of the day were horribly rusted and only marginally seaworthy. He knew that it was better to tear down the ships for scrap and use the scrap as part of a new shipbuilding effort. It was obvious that Hong Kong was too crowded to have a new facility for scraping old ships, and several years before the 1978 opening he had already proposed to higher authorities that they build the facilities in Shekou, which was located along the coast where there were lots of young men thirsting for employment opportunities.

When I interviewed Yuan Geng in Hong Kong, I realized that he was the perfect person for the job. He was originally from Bao'an County and during World War II had joined the Communist movement as a member of the East River Guerrilla forces in the area. After 1949, he had risen as a trusted official in Beijing in the Central Liaison Department, the branch of the Communist Party that had contacts with the outside world. He had frequently traveled to Hong Kong and had great contact with overseas Chinese from various parts of Southeast Asia. To scrap the ships, he needed the permission of the Ministry of Communications, and at the time the minister of communications was Zeng Sheng, also originally from Bao'an County, who had been Yuan Geng's commander and friend in the East River Guerrillas during World War II.

From his life in the office in Hong Kong, he had become familiar with the dynamic global economy and could see what China needed to become a dynamic market economy. It was easy to see that Yuan Geng was a poised leader with vision who commanded the confidence of superiors and the respect of those under him.

In 1987, in the facility in Shekou, aside from the area for scrapping the ships, was a small steel plant and a small aluminum plant and, in addition, one of the first foreign ventures in postreform China, the Pittsburg Plate Glass factory, which produced glass. Yuan Geng also had visions of a democratic government in Shekou, which were aborted because they were not supported by higher-level officials. He remained a leader in Shenzhen and maintained a liaison with high-level leaders in Beijing. In 1984, Yuan Geng benefitted from Deng's affirmation of the SEZs, and in 1992, he received Deng Xiaoping on his Southern Tour, during which Deng goaded the party leaders to continue vigorously his thrust for the Reform and Opening policy.

While Shekou was developing in the western part of Shenzhen, new construction was taking off in what was rapidly becoming the first center of the newly emerging Shenzhen City in the area around the train station. Some of the first modern buildings to be erected in Shenzhen after 1978 were hotels so that overseas Chinese and foreign businessmen could be comfortable as they came to invest in Shenzhen. Soon the attractiveness of Shenzhen spread to the surrounding rural counties, especially to Dongguan County (then less than one million, now a modern city with a population of more than ten million). Another early development along a lake north of the major urban area in Shenzhen was a modern luxury housing area where Shenzhen officials royally entertained high officials visiting from Beijing. In collaboration with Beijing pro-reform leaders who encouraged high officials to visit Shenzhen to build up support of the Reform and Opening policy, Shenzhen officials took advantage of the warm winter weather to invite high officials to visit during the seasons when northern cities were much colder. By providing great hospitality, Shenzhen leaders built up a group of high-level officials who would support them when questions arose about whether to continue Shenzhen's development. As some of China's first modern high-rise buildings and modern factories grew up, Party officials from Beijing and elsewhere were given tours of Shenzhen and went back as converts to the new market opening, which many of them had formerly opposed because they suspected it was tainted with capitalism.

Reform-minded officials in Beijing had helped Shenzhen in the early 1980s by sending in army construction crews that were demobilized in Shenzhen and were then available to become the core of modern construction companies. Hong Kong architects and construction companies, under the leadership of people like Gordon Wu, fresh from experience in leading the Hong Kong construction boom were ready to join in the construction business in Shenzhen. Major universities in various parts of China were encouraged to send people to the newly formed Shenzhen University. Departments of the central government and provincial governments bought properties in Shenzhen to keep posted on developments there and to make money as real estate prices appreciated. They sent officials there who enjoyed the attractive modern Shenzhen lifestyle and observed what could be done elsewhere, thereby transferring ideas back to their home units. Aspiring entrepreneurs congregated in Shenzhen to make their way, and rural dwellers first from areas in Guangdong came to find labor jobs. Skilled workers—who had originally come from Shanghai and other east coast cities and had been sent inland in the 1960s when Mao was developing the Third Front inland areas not vulnerable to possible attacks by Soviets and Americans—availed themselves of the opportunity to return to more urbane coastal areas. Young intellectuals who wanted to go to the West but did not have the opportunity flocked to Shenzhen, where they could enjoy more freedom and a better lifestyle than in many other parts of China. Local officials trying to make their way up the hierarchy went to Shenzhen to develop modern buildings, modern businesses, and to enjoy the greater freedom and better style of life than in other parts of China. In the midst of the gray area about how to move ahead, many officials in Shenzhen and elsewhere in Guangdong were willing to take risks even though they feared policy could again grow more conservative, as it had in previous lunges to the left. What happened in Shenzhen was partly a product of party and government leadership, but it was also the informal appeal of the modern, the open, the global cutting edge to many willing to take risks. By the late 1990s, foreigners were flying directly to Shanghai and Beijing without passing through Shenzhen. Hong Kong residents were taking through trains to Guangzhou without getting out of the train at Shenzhen.

Northerners coming down to Shenzhen were acutely aware of the influence of Cantonese language, Cantonese culture, and the influence from the modern Cantonese city, Hong Kong. But whereas Hong Kong was dominated by Cantonese people, Cantonese language, and Cantonese culture, Shenzhen was dominated by northerners and Mandarin. It became a melting pot—a cosmopolitan modern society where new ideas from abroad, as adapted by northerners and local Cantonese, could be tried out before they were spread elsewhere in China. Shenzhen was allowed to open a stock market that quickly became one of China's largest. Even Chinese residents of Shenzhen who, in the early years, were not allowed to travel to Hong Kong and the West could receive Hong Kong TV, not accessible elsewhere in China. First-time visitors to Shenzhen are still dazzled by the new buildings and elaborate stores that we can see there. Today Shenzhen no longer appears unique because so many other cities have learned from it. But in the 1980s and 1990s, the Shenzhen described by the authors of this volume was the pioneer, the cutting edge that helped shape modern urban life in all of China.

Shenzhen has not received the scholarly attention it deserves. At last we have a volume that has not only presented the basic facts of Shenzhen's development but given us a comprehensive picture of the political, economic, and social transformation of the fastest-growing city the world has ever seen.

Notes

1. *Canton under Communism* (Cambridge, MA: Harvard University Press, 1969).

2. *One Step Ahead in China: Guangdong under Reform* (Cambridge, MA: Harvard University Press, 1989). For a description of the early development of Shenzhen, see pages 125–60.

3. For a fuller description of Deng and the policies of the reform era, see my book *Deng Xiaoping and the Transformation of China* (Cambridge, MA: Belknap Press of Harvard University Press, 2011).

INTRODUCTION

Experiments, Exceptions, and Extensions

MARY ANN O'DONNELL, WINNIE WONG,
AND JONATHAN BACH

The production of policy through the production of "models"—model people, model factories, model villages—is a classic feature of socialist governance. Two famous models in China under Mao Zedong, Dazhai and Daqing Communes, were agricultural and industrial collectives said to showcase the collective spirit necessary to "surpass England and catch up with the United States." By the end of the Cultural Revolution, this rallying call sounded entirely hollow. But in the thirty years following, the People's Republic of China has thoroughly reversed that history of economic failure under the rule of a Communist Party, and it offers as startling proof of this achievement the city of Shenzhen.[1]

Shenzhen is a Potemkin village made real, a statistical wonder that inverts earlier models on their heads, cementing the legitimacy of the post-Mao Chinese party-state. With Shenzhen, the party-state has arguably achieved an economic liberalization far beyond what Western-led, neoliberal global institutions have managed in the postsocialist cities of Eastern Europe. As the Brookings Institute put it in 2013, "Shenzhen's economic model has to some extent become synonymous with China as a whole." How was this possible? And granting its obvious successes, should scholars and observers of China accept Shenzhen as a model for a new form of global Chinese capitalism run by single-party socialism and promote its redeployment for economic and political change throughout China, and even internationally? This volume unpacks these two questions by critically reexamining the city of Shenzhen itself through detailed ethnographic research, spatial and geographic history, and analysis of its political discourse, revealing the process by which the model was formed and made effective. The contributors investigate the socioeconomic itineraries of key actors and subjects; critical urban technologies of infrastructure, urban planning, and border administration; and the politicization of discourses such as reform, globalization, belonging, mobility, creativity, and health. The multidisciplinary contributions examine these shared processes of transformation across multiple scales and sites within the city, demonstrating how the Chinese party-state converted a zone of highly localized and controversial economic and political experimentation into just such a national and global model.

Shenzhen was founded as a Special Economic Zone (SEZ) in 1980, not long after American architectural theorists had promoted Las Vegas as a new model for architectural design and urban development.[2] These theorists had identified the liminal space of the Las Vegas strip as a site for radical reform of socioeconomic relations—a provocation unintentionally mirrored in the strip of land carved out of the rural southern borderlands by the post-Mao state in its politically strategic creation of Shenzhen. The dismantling of central planning policies with regard to China's SEZ also unintentionally echoed the Americans' call for an "architecture without architects"—a city in which the withdrawal of social planning, design, and control would allow the emergent and chaotic features of capitalistic environments and small-scale, everyday heroism to take shape. From the Chinese Cultural Revolution slogan, "Learn from Dazhai" to the new American postmodernist one, "Learning from Las Vegas," we see a broad and powerful "spatial" synthesis of culture and economy that has mobilized people and market forces, giving form to an ideology of informality, risk, and experimentation while also producing fast-paced material gains, intensive globalization, and endearing populism. This book's chapters account for this legacy of spatial exceptionalism, detailing its machinations and achievements and evaluating the consequences in space, subjects, and urban form. The book advances both China studies and the study of urban change and emergent global spaces more generally.

The "miracles" of economic prosperity and productivity evident today in Shenzhen's sinuous, solar lamp–lit highways and ostentatious postmodernist government buildings stand in contrast to the documentable human costs of exploitation, suffering, and complicit and illicit activity in Shenzhen. Shenzhen's "tale of two cities" is matched by divisions among current scholars. On the one hand, those who emphasize the achievements of China's post-Mao economic reinvention tend to see Shenzhen as a site of veritable success stories—of progressive political movements, of the "Shenzhen speed" of modernization, of peasants who have made their fortunes, and of migrants who have found a livelihood and a home.[3] On the other hand, those focused on the thorny problems of global capitalism in an authoritarian political system regard Shenzhen as a site of its worst abuses—the massive exploitation of migrant labor and the host of illicit global trade networks (counterfeits, piracy, corruption, technology theft, and organ trade) rumored to be centered in Shenzhen.[4] The contributors to this volume untangle the utopian and dystopian strains running through representations of this city of more than fifteen million by exploring how the post-Mao China appropriation of capitalist logics led to a dramatic remodeling of the dominant institutions and discourses of the socialist planning era. In doing so, we bring the extensive literature of China studies into dialogue with urban studies and its recent rethinking of the "spatial turn" that has come to define the field.

As totalizing as such transformations in China may seem, however, they have not threatened the socialist cultural imaginary or indeed the legitimacy of the Chinese Communist Party. On the contrary, this volume reveals the surprising potency with which the socialist party-state has reconstituted Shenzhen "the special zone" as Shenzhen "the model city." In 1992, in the wake of the Tiananmen Square disaster, Deng Xiaoping's official tour to Shenzhen solidified reform policies that would be expanded to other zones along the coast and eventually to other Chinese cities. By its thirty-year anniversary, Shenzhen had become the poster city for the success of official planning and policies. Model leaders, model workers, model villages, model industries, model institutions, and model governance are evident everywhere in Shenzhen today.

The contributors to this volume therefore broadly concur with the dominant Chinese political discourse that Shenzhen serves as a pivotal case study from which important lessons can be learned. Our analysis differs from the official story of the "Shenzhen Miracle," however, in our criticism of how it elides the actual factors and agents of reform in that miracle. While the official perspective lauds the model as the logical result of coherent policies and plans, we argue that nothing of the sort took place. In fact, the model emerged out of a period of illicit (and often outright politically unapproved) experimentation, which itself allowed for a series of exceptions to the model's own policies. These results were later cannily reinscribed ex post facto into officially sanctioned (and newly invented) narratives, and eagerly extended into other cities, industries, and continents. As researchers whose fieldwork consistently observed this powerful form of policy experimentation and political model making, we collectively call for a level of theoretical inquiry addressing the contextualized histories below the urban scale while critiquing the regional, national, and even international frameworks that the "city" is so often mobilized to explain. As a result, we highlight the constantly shifting combinatorics of socialism and capitalism in Shenzhen's history, detailing in this Chinese case what David Stark has called "playing capitalism with communist pieces."[5] Likewise, the analyses brought together here prioritize neither party-state–led policies nor the raw hand of the global market, focusing instead on the processes by which the construction of the Shenzhen zone, then model, emerged. Our case studies bring to the fore numerous individuals and organizations that have been forgotten in the dominant history of Shenzhen, exploring how agents of reform were transformed into model subjects of new order. Together the chapters in this volume show how the logic of socialist fabulation and the logic of capital have come together in a new kind of myth making.

Globalization as Arbitraged Difference

Shenzhen is often referred to as China's "window to the world," as epitomized by its well-known theme park of the same name, built in Shenzhen's Overseas Chinese Town (OCT) in 1994. The architectural metaphor of *window*, like Georg Simmel's famous invocation of the metaphors of *bridge* and *door*, assumes a dualistic ontology that posits a world of disconnected objects connected by humans.[6] Yet the window metaphor is also, as Georges Teyssot writes, an optical apparatus, an eye that gazes at the same time as it frames the gaze.[7] At the beginning of Shenzhen's existence, the state sought to frame the gaze through a metaphoric and theme-parked window, but as the decades progressed, the window became less centralized, supplanted by many smaller windows—not the least of which consist of electronic screens held in the palm of the hand (what Teyssot calls "nomadic devices"), famously made in the Shenzhen Foxconn and Huawei factory campuses.[8] These devices, variously copied, tweaked, and even improved upon through *shanzhai* (pirating) practices in Shenzhen's Huaqiangbei area, are transforming Chinese civil society.[9] In a sense, the stock metaphor of a frontier window to describe Shenzhen has been overrun. It is no longer access, but space—its production, regulation, invention, intensification, commodification, and reinvention—that makes Shenzhen viable.

The spatial politics of Shenzhen is in turn concerned with one overriding issue: the production of value through the regulation of movement. Shenzhen's spatial politics center on facilitating and controlling movement, especially movement across borders, where value "steps up" and "steps down."[10] The people of Shenzhen have developed extraordinarily complex sociotechnical machineries to regulate, evade, evoke, and provoke the movement across its bordered spaces, whether of money, products, people, labor, or ideas. Borders and the values they create condition nearly the entire urban and rural form of Shenzhen. Numerous cases throughout this book reveal how seemingly inept governance—or at best, a certain looseness in the policing of boundaries—is in fact the crucial mechanism of Shenzhen's dynamic growth. Our analyses show that for boundaries to function as value-producing mechanisms, they must be both ideologically firm and functionally blurry.

If the purpose of boundaries is to create value out of the interplay of mobility and immobility, the most powerful and pervasive of these boundary mechanisms in China is not a cartographic border at all but the *hukou*, or household registration system. This long-standing institution continues to designate Chinese born in rural villages as peasants (with peasants' rights of collective property), while Chinese born in cities enjoy privileges within their particular city of birth. Historically, *hukou* determined one's place of residence and (legitimate) work, effectively creating two mutually isolated, but nevertheless hierarchically integrated, societies.[11] In the post-Mao era, as *hukou* constraints on mobility between rural and urban areas have loosened without a concomitant restructuring of welfare benefits and property rights, the system has created an entire class of illicit workers—namely, those rural peasants who have left their home villages to work in the cities without appropriate *hukou*. Such a move may enhance their economic survival but limits their social mobility because educational opportunities, health care access, police protection, and property rights are still primarily determined by *hukou*. Moreover, *hukou* discrimination has deeply pervasive effects across generations. Because *hukou* also determines where a child can be educated and ultimately take the national college entrance exam (*gaokao*), migrant workers often leave their child(ren) at home with aging grandparents, reuniting with them only during the Chinese New Year holiday.

When the *hukou* system met the production of Shenzhen's spatial boundaries and the zone's exceptional and experimental economic policies, the result was a massive, unsanctioned surge of internal rural-to-urban migrants. If the *hukou* system was a form of what Agamben would call an inclusive exclusion, or being included in the legal order only to be more effectively excluded, the new and growing population of rural migrants in Shenzhen could be thought of as an exclusive inclusion—that is, excluded by the legal order only to be more effectively included in the economic life of the city.[12] Rural migrants, however, were not the only subjects caught in a precarious relation to the city and citizenship, for the central government initially required all Chinese (regardless of *hukou*) to obtain permits to legitimately enter the SEZ, which lay between the mainland and the wider world of Hong Kong and beyond. At first this permit to cross into the Shenzhen SEZ was both difficult and expensive to obtain. Over time it became easier to secure a border pass, and eventually the internal border opened, though border checkpoints, selectively enforced, remain in place. Enforcement not only varies over time and from checkpoint to checkpoint; many well-known, illicit border-crossing schemes continue to be tolerated. Precise dates in this history vary across the city because the process has been functionally opaque and selectively implemented, and in practice the boundary was always porous. These examples of selective control over borders and residency status, consistent throughout Shenzhen's history, provided legitimacy for the central government while they simultaneously created conditions that escaped or undermined its authority. The creation and navigation of the resulting ambiguity are central to understanding Shenzhen as a lived reality, an economic phenomenon, and a political project.

Over the course of the zone's first thirty years, the tiers of difference created by its borders remain constantly in a state of renegotiation alongside the spatial redefinition of the city. One effect was to position Shenzhen at the forefront of reforms to the *hukou* system. Earlier than any other city in China, Shenzhen gradually dismantled the exclusive privileges of subsidized housing, education, property, and medical care for urban-defined residents, even promulgating laws that made *hukou* and "long-term residents" equivalent legal statuses within the municipal apparatus.[13] This went hand in hand with the formation of new governance spaces inside the city. Shenzhen witnessed the steady creation of new administrative districts and subdistricts carved out of the historically rural Bao'an County and assigned municipal status as urban districts (see table 1). "Urbanization" in Shenzhen has therefore included a process of "rural urbanization" through which the city has grown by absorbing rural space, only to find the rural coming to largely define its unique form of urbanization, as explored in Huang's discussion of the Overseas Chinese Town (chapter 3), Bach's discussion of Tianmian (chapter 7), and Wong's discussion of Dafen Oil Painting Village (chapter 9).

Through separate case studies, this volume shows how three intertwined forms of spatial production—borders, citizenship status, and rural urbanization—mediate its political and vernacular geography of arbitraged difference. Intentionally or not, Shenzhen as model and reality has enabled the production of specific labor mobilities and immobilities in the form of rural-urban mass migration under conditions of quasi-legality. These are combined with a cosmopolitanism that includes international migration. At each of Shenzhen's boundaries, who and what is mobile, and who gets to control this mobility, is hierarchized. Value here extends beyond the cost of materials, products, and factory labor to places like Shenzhen's "second wives villages"—border communities where the "second" families of Hong Kong men live—that speak to how spatial borders amplify desires, inequities (e.g., in gender), and forms of acquisition. As Willa Dong and Yu Cheng's chapter on female sex workers shows, frisson at the borders creates sites for conditions of arbitrage—for commodities, properties, people, moralities, and legalities.

INTRODUCTION 7

But this arbitrage takes place both at and beyond the border, as Max Hirsh details in his chapter on Shenzhen's so-called international airport as an infrastructure that extends across several borders and international waters. We also see this arbitrage in the decline of Shenzhen's "big brother" across its southern border, Hong Kong, as an aspirational model for Shenzhen. The central government is planning for the eventual administrative integration of the two cities into a single metropolitan area—a prospect only imaginable when a viable uniformity between the two cities appears reachable. As the economic gap between the two cities has narrowed, however, the perceived moral gaps have widened. Here we see the overlap of economic and moral value, a theme running through this book. Hirsh describes, for example, the bizarre practices separating Hong Kong International Airport workers from their Shenzhen counterparts at the site of border contact, epitomizing anxieties that run through everyday Shenzhen–Hong Kong interactions. Shenzheners treat Hong Kong as the retailer of regulated goods (even the same ones they could buy in Shenzhen), while Hong Kongers see Shenzheners as vulgar country cousins not because of their impoverishment, as was the case in the 1980s, but because of their overt materialism. This book examines the intersections of economic and moral value, offering insights into how Shenzhen came to function as a key imaginary economic frontier of Chinese globalization and economic reform from Deng's China through today.

In his work on Russia, Stephen Collier identifies the total planning of the Soviet city as essential to what he calls the "Soviet social," and he explores how socialist modernity is reassembled in the neoliberal "post-Soviet social."[14] Shenzhen, as an urban morality tale of successful globalization and socialist transition for China, presents an interesting parallel imbrication of governmental practices, technologies of power, and the city-as-model in the particularly Chinese search for modernity under conditions of market socialism. Embedded and echoed in this search for modernity are the practices and legacies of other encounters with the modern—colonial, socialist, and now, for lack of a better word, neoliberal modernity—sometimes eliding each other's histories, sometimes accentuating their differences and similarities.[15]

Value Village

The spatial politics delineated above and the analysis we make of it throughout this volume highlight one of the central and most distinct features of Shenzhen's urban space: its "urbanized villages." This is one aspect of Shenzhen that permeates nearly all the fieldwork represented throughout this volume, building significantly on earlier scholarship on Shenzhen.[16]

In 1982, the year that Shenzhen's first comprehensive plan was promulgated, the central government amended the Chinese Constitution to define collective (i.e., rural) property rights, without, however, actually defining a collective. This omission effectively freed villages throughout the country, including in Shenzhen, to engage in production outside the national plan. The only condition was that, although the village collectives maintained their "land use rights," they could not "appropriate, buy, sell or lease land, or unlawfully transfer land in other ways" (see figure 1).[17] The creation of Township and Village Enterprises (TVEs) enabled villages and townships to incorporate as stockholding companies and engage in joint ventures with foreign firms.[18] By the mid-1980s, TVEs accounted for more than 26 percent of China's gross national product.[19] Under this ad hoc assemblage of "village" and "enterprise," villagers became stockholders and village chiefs CEOs, even as the patriarchal social structure of village power remained largely unchanged. These enterprises created firms and factories—sometimes village owned, sometimes leased—and produced consumer goods, textiles, and building materials, quickly becoming an important sector within China's economy, though not always captured in official statistics. The incorporation of the rural into the urban through the expansion of city boundaries into formerly rural areas, as well as the reproduction of rural kinship communities through migration into cities, has become an important area of China studies. Pathbreaking work by Li Zhang, Helen Siu, Pun Ngai, and Ching Kwan Lee, among others, demonstrates the centrality of the village structure for understanding China's economic and social change in urban settings.[20]

Nowhere is the contemporary rural-urban imbrication more evident than in Shenzhen. In Chinese, such spaces are most often referred to as *chengzhongcun*, generally translated as "village-in-the-city," or "urban villages." We argue the expression is best translated as "urbanized villages," which captures how these spaces co-evolved with the development of the city itself. The "city" has progressively enclaved the "village," surrounding it by high-rises, shopping malls, and gated housing estates. In some instances, actual walls have been erected between urbanized villages and their neighbors. Shenzhen urban planners and architects have ironically described this process as "the city surrounds the countryside," inverting that Maoist pro-peasant slogan, "the countryside surrounds the city," a discursive transformation that Mary Ann O'Donnell analyzes in chapter 5.[21] As de facto exceptional spaces within an already exceptional SEZ, Shenzhen's urbanized villages maintain a mix of rural privilege and urban dynamism that has allowed them to mediate licit and illicit mobility in the city and exert significant autonomy within the limits of the village. They have become a zone within a zone, playing a critical role in shaping the city.

The diversity of urban villages examined in the following chapters suggests both the practical limits to, and the experimental possibilities of, urbanization outside state models. On the one hand, TVEs located along the rail lines and highways have focused their investment strategy on assembly manufacturing, which requires low capital investment to build the factories and an educated, but docile, labor force. In this sense, Shenzhen TVEs have actively participated in the localization of global production chains. On the other hand, many villages located along the Shenzhen–Hong Kong border have focused on providing entertainment, ranging from spas to sexual services for mobile men. These TVEs have benefited from the state's limited ability to regulate activity within village boundaries and have become the subject of stigmatization. Importantly, all Shenzhen villages have developed as neighborhoods for the city's working poor and recent migrants, many of whom labor in the official economy outside village boundaries. These neighborhoods not only provide low-cost housing for over half the city's population but also anchor Shenzhen's extensive gray economy, composed of small shops, labor-intensive services, and cheap food. The construction, building, investment, labor management, and social change that created Shenzhen has therefore allowed the canniest villages to produce surplus profit for both capital reinvestment and individual enrichment while simultaneously creating new exploited classes of laborers.

Initial efforts to integrate urbanized villages into the municipal apparatus came only after more than a decade of urban growth. The first rural urbanization campaign of 1992–96 brought the inner district collectives into the municipal apparatus as "neighborhoods," the lowest level in the municipal hierarchy. During this period, the central government tasked state-owned enterprises with negotiating the transfer of land use rights from the collectives to the municipality, which in turn sold these rights to various companies. This double exchange of land both provided investment capital to the villages and financed the construction and development of the municipal bureaucracy, including the city's infrastructure. Through this process, Shenzhen became one of the earliest and most vibrant examples of what scholars call "vertical accumulation," a reference to the process by which real estate speculation and infrastructure development together produce vast sums of money and reshape cityscapes across East Asia.[22]

The administrative status of the villages and their enterprises took on even greater importance in the 2000s, when Shenzhen increased its efforts to bring its municipal administrative apparatus in line with other national-level cities. Through an administrative sleight of hand in 2004—and with much fanfare—Shenzhen became China's first "city without villages." The municipal apparatus had incorporated former village collectives while the state apparatus had transformed former villagers into "city citizens," thereby ending the Maoist system of collective ownership of land and the means of production in rural areas within Shenzhen. Though Shenzhen was now officially "without villages," TVEs continued to exist and the newly incorporated villages now called themselves "new villages," raising landmark arches labeling themselves as such. As many of the chapters in this volume make clear, this remarkable transformation of Shenzhen village lifeways has strengthened, rather than diminished, village-based cultural identities. The persistence of village identity has created serious political, economic, and ideological challenges for both Shenzhen officials and urban migrants. One of this volume's claims is that Shenzhen's urbanized villages, through their status as outside yet integral to mainstream Shenzhen society, represent an important figuration of the Chinese nation as a locus of new forms of value production.

The following discussions of urbanized villages and village corporations explore competing understandings of how rights to the emergent city should be (re)distributed and regulated. During the early years of Shenzhen development, these functions were split between TVEs and the municipal apparatus and its industrial proxies. The 2004 integration of Shenzhen's collective enterprises into the municipal apparatus, however, reframed the problem from one of contradictions between the state and the collective to one *within* the state itself, a discursive and regulatory shift that recast Shenzhen's urbanized villages as a "problem" of urban planning rather than the target and beneficiary of systemic reform.

Overview of the Book

This volume details the progression of Shenzhen from experiment to official model. Taken together, the chapters contrast an early "experimental" mood, during which the state was often an absent presence, with the most recent acts of urban consolidation, when we increasingly see how the party-state apparatus increasingly intervenes in microlevel social change, appropriating the city's stunning economic successes into larger national political narratives. The socialist model sutures these two histories even as it sutures the history of Shenzhen to the state's national project to modernize.

We chronicle this shift over three periods: The first period of experimentation begins with the founding of Shenzhen and the SEZ in 1979 and ends with the administrative redesignation of all villages within the SEZ districts in 1992. The second period of uncertain exceptions from norms begins in 1992, while the third period of extensive consolidation of the "model" began in 2004, when all the villages of the city's outer districts were also officially urbanized.

INTRODUCTION

By 2010, the national narrative of the "model" had become fairly visible, and with the official decommissioning of Shenzhen's internal boundary, all of the city became, in effect, a "Special Economic Zone" even as that exceptionalism was no longer functional. The book is thus divided into three parts: experiments, exceptions, and extensions, which reflect both chronological periods but also material administrative changes on the ground.

Part 1, "Experiments (1979–92)," begins with the ideological separation of politics and economics that positioned the "zone" as a desirable target of governmental strategy. In chapter 1, Jonathan Bach examines the historical and global context of Shenzhen as a space of economic privilege, showing how "zonal logics" have worked to turn the exception into the rule. He situates the city's emergence against the backdrop of two genealogies of the SEZ—as a site of economic planning that enhances the accumulation of capital and as a version of an imagined modern, rational city with its inherent civilizing mission—and attempts to expunge whatever does not fit the model. In such experimental places, Bach argues, the SEZ requires both political endorsement and disavowal in order to create the desired conditions for economic enterprise. In chapter 2, Mary Ann O'Donnell provides concrete examples of this shifting relationship between economics and politics by detailing the thwarted and erased political reforms introduced by three of Shenzhen's most important early leaders. *Experimentation* means, in the first instance, strategic failure, which was on full display in the early years of Shenzhen's construction. Through a political history of the Shenzhen zone, O'Donnell tracks the ability of Mao-era Communist Party discourse to adopt almost complete reversals in socialist ideology, as well as the strikingly experimental modification and interpretation of Deng-era socialist language by early Shenzhen leaders, who briefly became model heroes for a new city and era. She traces their success and downfall and highlights the importance of the process of "modeling" in the Shenzhen experiment, as discussed in subsequent chapters that explore model roles, spaces, and individuals as they become visible in official and public discourses.

If Shenzhen's early years proved paradoxical for politicians, many of these same politicians forced the new city's urban planning apparatus into strategic failures as a way of acknowledging difficult realities on the ground and the need to diverge from national policy. In chapter 3, Weiwen Huang examines the role of urban plans and administrative planning in the development of the SEZ's first urban areas, showing how the contradictory geography of immense density and sudden sparseness must be understood as a unique product of Beijing urban planning perspectives, the reality of migrant workers, and the SEZ's modes of development. In chapter 4, Eric Florence brings this part to a close by exploring the shift from model leader to model worker in official and public discourses as Shenzhen's new society begins to consolidate in the 1990s. Anticipating more far-reaching changes in national Chinese perceptions of rural-to-urban workers, Florence details how party-led campaigns, state-run media, and migrant worker–written narratives operated together to forge a new "consciousness" through concepts such as the Shenzhen "spirit," the feeling of urban "belonging" for the dislocated worker, and the city as a site for the realization of personal "dreams." In trying to construct a narrative of rural migrant workers to fit within the dominant imaginary for Shenzhen, Florence shows how the party-state constructed rural migrants as subjects for and of a modern international city and successful model for economic development. In the process, the discourse around Shenzhen's rural migrants served as a harbinger for the ideological revamping that later produced a new role for the Party as it sought to adapt to China's far-ranging socioeconomic transformations.

But Shenzhen produces within its already experimental and exceptional space additional spatial, economic, and administrative zones of exceptions. The book's second group of essays, "Exceptions (1992–2004)," focuses on how the models discussed in the first part create and coconstitute exceptional spaces, from the borders themselves to the formerly rural areas now located within the city. Exceptional spaces are crucial to the city's economic success and regional and transnational competitiveness, yet they also create spaces of exclusion and social marginalization, including the urbanized villages, border spaces, and workers—specifically migrant sex workers—that fall outside the accepted discourse. The cases in this part pivot on the middle phase of Shenzhen history, after the experiments have been made but before the city has become a national and international model. These chapters illuminate the drivers and costs of transformation, as well as the complexity and uniqueness of local factors in a city often held up as a model for other countries.

In chapter 5, Mary Ann O'Donnell introduces the complex political and cultural geography of urbanized villages as the fulcrum of Shenzhen's urbanization. She shows how the government inverted Mao-era socialist language and values in the reform era, with consequences for the spatial transformation of China's cities, and resulting in the exceptional space of the urban village as a flexible and responsive spatial typology. While rural migration to centuries-old metropolises such as Beijing and Shanghai dominate the Western image of migration within China, migrants and villages in Shenzhen, a new city, have a different starting point in their relation to the urban. Shenzhen's villages, which have housed the migrants who provide its skilled administrative, technical, commercial, and design employees, continue to assert a "village" identity

INTRODUCTION

while forming a vernacular geography of "modernized," "urbanized," and "everyday" spaces that shape Shenzhen's rapidly changing built environment.

Following O'Donnell's framing of Shenzhen as a city defined by both urbanization and ruralization, in chapter 6, Emma Xin Ma and Adrian Blackwell provide a history of the spaces that physically delineate Shenzhen's political geography: the Sino-British border and the Second Line (or internal boundary). Their account of these borders as "technologies of differentiation" leads to a discussion of the zones of exception that arise when spaces (such as villages or streets) straddle administrative attempts to separate and control. Like the plans discussed in Huang's chapter, these attempts consciously enact their own failure, thereby allowing for proliferation of gradations of the licit and the illicit. In analyzing the problem of the illicit—so persistent throughout Shenzhen's history—Ma and Blackwell stress the central role of the border(s) in determining labor and capital movement.

Shenzhen, it appears, is increasingly defined by its seemingly infinite ability to create gradations—of licit and illicit, formal and informal, success and failure, rural and urban. In chapter 7, Jonathan Bach unravels the tensions between official and village approaches to these gray areas of quasi-informal and quasi-legal communities, examining how villages make use of their rural vestiges to coproduce the city. He considers in particular the fraught transition from villager to urban citizen and the changing cultural imaginaries this entails. Bach describes the call and response between village and city as they work through policies of expropriation, practices of appropriation, and the belated "rediscovery" and redevelopment of villages as integral to the city's past and future.

While the city government is eager to make "citizens" out of "peasants," as described in Bach's chapter, Willa Dong and Yu Cheng take us into the lives of those most marginalized of Shenzhen's migrants, female migrant sex workers, whose day-to-day existence is shaped by the socioeconomic inequality and structural violence of rural-to-urban migration, as well as by a version of the desire for urban belonging discussed in Florence's chapter. In chapter 8, Dong and Cheng considered these women's mental health by conducting in-depth interviews to understand their behaviors, coping strategies, anxieties, and needs, while also disclosing the workings of nested spaces of exceptions and the articulation of Shenzhen's geography and history in these workers' lives as they seek to secure income in barely concealed illicit spaces inside their urban villages.

The third and final part, "Extensions (2004–Present)," follows the Shenzhen model from its early construction of model leaders, workers, villages, and citizens to its contemporary form of city as model, boasting model industries, infrastructures, and professions. This entails mutually referential negotiations with the global—Shenzhen becomes a model for the world through modeling itself after international examples. The chapters in this part present cases from creative industries, public health, and transport infrastructure to underscore Shenzhen's increasing concern with the visibility of its "model" and that model's mobilization within the city's discourse of exceptionalism. Ironically, we argue, the more Shenzhen presents itself as a model for others, the more it erases or marginalizes the diverse local histories and the very spaces, people, practices, and processes that were central to its development.

In chapter 9, Winnie Wong explores the fusion between the notion of the model urban village presented by Bach in chapter 7 and the rise of a specific model industry, the creative industry, which later becomes part and parcel of the ideology of the "China Dream." Echoing Florence's discussion of the model worker, Wong analyzes the artist as worker in the infamous Dafen Oil Painting Village, raising new questions about subject creation in an era marked by China's fetishizing of its cultural and creative industries. This fascination with creative industries is not wholly capitalist; it can be situated as part of the international legacy of socialist commitments to universal creativity and worker justice. Wong delineates the crucial function of socialist language and politics in the elevation of this "model" while also exposing how creative industry propaganda departs radically from the community and trade it purports to depict. Such "models," she argues, allow officialdom to imagine its own, already approved "success" as the Shenzhen brand moves into its prime. The historical distinctions between the legal and illegal, authentic and fake, original and copy that once created a productive tension in Shenzhen's experimental days are now increasingly elided into a self-perpetuating apparatus of discursive appropriation.

If the rural migrant looks to the urban professional as a model, the urban professional often looks abroad. In chapter 10, on public health, Katherine A. Mason contrasts the professionalization of urban-to-urban migrants in Shenzhen with priorities of international funding structures and global health discourses. She shows how a successful implementation of an international model in Shenzhen actually brought about a public health crisis in 2009 when American, rather than Hong Kong, standards of global health came to define Shenzhen's governmental goals in addressing H_1N_1. To understand the contradictory impact of public health models in Shenzhen, Mason also examines the earlier SARS crisis of 2003—which inverted the relationship of Shenzhen and Hong Kong thanks to the successful performance of professional public health action in Shenzhen—comparing it to the different approach to

H1N1 several years later. Mason turns to an analysis of complex local needs to explain why success in adapting global standards was not sustainable. Lastly, she tracks a return among Shenzhen professionals to moral ideals of migrant worker social justice, engaging themes also raised in the chapter by Dong and Cheng.

The part's final chapter, by Max Hirsh, circles back to the dreams of smooth, frictionless globalization that informed the fantasies of economic zones discussed by Bach in chapter 1 of the volume. Shenzhen's infrastructure is the material scaffolding for the hopes of its denizens, from white-collar professionals to villagers and migrant workers. Nowhere is this infrastructure more symbolic, serving as both cultural capital and propaganda, than in Shenzhen Bao'an International Airport. This airport, however, exists in practice as just one element of a regional infrastructure whose other elements include those co-located in less glamorous spaces of urban villages and shopping malls. Looking at the promise of global mobility inherent in Shenzhen's ambitions to be a "world city," Hirsh investigates the development of informal check-in terminals that allow passengers in Shenzhen to check in for flights departing from Hong Kong International Airport. Located in storefronts, shipyards, and border crossings, and operated by local bus and ferry companies, these terminals have extended the infrastructure of aerial mobility into the everyday urban fabric of Shenzhen. Hirsh also argues that constraints on Shenzhen's ambitions produce unexpected forms of spatial and typological innovation in the urban landscape, such as a cross-boundary bus system that radically rearranges the city's connections to the outside world. Hirsh's chapter shows how Shenzhen's dual external borders, as discussed by Ma and Blackwell, now appear as an extension into the everyday space of the city, unmasking one of the city's most visible models as dependent on the fine-grained texture of the ordinary that it seeks to erase.

Taken together, these essays provide a historically sensitive, ethnographic, spatial, and discursive study of a city that manifests the structural changes of post-Mao China like no other. The chapters form a historical overview of Shenzhen's development from a Cold War borderland to Hong Kong's backyard factory, from a patchwork of urbanized villages to a regional metropolis with global aspirations and impact. They provide critical and multidisciplinary assessments of Shenzhen's viability as a model of Chinese urbanization, situating both its successes and failures within broader regional, national, and transnational problems. We pay special attention to the functional contradictions that are everywhere in Shenzhen's history as we document how villages became cities (while retaining their village power structures), workers became heroes (while deriding the hero status), the global became local (though most transnational immigrants returned), and the migrant became cosmopolitan (though a very nationalistic citizen). Contributors to this volume do not take issue with the Shenzhen "model" for being somehow false—indeed, on the whole we highlight its many achievements—but we do raise skepticism about the model's uncritical incorporation of events that were often produced after the fact, with little recognition of the main actors. Given the appeal of such a model as the basis for reforms in second- and third-tier cities throughout China, and possibly as a roadmap for the development of Chinese influence in Africa, Latin America, and India, the processes by which these contradictions are sustained must be understood.

The book's title, *Learning from Shenzhen*, intentionally refers to both the exhortations of Mao-era models (Learn Agriculture from Dazhai! Learn Industry from Daqing! Learn Values from Lei Feng!) and Robert Venturi, Denise Scott Brown, and Steven Izenour's landmark architectural pattern book, *Learning from Las Vegas*.[23] While the lessons of Las Vegas did not comprehensively transform American cities, the vast changes that first took root in Shenzhen have thoroughly transformed Chinese society. Although this is not a book about Shenzhen's architecture, we consider architects and urbanists among the stakeholders in China's historic experiment who can—and should—speak to the interactions we explore between individual agents and government apparatuses, and between foreign or migratory workers and local, legitimized citizens. For students and scholars of Chinese cities, as well as those interested in new directions in China studies, global studies, and cultural and social analyses of modernity in Asia and beyond, this book charts the production of a spectacular form of globalization through a fine-grained analysis of how Shenzhen came into existence. Other Chinese cities, particularly Beijing, Tianjin, or Shanghai, have longer and more politically significant international and cosmopolitan interactions, and other city regions such as Xiamen, Chongqing, and the Jiangsu Delta have been equally influenced by foreign investment and ideas. But Shenzhen remains unique within China's post-Mao history as a place where the central government's early tolerance of informal and illicit practices, loosely confined within variably administered spaces, created an excess of capital, knowledge, and imaginaries—including the world city and world citizen—that extended beyond Shenzhen into the rest of China. In this sense, this volume addresses how Shenzhen contributes to the ongoing reconfiguration of the meaning of collective life in today's China along with the reconfiguration of global economic space, for which Shenzhen is both emblematic and enigmatic.

Notes

1. On policy making, models, and experimentation from the Mao to the Deng eras, see Sebastien Heilmann, "Policy-Making through Experimentation: The Formation of a Distinctive Policy Process," in *Mao's Invisible Hand: the Political Foundations of Adaptive Governance in China*, ed. Sebastien Heilmann and Elizabeth Perry (Cambridge, MA: Harvard University Asia Center, 2011), 62–101.

2. Robert Venturi, Denise Scott Brown, and Steven Izenour, *Learning from Las Vegas: The Forgotten Symbolism of Architectural Form* (Cambridge, MA: MIT Press, 1977).

3. See, for example, Hong Chen, *Shenzhen zhongda juece he shijian minjian guancha* (Wuhan, China: Changjiang wenyi, 2006); Tu Qiao, *Yuangeng Zhuan/Gaige Xianchang (1978–1984) [The Story of Yuangeng/The Scene of Reform]* (Beijing: Writers' Publishing House, 2008); Rem Koolhaas et al., *Mutations: Harvard Project on the City* (New York: Actar, 2000); Ezra Vogel, *Deng Xiaoping and the Transformation of China* (Cambridge, MA: Harvard University Press, 2011); Winnie Wong, *Van Gogh on Demand: China and the Readymade* (Chicago: University of Chicago Press, 2014).

4. New scholarship in this area on Shenzhen includes Jenny Chan, "A Suicide Survivor: The Life of a Chinese Migrant Worker at Foxconn," *Asia-Pacific Journal* 11, no. 31 (2013): 1; Jenny Chan, Ngai Pun, and Mark Selden, "The Politics of Global Production: Apple, Foxconn and China's New Working Class," *Asia-Pacific Journal* 11, no. 32 (July 2013): 2. Larissa Heinrich's work on *Bodyworlds* attempts to document rumors connecting the body trade to Shenzhen. See her "The Dissection Controversy," in *The Harvard Illustrated History of Chinese Medicine and Healing*, ed. Linda Barnes and T. J. Hinrichs (Cambridge, MA: Harvard University Press, 2012). In American popular culture and media, the case of the dramatist Mike Daisey inventing tales of labor exploitation at the Foxconn factory on the radio program *This American Life* illustrates the extent to which Shenzhen is readily accepted as a site of dystopian industrialization. See James Fallows, "The Sad and Infuriating Mike Daisey Case," *The Atlantic Monthly*, March 17, 2012.

5. David Stark, *The Sense of Dissonance: Accounts of Worth in Economic Life* (Princeton: Princeton University Press, 2009).

6. Georg Simmel, "Bridge and Door," *Theory, Culture, and Society* 11, no. 1 (repr.; 1994 [1909]): 5–10.

7. Georges Teyssot, *A Topology of Everyday Constellations* (Cambridge, MA: MIT Press, 2013).

8. Teyssot, *Topology of Everyday*, 276.

9. On *shanzhai* in relation to Shenzhen, see Cara Willis and Jack Linchuan Qiu, "Shanzhaiji and the Transformation of the Local Mediascape in Shenzhen," in *Mapping Media in China: Region, Province, Locality*, ed. W. Sun and J. Chio (Hoboken, NJ: Taylor and Francis, 2012), 109–25; and Silvia Lindtner, "Making Subjectivities: How China's DIY Makers Remake Industrial Production, Innovation & the Self, " in "Political Contestation in Chinese Digital Spaces," special issue, *Journal of China Information* 28 (2014): 145–67.

10. Josiah Heyman, "Ports of Entry as Nodes in the World System," *Identities: Global Studies in Culture and Power* 11 (2004): 303–27.

11. Helen F. Siu, "Grounding Displacement: Uncivil Urban Spaces in Postreform South China," *American Anthropologist* 34, no. 2 (2007): 329–50; Dorothy Solinger, *Contesting Citizenship in Urban China: Peasant Migrants, the State, and the Logic of the Market* (Berkeley:

University of California Press, 1999); Kam Wing Chan and Li Zhang, "The Hukou System and Rural-Urban Migration in China: Processes and Changes," *China Quarterly* 160 (1999): 818–55; Pun Ngai, "Women Workers and Precarious Employment in Shenzhen Special Economic Zone, China," *Gender and Development* 12, no. 2 (2004): 29–36; Pun Ngai, *Made in China: Women Factory Workers in a Global Workplace* (Durham, NC: Duke University Press, 2005); Tiejun Cheng and Mark Selden, "The Origins and Consequences of China's *Hukou* System," *China Quarterly* 139 (1994): 644–68; Li Zhang, *Strangers in the City: Reconfigurations of Space, Power, and Social Networks within China's Floating Population* (Stanford: Stanford University Press, 2001).

12. See Giorgio Agamben, *State of Exception*, trans. Kevin Attell (Chicago: University of Chicago Press, 2005).

13. Linda Wong and Hue Wai-Po, "Reforming the Household Registration System: A Preliminary Glimpse of the Blue Chop Household Registration System in Shanghai and Shenzhen," *International Migration Review* 32, no. 4 (1998): 974–94. See also Kam Wing Chan and Will Buckingham, "Is China Abolishing the Hukou System?" *China Quarterly* 195 (2008): 582–606.

14. Stephen J. Collier, *Post-Soviet Social: Neoliberalism, Social Modernity, Biopolitics* (Princeton: Princeton University Press, 2011).

15. On the ambivalent relation of postcolonial studies to socialist and post-socialist China, see Tani E. Barlow, "Colonialism's Career in Postwar China Studies," *Positions* 1, no. 1 (1993): 224–67. On China's (post)socialist modernity and its relation to the imperialist past, see also Ann Anagnost, *National Past-Times: Narrative, Representation, and Power in Modern China* (Durham, NC: Duke University Press, 1997); Lisa Rofel, *Other Modernities: Gendered Yearnings in China after Socialism* (Berkeley: University of California Press, 1999); and Yukiko Koga, *Inheritance of Loss: The Political Economy of Redemption after Empire* (Chicago: University of Chicago Press, 2016).

16. Urbanus, *Village/City City/Village* (Beijing: CEPP, 2006); Yan Song, "Housing Rural Migrants in China's Urbanizing Villages," Lincoln Institute of Land Policy, *Land Lines* (2007): 1–7; Ya Ping Wang, Yanglin Wang, and Jiansheng Wu, "Urbanization and Informal Development in China: Urban Villages in Shenzhen," *International Journal of Urban and Regional Research* 33, no. 4 (2009): 957–74; Weiwen Huang, "Chengshi Guihua Yu Chengzhongcun, Shei Lai Gaizao Shei? [Urban Planning and Urban Villages: Who's Changing Who?]" *Zhuqu* 45 (2012a); Hang Ma, "'Villages' in Shenzhen: Persistence and Transformation of an Old Social System in an Emerging Mega City" (PhD dissertation, Faculty of Architecture, Bauhaus University, Weimar, Germany, 2006); Laurence J. C. Ma and Fulong Wu, eds., *Restructuring the Chinese City: Changing Society, Economy and Space* (New York: Routledge, 2005); Shenzhen Urban Planning Bureau, *Shenzhen Chengzhongcun (Jiucun) Gaizao Zongtu Guihua Gangyao [Shenzhen Urban Villages (Old Villages) Renovation Comprehensive Plan]* (Shenzhen: Shenzhen Government Publication, 2006); Chris Webster, Fulong Wu, and Yanjing Zhao, "China's Modern Gated Cities," in *Private Cities: Local and Global Perspectives*, ed. G. Glasze, C. J. Webster, and K. Frantz (London: Routledge, 2005), 153–70; James Jixian Wang and Jiang Xu, "An Unplanned Commercial District in a Fast-Growing City: A Case Study of Shenzhen, China," *Journal of Retailing and Consumer Services* 9, no. 6 (2002): 317–26.

17. Article 8, Constitution of the People's Republic of China (adopted December 4, 1982), accessed February 26, 2013, http://english.people.com.cn/constitution/constitution.html.

18. See Chenggang Xu and Xiaobo Zhang, "*Township-Village Enterprises Revisited*" (Institute for Food Policy and Research, Washington, DC, discussion paper no. 854, 2009); James Kai-Sing Kung and Yi-Min Lin, "The Decline of Township-and-Village Enterprises in China's Eco-

nomic Transition," *World Development* 35, no. 4 (2007): 569–84; and Wei Zou, "The Changing Face of Rural Enterprises," *China Perspectives* 50 (2003), accessed August 7, 2015, http://chinaperspectives.revues.org/773.

19. David Zweig, "'Developmental Communities' on China's Coast: The Impact of Trade, Investment, and Transnational Alliances," *Comparative Politics* 27, no. 3 (1995): 253–74.

20. Ching Kwan Lee, *Against the Law: Labor Protests in China's Rustbelt and Sunbelt* (Berkeley: University of California Press, 2007); Li Zhang, *Strangers in the City: Reconfigurations of Space, Power, and Social Networks within China's Floating Population* (Stanford: Stanford University Press, 2001); Siu, "Grounding Displacement"; Pun, *Made in China*. See also Stefan Al, ed., *Villages in the City: A Guide to South China's Informal Settlements* (Hong Kong: Hong Kong University Press, 2014); and Bruno De Meulder, Kelly Shannon, and Yanliu Lin, *Village in the City* (Zürich: Park Books, 2014).

21. See Robin Visser, *Cities Surround the Countryside: Urban Aesthetics in Postsocialist China* (Durham, NC: Duke University Press, 2010).

22. Hyun Bang Shin, "Vertical Accumulation and Accelerated Urbanism: The East Asian Experience," in *Urban Constellations*, ed. Matthew Gandy (Berlin: Jovis Verlag, 2011), 48–53; You-tien Hsing, "Socialist Land Masters: The Territorial Politics of Accumulation," in *Privatizing China: Socialism from Afar*, ed. Li Zhang and Aihwa Ong (Ithaca: Cornell University Press, 2008), 57–70; Zhang Li and Aihwa Ong, *Privatizing China: Socialism from Afar* (Ithaca, NY: Cornell University Press, 2008); and Ananya Roy and Aihwa Ong, eds., *Worlding Cities: Asian Experiments and the Art of Being Global* (Malden, MA: Wiley-Blackwell, 2011).

23. Robert Venturi, Denise Scott Brown, and Steven Izenour, *Learning from Las Vegas: The Forgotten Symbolism of Architectural Form* (Cambridge, MA: MIT Press, 1977).

PART I

Experiments (1979–92)

1

Shenzhen: From Exception to Rule

JONATHAN BACH

Introduction

Writing of the spectacular architecture that fills the skylines of Shenzhen and other Asian cities, Aihwa Ong borrows the term *hyperbuilding* from the architect Rem Koolhaas to describe how the new Asian city creates stunning urban infrastructures that simultaneously serve national and global aspirations.[1] The pursuit of the national and the global, she emphasizes, cannot be divorced from each other—the (re)construction of the nation involves a positioning of the nation on the global stage. The origins of today's Asian cities of hypermodernity lie in an unprecedented reconfiguration of national and global economic space in the second half of the twentieth century, from which Shenzhen emerged as one of the leaders of a fast urbanism that came to define the image of the new Asian city.

For China, emerging from the political chaos and economic paralysis of the Cultural Revolution of 1966–76, the creation of Special Economic Zones (SEZs) linked the national and the global in an ongoing speculative project that indelibly transformed China's identity, economy, and urban landscape. Shenzhen was the largest and most successful of these early endeavors, and this chapter shows how Shenzhen emerged from the interplay between national and global dynamics and entered into the global circulation of models of economic and urban development. Shenzhen arose specifically against the backdrop of two genealogies of the SEZ—the zone as a site of economic planning that enhances the global circulation of goods and the accumulation of capital, and the zone as a version of the imagination of the modern rational city with an inherent civilizing mission. By internalizing and expanding on these origins, Shenzhen became a prototype for China and beyond as a model (or antimodel) for a unique blending of economic and urban innovation and development.

Harnessing the Logic of the Zone

AN EXPORT ZONE BY ANY OTHER NAME?

Shenzhen originated from the attempts of the post-Mao leadership in the late 1970s to undo the economic paralysis of China's economy during the Cultural Revolution of 1966–76. A major challenge lay in acquiring badly needed foreign capital and technology from the class enemy without appearing to betray socialist principles. An economically attractive, though ideologically problematic, model appeared in Taiwan and South Korea, both firmly within the "imperialist" camp, who themselves sought ways to gain capital and technology with significant success. In 1966, while the Cultural Revolution was getting under way in the People's Republic, Taiwan established an export processing zone at Kaohsiung that grew at an astounding rate—just under 50 percent for the first four years and around 26.5 percent on average in the period up to 1979.[2] As the Cultural Revolution was winding down, South Korea established its first zone in Masan in 1974, which was widely successful in precisely the areas that China was concerned with: foreign exchange earnings, the introduction of new technologies, and the training of workers and managers in these new technologies. This had a ripple effect on other parts of the South Korean economy through employment, services, and materials.

Taiwan and South Korea, in turn, had looked toward the colonial entrepôts of Singapore and Hong Kong as inspiration for new ways to attract foreign direct investment (FDI), and therein lay a significant ideological problem. Singapore had only recently acquired formal independence in 1963, and Hong Kong remained a British colony (until 1997). South Korea and Taiwan were firmly within the American economic and military orbit. Were economic zones just another form of concession to Western powers—a modern version of the hated colonial treaty ports forced on China after it lost the Opium War in 1842—or was there a way to use the logic of the zone to China's advantage without compromising its principles?

These kinds of considerations faced Deng Xiaoping, Vice-Minister Gu Mu, and other leaders as they gathered throughout 1979 to discuss the best way to relink China to the outside world. Both Deng and Gu were supporters of creating special zones, and they ultimately carried the day. The idea of a "special district," first introduced in March 1979, coincided with Deng's desire to use the districts to not only attract foreign money but also gain room for broader reform attempts that would surely be too controversial if attempted in established cities. Seeking to justify the idea of an experimental space, Deng is said to have exclaimed, "Yannan was also a zone!," referring to the

Communist Party's mountain headquarters during the war. After a debate about what name would be least ideologically delicate, the decision was made to announce four SEZs that year—Shenzhen, Zhuhai, Shantou, and Xiamen. Of these, Shenzhen, located on the border of Hong Kong, became the largest and most well known outside of China.

The term SEZ struck a rhetorical balance between reassurance and boldness. These were not capitalist export processing zones (EPZs) as in Taiwan and South Korea but special *economic* zones that would, as Deng once put it, not be capitalist enclaves but "develop the productive forces under socialism."[3] They were not located in established cities but strategically placed to attract investment from overseas Chinese in Hong Kong and Taiwan, where their peripheral locations also conveniently gave the government a sense that they could be erased if they failed. They would not be limited to factories but would become model cities with diversified industries, including tourism and real estate development. In a somewhat updated version of learning from the industrial city of Daqing, the hope was that someday the rest of China would learn from the SEZs.

THE RISE OF THE EXPORT STATE

In creating these zones, China was taking part in the midphase of a phenomenon that had become an integral part of the global economy by the twenty-first century. Forms of economic enclaves have long been part of the history of trade: from the extensive trading network of Hanseatic cities from the thirteenth to the seventeenth century in Europe, to the Japanese island of Dejima where Dutch traders were confined from the seventeenth to the nineteenth century, to the Ming- and Qing-era port of Guangzhou (Canton) where European trade was concentrated and then confined from the 1700s until the British invaded to end the system, to the aforementioned colonial treaty ports from Hong Kong to Gibraltar. The contemporary concept of the economic zone, defined as a spatial designation where one country alters its laws to give preferential treatment to foreign investors and manufacturers, gained its current contours primarily in the postwar period. The zone as a contemporary phenomenon was spurred into being by mobile capital seeking the relatively cheaper wages of immobile labor. This process became greased by new developments in transportation (faster, more capacious) and an increased desire by developing countries to attract more investment to transform their economies.

Up until the mid-1960s, there had been scattered attempts by a small number of countries to attract investment by tempting investors with a different set of rules to allow them to take advantage of cheaper labor, lower or no taxes, and exemption from various regulations. Puerto Rico, for example, was promoted in 1948 as a de facto SEZ *avant la lettre* with tax and duty exemptions for US companies who produced there. In the mid-1960s, Mexico also worked out special incentives for companies to produce in a strip along their northern border in *maquiladora* factories, mainly to allow US companies access to cheap Mexican labor after the end of a "guest worker" program called Bracero that had enabled migrant labor in the United States. But it was Shannon, Ireland, that in 1958 hit upon the formula that now exists in approximately 3,500 variations around the world. Transatlantic flights to Europe used to stop in Shannon to refuel. Once long-haul flights no longer made this necessary, Shannon sought to keep air traffic by creating a geographic special zone that brought global transportation links, various investment incentives, and industries together in one legally exempted and proscribed space.[4]

By the time China created its SEZs, the idea of exports as a growth model for developing countries had taken hold. Developing countries were increasingly eager to adopt zones because they had grown disillusioned with the widespread strategy for economic growth of the early postwar era known as "import substitution," where countries aimed to grow and protect their own industries for domestic production and made imports prohibitively expensive, among other measures. Often faced with an excess of labor and lack of foreign exchange revenue, starting in the 1960s countries like India, Mauritius, the Philippines, and Kenya began to create EPZs that incorporated elements of Shannon's model, where foreign companies could employ local workers, goods could enter and leave with minimal regulation, and foreign exchange could be earned. These zones, however, had a relatively limited economic impact. It was only when Taiwan and South Korea began to reconceptualize the zone as a means of national development that the idea of export processing as a path to growth caught fire. The "Asian Dragons," or "Asian Tigers" (Singapore, Hong Kong, South Korea, and Taiwan), owed their meteoric rise to export-led growth, and this was not lost on China. The developing world was shifting to an export-oriented phase of industrialization that would later prove to be central to China's own strategies for economic growth.

The flip side of national growth driven by exports was the massive flight of manufacturing from the developed countries—primarily the United States and, to a lesser extent, Western Europe—to low-wage factories in Asia. From its peak of 19.5 million manufacturing jobs in 1979, for example, the United States lost 5.7 million jobs by 2007 (the year before the global economic downturn).[5] Not all jobs went overseas, since automation and a rise in productivity also reduced workplaces, but the majority did, and most of them went to East

Asia. In a comparable period (1980 to 2005), East Asia gained a whopping 42 million manufacturing jobs (from 27 to 69 million), an indication of not only the overall growth in manufacturing but also the incredible competitiveness of the region when manufacturing jobs declined everywhere else except Eastern Europe and India.[6]

The rise of "exportism," as Ngai Ling Sum has called this phenomenon, as a new form of development, along with the shift in manufacturing away from the industrialized countries, is part of an epochal rescaling of national economic space in which zones came to play a key role.[7] As Western governments sought to ease regulations from the late 1970s onward, corporations began to relocate manufacturing in large numbers to the most competitive locations. This trend was emboldened by Western government policies of deregulation and privatization that became dominant in the 1980s, starting with Britain's Prime Minister Margaret Thatcher and US President Ronald Reagan.

Yet the political context alone would have been insufficient for such a global rescaling of space had it not been for major changes in technology, especially long-distance transport and computerization. The container ship, which had only been invented in 1956 and standardized in 1961, allowed freight costs to drop from nearly 25 percent of product cost in the 1950s to a tiny fraction of that today.[8] The earliest of these ships could carry five hundred to eight hundred containers—today the largest carry eighteen thousand.[9] Together with a corresponding rise in the availability of air freight, these technological innovations allowed for the rise of "flexible" production methods such as "just-in-time" production that tailored production to consumer demand.

China's experiment with SEZs thus came as the global economy was shifting to internationalized production for ideological, structural, and technological reasons, where companies increasingly distributed their processes across the globe in a complex choreography of design, labor, production, transport, and assembly. When Shenzhen came into existence in 1979, then, it became part of an already rapidly expanding global system of economic zones.

FROM EXCEPTION TO EXPANSION: THE ZONE AS INTERNAL MODEL

Each host country selectively applies their domestic law to their economic zones in order to generate investment, primarily by enhancing the production and circulation of commodities. In doing so, they are taking part in the logic of sovereign exception, where states effectively section off a part of their country and give it an intermediate status between home and abroad. Political and economic space becomes separated within the boundaries of the nation-state, and companies and workers alike receive different status within the zone than in the rest of the country.

As Ronen Palan has shown in his important work on offshore spaces and tax havens, the ability of states to parcel their territory and market it for profit is inherent in the very structure of the modern international system even as it can seem to contradict the sacred status of the sovereign nation as an organic whole.[10] The sovereign state system achieved near universal status after decolonization in the wake of the Second World War. It is premised on a doctrine of noninterference in other states' internal affairs. This gives, in principle, a state the freedom to determine whether laws apply equally or differentially within its borders. For much of the history of modern capitalism, especially industrialization in Western Europe and North America, the global economic system functioned effectively within states where political and economic spaces overlapped. What is colloquially called "globalization," however, has separated political and economic space and spread both production and capital accumulation across the world. By moving across borders, including in and out of zones, the value of materials and goods can be "stepped up" or "stepped down" at each boundary crossing so that it becomes cheaper to make a single computer with parts designed, sourced, made, assembled, and packaged in a half-dozen countries by dozens of companies than to locate these processes all within one country.

Zones played a significant role in enabling this international production, functioning as a kind of spatial adaptation of the sovereign state system to global forms of capital accumulation. All zones offer variations on the same theme: a different regulatory regime than in the rest of the country, usually a confined geographic area that provides better infrastructure and good transport, and zone governance dedicated to business as the primary denizen of the zone. The zone thus became a privileged economic space for manufacturing, and later also services, because it became a space of exception to the tax, labor, and customs laws of countries worldwide, allowing a parallel system of global production to emerge that today employs 130 million people directly and indirectly.[11]

The historical synchrony of political and economic space that is being undone today has its origins in not only economic rationales but also the powerful idea that all citizens deserve equal treatment regardless of their location within the country. As Aihwa Ong has argued, the slicing of sovereignty into different zonal spaces—what she calls "graduated sovereignty"—allows states to reclassify people as well as goods.[12] With millions of workers, entrepreneurs, and professionals, the zone becomes a space for states to find new ways to not only produce goods and attract foreign exchange investments but classify and manage populations. This biopolitical aspect of the zones takes different forms according to the pressures in the respective country. In Shenzhen, for example, in addition to economic issues, the growth of the zone raised the question of how rural populations and millions of migrants can be integrated into the national urban system and what sort of politics can emerge from and through the management of socioeconomic diversity.

From the beginning, Shenzhen and its sister SEZs were meant not merely to process exports but to act as thresholds through their exceptional status. The zone is neither fully home nor abroad—it is at the same time both inside and outside the system. This is the zone's particular strength, what allows the zone to let goods, money, people, and ideas circulate in ways that might not otherwise be possible. In Shenzhen, as in other cases, the zone as threshold serves two liminal functions that propel it from a site of production to a site of transformation. First, the zone serves as a *spatial* threshold that mediates between China's economic space and that of other countries—as the "window to the world," as the leadership called it—through which one can look both in and out. Second, the zone serves as a *temporal* threshold between stages of development—the China that was and the China that will be. Deng and his supporters were explicit about the SEZs serving as places to test out reforms that could later be extended to the whole country. By dint of this threshold function, the hope was that the zone would indeed have a magical effect of sorts, transforming the rest of the country in the process. Further, they were to serve as a link to not only global capital in general but overseas Chinese in particular, thus incorporating and reshaping networks and trajectories of Chinese beyond China.

From the very beginning, Shenzhen was to fulfill this threshold function in two ways: as a model for the rest of China and as an example for the rest of the world to see China's capability and commitment to reform. This was an ambitious task for which half measures would not work; the central government rejected Shenzhen's original master plan from August 1980 as too modest. Beijing insisted on not just a small city of 49 km^2 supporting a larger industrial area (as in the original plan) but a major industrial city with the full range of urban functions from tourism to commerce and residential housing (the urban area is now more than 400 km^2 in size).[13] Shenzhen became a kind of incubator, relatively protected from provincial political constraints and the established elite interests that dominated older Chinese cities. It thus became the site of many of China's fabled "firsts" that are now standard across the country: the first stock exchange, the first privatization of state-owned enterprises, the first novel foreign exchange transaction systems, the first radical reforms in the commercialization of housing, the first sale of industrial land, the first labor-contract systems that aligned wages with a market system and created a professional class of entrepreneurs, and many more.

The core element of the economic zone—preferential treatment for investment involving a mixture of tax incentives, protection of property, and extended land use rights—has also spread across the country. The once controversial and "special" zone has become the national model for municipalities to attract FDI and integrate with the global economy. An ever-larger number of zones of different shapes, sizes, locations, and nomenclatures followed Shenzhen and its three sister SEZs of 1979 and 1980. The 1980s saw the opening of fourteen port cities in 1984 for foreign investment, followed in 1988 by the initially less successful addition of Hainan Island as an SEZ and the more successful "opening" of seven coastal areas throughout the decade.

In the 1990s into the 2000s, the zone as a development model took off on a national scale. The creation of Pudong in Shanghai in 1990 was followed by the creation of fifteen economic border zones since 1992. China joining the World Trade Organization in 2001 led to the steady addition of a dizzying array of national- and provincial-level zones. By 2013, for example, there were at least fifty-three National High-Tech Development Zones and forty-seven National Economic and Technological Development Zones.[14] These are no longer grouped exclusively along the coast or borders but extend into the interior of the country. Jin Wang counts 222 national-level and 1,346 provincial-level zones across China's municipalities.[15] In 1978, she notes, there were no SEZs within any municipality, and the initial SEZs of 1979 and 1980 were purposely located outside existing cities. By 1990, however, nearly a quarter of all Chinese municipalities had an SEZ-like zone within its borders, and by 2008 zones were present in nearly all Chinese cities (92 percent). If the goal had been to turn the exception into the rule, the original SEZ experiment had certainly bore fruit. Jin Wang puts this in a dramatic visual form, showing what could be called the zonification of China (see figure 2).

The Zone as a Modern Urban Space

PRODUCING VALUE(S)

As Shenzhen grew, it had to find its place within two parallel processes: a global shift from labor-intensive manufacturing to knowledge-based economies, and increased competition from the ever-growing array of zones within China vying for both domestic and foreign attention. The diffusion of Shenzhen's market model across China means that the city can no longer lay unique claims to the sectors that once made it a destination for workers and investors alike. It faces serious domestic competition from Shanghai, Tianjin, Chongqing, and many other cities and regions, and increased international competition from both low-wage countries like Vietnam and Bangladesh and high-end centers such as Seoul, Singapore, and from its neighbor, Hong Kong.

For Shenzhen to continue to produce, economic value required continuous adaptation of what Shenzhen was able to offer. As labor costs increased, it became less profitable to maintain small-scale factories. Already by 1985, low-skill manufacturing began to give way to more highly skilled demands from the emerging high-tech sector. High tech came to dominate Shenzhen, with flagship Chinese companies such as Huawei (telecommunications) and Tencent (Internet) and with globally famous (or notorious) branches of companies such as Foxconn making components for Dell, Hewlett Packard, and Apple. High tech is still the mainstay of Shenzhen, accounting for about 60 percent of its total industrial output, but is itself being retooled to focus on the "new" industries of the twenty-first century, key among them e-commerce, non-carbon-based energy sources (e.g., solar, wind), and the biomedical sector, including stem cells and biomedical equipment. Similarly, Shenzhen's vast infrastructure for transport and storage—key to the institutionalization of zones as nodes in the global economy—augments its size as the world's fourth-largest container port with a new focus on services and back office work. Financial services (especially fund and venture capital) and creative industries (especially design) round out Shenzhen's economic profile on the world stage.

Its ultimate value, however, is not to be measured in GDP (a respectable US$25,038 per capita in 2014), exports (more than US$245 billion in 2015), or the fifty-three top Chinese companies headquartered in the city alone. Rather, it is in the perception of Shenzhen within China as a world-class city with a mixture of spectacular architecture, "civilized" citizens, clean streets, and an entrepreneurial spirit in line with the city slogan: "Dare to Become the World's First." In short, Shenzhen wishes to be at the pinnacle of modernity. The modernist dream of the city as the ultimately rational, civilizing force in human development is the other legacy of the zone, one as important as its origins in the postwar export economy. This modernist fantasy comes from a long tradition in (mostly Western) philosophy that Stephen Toulmin characterizes as the dream of Cosmopolis: a rationally ordered society where nature and society fit into precise categories and interact productively according to an unerring logic.[16] The modernist-planned city was thought to give rise to this ordered society and has a long history of seeing the fresh start as its essential ingredient, from American utopian communes in the nineteenth century to Soviet total planning cities in the twentieth century to contemporary gated communities today.[17]

This is the fantasy of the ideal modernist city as a clean slate, a *tabula rasa*. The fantasy of the perfect city as a tabula rasa sits deep within Shenzhen, which seems to take to heart the playwright and poet Bertolt Brecht's exhortation in his 1926 *Handbook for City Dwellers* to "erase the traces!"[18] Invariably, Shenzhen is presented in media, promotional materials, and conversation as a city with no history, arising from a proverbial small fishing village or, somewhat more accurately, a small border town. Its historical predecessors, Xin'an (which encompassed both present-day Hong Kong and Shenzhen) and Bao'an counties, become mythologized and temporalized.[19] Its former villages, still physically and psychologically present as traces of a rural past turned urban anomaly, disappear in official representations of the city.

This elision of the past is, in part, what enables Shenzhen to present itself as a unique space that can redeem the past precisely because it is unencumbered by it. Redemption occurs through a focus on the present where, in the words of the city's popular slogans, "practical work brings prosperity" and "time is money, efficiency is life." This emphasis on the pragmatic application of grit and entrepreneurial spirit not only redeems the "lost" decades of the Cultural Revolution but also helps settle the larger score of being subjected to colonialism. During the city's thirtieth anniversary, for example, a common catchphrase touted how Shenzhen accomplished in thirty years what it took Western society three hundred years to achieve. This is the heroic, even miraculous Shenzhen, which former Chinese President Hu Jintao referred to as "a miracle in the world's history of industrialization, urbanization and modernization."[20]

MODELS, MIRRORS, AND MASTERS

This miracle, however, always looked over the border at Hong Kong. As Mary Ann O'Donnell writes, Hong Kong was simultaneously Shenzhen's future (what it *would be* like in years to come) and its spectral past (what Shenzhen *might have* been like under other circumstances): "Hong Kong's postwar history becomes the past that Shenzhen would have had if not for a cruel twist of socialist fate."[21] This is highly relevant to understanding Shenzhen in its global context. While initially it was the EPZs of Taiwan and South Korea that sparked the idea of Shenzhen, it was always Hong Kong that loomed as the "big master" that provided both the rationale for the zone's location and the lion's share of its foreign investment. It also provided the type of image Shenzhen wanted to cultivate on the world stage. In essence, Shenzhen sought to use the zone models from Taiwan and South Korea to provide the means to become a major city like Hong Kong, and ultimately, perhaps, to overshadow it.

If Shenzhen was powered in large part by its initial abundance of low-skilled labor streaming from the rural provinces of the People's Republic across the "Second Line" (the city's internal border) into the zone, it was also powered by a mirror image across the external border to Hong Kong and beyond of what Constance Clark calls the "supranational connections of kin and capital": "We cannot understand Shenzhen," Clark writes, "without conceptualizing it as part of a broader network connecting Guangdong Province, Hong Kong, and parts of Southeast Asia, linked through flows of capital, kinship, information, and labor."[22] The tens of thousands of refugees who fled to Hong Kong, Macao, and Taiwan after 1949—together with overseas Chinese in Singapore, Indonesia, and elsewhere—had created a deep and wide pool of money and knowledge of which the People's Republic was all too aware. Prior to the establishment of the SEZs, write Peck and Zhang, "actually existing 'Chinese capitalism' was a strictly offshore, extra-territorial phenomenon, more or less coterminous with the vibrant, diasporic economies of the 'overseas Chinese,' but entirely absent from the People's Republic of China itself."[23] The SEZs were a way to reterritorialize at least the money, if not also some of the talent, from these diasporic economies, and in so doing also lay further claim to the historical ties to Hong Kong, Macao, and perhaps eventually Taiwan as well.

This broader network, especially the ties to Hong Kong, was what enabled the market reforms that have come to define Shenzhen to be originally "assembled bit by bit" from Hong Kong.[24] The most radical early reforms, as Jun Zhang details, occurred under pressure from Hong Kong investors who convinced Shenzhen officials to introduce contract-based labor and Hong Kong–style land management regimes with land auctions and use rights and to copy the Hong Kong stock exchange.[25] Hong Kong investments dominate Shenzhen both overall and cumulatively, even if after Hong Kong's return to China in 1997 such investments were no longer technically "foreign." Since the 1997 Handover, this relation has changed significantly. Today, Shenzhen can have the luxury of imagining that it is more essential to anchoring Hong Kong's future in China than Hong Kong is for Shenzhen's future rationale. Shenzhen's 2030 urban development strategy now includes Hong Kong in *its* own plans to become a world city.

In the vision for 2030, *ShenKong* (Shenzhen plus Hong Kong) appears as one extended urban area and common capital market with twin hubs for finance, trade, and shipping linked by seamless transportation and high-tech and service industries.[26] This does not mean, however, that Hong Kong is no longer a model. The *Shenzhen Daily* reports that leading up to the 2030 urban development strategy, "the city sent 10 delegations comprising hundreds of officials to learn from the many different aspects of Singapore and Hong Kong in urban planning, environmental protection, housing management and traffic control. The plans were drawn up after rounds of discussions by the visiting officials who were tasked with specific research in Singapore and Hong Kong."[27]

The role of Singapore—Shenzhen's "second master" after Hong Kong—is significant here because, as Zhang argues, Singapore's importance increased proportionally to Shenzhen's concerns with governing the very capitalism it has created.[28] Already in the 2000 master plan for Shenzhen, the rhetoric was to emulate Hong Kong's efficiency but Singapore's "environment"—that is, its political environment. With its blend of authoritarian capitalism and knowledge-based economy, Singapore has been raised to the status of premier role model in recent years by both the Shenzhen and the Beijing governments.

However, as Zhang points out, if Singapore's authoritarian capitalism is an attractive model, then there are serious divergences from Shenzhen that seem to move in the opposite direction. For example, in Singapore, the government owns and manages most of the housing, something the Shenzhen government has been trying to move away from. Further, the ability of Singapore's government and its planners to exert almost absolute control over land and development is something the Shenzhen government can dream of, but it is complicated by the legacies of rural and urban legal distinctions, explored in detail in other chapters in this book, that make much of Shenzhen's ownership opaque and complex and therefore difficult to control.[29]

"SHENZHEN" AND THE SEZ IN GLOBAL CIRCULATION

Yet while Shenzhen looked to Hong Kong and Singapore, it became a frequently invoked ideal for other countries that dreamt of a similarly "miraculous" effect from SEZs. In India, Shenzhen often appears as a paradigmatic case to study and learn from, while Africa looks increasingly toward the Chinese SEZ model for its own development, which China seeks to promote. As Bräutigam and Xiaoyang write, "Over time, China's national and provincial governments, and its zone developers, have acquired considerable expertise in planning, developing and operating various kinds of industrial parks. At least three of these experienced Chinese developers are themselves actively involved in the African zones and provide a learning resource."[30]

Whether the SEZ model perfected in the context of the East Asian development state can be "exported" to Africa or adapted elsewhere is the subject of considerable debate and skepticism. This has not dimmed the enthusiasm for zonal models for economic development, though they have become increasingly fraught with anxieties of exploiting workers or reducing wages rather than raising them. In 2005, India passed a highly controversial law approving more than five hundred SEZs as the basis for exports and foreign exchange earnings that critics saw as a massive land grab by developers. In the same year, Russia passed a law creating twenty-four SEZs. In Europe, the anxiety about zones focuses on what is seen by critics as an attempt to roll back the social consensus regarding labor and welfare. For example, in Poland there are seventeen SEZs that labor unions find threatening as a model for the rest of Europe. Comments such as this one by the president of the Federation of German Industry in 2011, who proposed that to address its economic collapse, "Greece should become a kind of special economic zone within the euro zone, equipped with necessary and permitted financial aid, but also with the presence of foreign European Union officials," added to critics' concerns that the SEZ model would be imported to the developed world as a response to financial crisis in order to undermine the gains of social democracy.[31]

Shenzhen serves in this increasingly global discourse as both a model and antimodel. The World Bank holds up Shenzhen as "an illustration of the effectiveness of the SEZ model in the Chinese context."[32] Right-wing commentators in the United States referenced Shenzhen when fiercely attacking plans (since halted) for the Chinese machine tool company Sinomach to build a fifty-square-mile SEZ in 2011 south of Boise, Idaho, that would have been a "self-sustaining" city, including manufacturing, housing, and retail. In an ironic reversal of Chinese hard-liners' opposition to Deng's plans to create Shenzhen, right-wing American hard-liners raised the specter of colonialism. Arguing against Sinomach's plans in the United States, one blogger wrote, "That's exactly how India came to be ruled by the British."[33]

Yet while "Shenzhen" indexes the idea of a successful development model as an export zone for others around the world, Shenzhen sees itself no longer as a zone but as a city that seeks to position itself in the global urban order by "catching up" with and surpassing "the best cities in Asia, including Hong Kong, Singapore and Seoul" to become "the cleanest and most ecologically sound city on the mainland."[34] Since its ninth five-year plan (1996–2000), Shenzhen has shifted the emphasis of its place in the international citational economy to becoming a "world class city," completing a shift in its self-presentation from a space for economic production to a space for urban experimentation that channels classically modern dreams into the rhetoric of sustainable, technologically sophisticated megacities.[35]

Conclusion

"THE CLEANEST AND MOST ECOLOGICALLY SOUND CITY ON THE MAINLAND"?

We have seen how Shenzhen emerged from a shift in global economic and political space in the mid to late twentieth century that elevated the export zone into a starring role for both production and development. Similar to Taiwan and South Korea, China was able to use SEZs to radically transform and globalize its economy and society. Taken as part of a larger systemic transformation, these zones were a remarkably successful attempt to combine jurisdictional tactics and technological adaptations to produce not just new spaces for the production of commodities but new foundations for rapid national economic growth and new spaces for urban experiments. Because of the ideological constraints that required Shenzhen to always be more than just another export processing zone, Shenzhen became one of the earliest prototypes of an instant city enabled by the postwar internationalization of production. Out of Shenzhen grew what architect Keller Easterling has called the "new urban paradigm" of the zone, a kind of urbanism I have called elsewhere the "ex-city" to describe an urban space fashioned out of a powerful mixture of exports, exception, and exhibition.[36]

The journalist Greg Lindsay pays Shenzhen a backhanded compliment when holding it up as the antimodel for the more recent ex-city, South Korea's New Songdo City.[37] This sleek yet still sparsely occupied SEZ is contrasted by Lindsay against Shenzhen, which for him occupies the space of the "archetypical Asian city... unplanned and uncontrollable... representing 20th-century urbanism at its worst." The strength of his claims are less interesting than the way Shenzhen here comes to occupy the role as the model against which the newest zone on the block, New Songdo City, both owes homage to and seeks to differentiate itself from. Shenzhen, it seems, even when it is being pilloried, has become a key site for what Ananya Roy calls the "material and discursive practices of inter-referencing through which cities are made and inhabited."[38] Shenzhen is part of the global discourse on what constitutes contemporary urbanism and capitalism in Asia and beyond, and how the infrastructure of the city relates to its economic and cultural claims. In this role, it is already the very world city it wants so desperately to become.

Notes

1. Aihwa Ong, "Hyperbuilding: Spectacle, Speculation, and the Hyperspace of Sovereignty," in *Worlding Cities: Asian Experiments and the Art of Being Global*, ed. Ananya Roy and Aihwa Ong (Malden, MA: Wiley-Blackwell, 2011).

2. Jing-dong Yuan and Lorraine Eden, "Export Processing Zones in Asia: A Comparative Study," *Asian Survey* 32, no. 11 (1992): 1026–45.

3. Jamie Peck and Jun Zhang, "A Variety of Capitalism . . . with Chinese Characteristics?," *Journal of Economic Geography* 13, no. 3 (2013): 357–96.

4. Thomas Farole and Gokhan Akinci, *Special Economic Zones: Progress, Emerging Challenges, and Future Directions* (Washington, DC: World Bank, 2011).

5. Justin R. Pierce and Peter K. Schott, *The Surprisingly Swift Decline of U.S. Manufacturing Employment* (Washington, DC: US Census Bureau, 2012).

6. Richard Newfarmer and Monika Sztajerowska, "Trade and Employment in a Fast-Changing World," in *Policy Priorities for International Trade and Jobs*, ed. Douglas Lippoldt (Washington, DC: Organization for Economic Cooperation and Development, 2012).

7. Ngai-Ling Sum, "Theorizing the Development of East Asian Newly-Industrializing Countries: A Regulationist Perspective," in *Regulation Theory and the Crisis of Capitalism*, ed. Bob Jessop (Cheltenham: Edward Elgar, 2001); Neil Brenner, *New State Spaces: Urban Governance and the Rescaling of Statehood* (Oxford: Oxford University Press, 2004).

8. Marc Levinson, "Container Shipping and the Economy," *TR News*, no. 246 (2006): 10–12.

9. Jean-Paul Rodrigue, *The Geography of Transport Systems* (New York: Routledge, 2013).

10. Ronen Palan, *The Offshore World: Sovereign Markets, Virtual Places, and Nomad Millionaires* (Ithaca, NY: Cornell University Press, 2003).

11. Thomas Farole, *Special Economic Zones in Africa: Combining Performance and Learning from Global Experience* (Washington, DC: World Bank, 2011).

12. Aihwa Ong, *Flexible Citizenship: The Cultural Logics of Transnationality* (Durham, NC: Duke University Press, 1999).

13. Mee Kam Ng and Wing-Shing Tang, "The Role of Planning in the Development of Shenzhen, China: Rhetoric and Realities," *Eurasian Geography and Economics* 45, no. 3 (2004): 190–211.

14. Consulate-General of the Peoples Republic of China in San Francisco, "News on Startups in China," accessed June 2, 2013, http://www.chinaconsulatesf.org/eng/kj/zyxx/t43953.htm.

15. Jin Wang, "The Economic Impact of Special Economic Zones: Evidence from Chinese Municipalities," *Journal of Development Economics*, no. 101 (2013): 133–47.

16. Stephen Toulmin, *Cosmopolis: The Hidden Agenda of Modernity* (New York: Free Press, 1990).

17. Jonathan Bach, "Modernity and the Urban Imagination in Economic Zones," *Theory, Culture & Society* 28, no. 5 (2011): 98–122.

18. The German phrase "Verwisch die Spuren" can also be translated as "cover your tracks." Bertolt Brecht, "Lesebuch für Städtebewohner," poem, accessed August 7, 2015, http://www.usc.edu/libraries/archives/arc/libraries/feuchtwanger/exhibits/Brecht/Lesebuch.html.

19. Mary Ann O'Donnell, "Becoming Hong Kong, Razing Baoan, Preserving Xin'an: An

Ethnographic Account of Urbanization in the Shenzhen Special Economic Zone," *Cultural Studies* 15, no. 3/4 (2001): 419–43.

20. Chen Hong, "President Hails Shenzhen SEZ a World 'Miracle,'" *China Daily*, September 7, 2010, accessed August 7, 2015, http://www.chinadaily.com.cn/china/2010-09/07/content _11264644.htm.

21. O'Donnell, "Becoming Hong Kong," 419–20.

22. Constance Clark, "The Politics of Place Making in Shenzhen, China," *Berkeley Planning Journal* 12, no. 1 (1998): 103–25.

23. Peck and Zhang, "A Variety of Capitalism," 4.

24. Jun Zhang, "From Hong Kong's Capitalist Fundamentals to Singapore's Authoritarian Governance: The Policy Mobility of Neo-Liberalising Shenzhen, China," *Urban Studies* 49 (2012): 2860.

25. Zhang, "Hong Kong's Capitalist Fundamentals," 2860.

26. Xiangming Chen and Tomas de'Medici, "The 'Instant City' Coming of Age: Production of Spaces in China's Shenzhen Special Economic Zone," *Urban Geography* 31, no. 8 (2010): 1141–47.

27. Han Ximin, "Shenzhen to Par with Top Cities," *Guangdong Impression*, April 9, 2008, accessed August 7, 2015, http://gocn.southcn.com/english/impression/200804090057.htm.

28. Zhang, "Hong Kong's Capitalist Fundamentals," 2856–60.

29. Zhang, "Hong Kong's Capitalist Fundamentals," 2865–66.

30. Deborah Bräutigam and Tang Xiaoyang, "African Shenzhen: China's Special Economic Zones in Africa," *Journal of Modern African Studies* 49, no. 1 (2011): 48.

31. Hans-Peter Keitel, quoted in Tomasz Konicz, "Europa als Sonderwirtschaftszone," *Gegenblende* 17 (September/October 2012), accessed June 3, 2013, http://www.gegenblende.de/-/XCK.

32. Farole, *Special Economic Zones in Africa*, vii.

33. "China Plans Special Economic Zones in the US," *Want China Times*, June 13, 2011, http://www.wantchinatimes.com/news-subclass cnt.aspx?id=20110613000070&cid=1102.

34. Han Ximin, "Shenzhen to Par with Top Cities."

35. Carolyn Cartier, "Transnational Urbanism in the Reform-Era Chinese City: Landscapes from Shenzhen," *Urban Studies* 39, no. 9 (2002): 1513–32.

36. Keller Easterling, *Enduring Innocence: Global Architecture and Its Political Masquerades* (Cambridge, MA: MIT Press, 2005); Bach, "Modernity and the Urban Imagination," 109. See also Keller Easterling, *Extrastatecraft: The Power of Infrastructure Space* (New York: Verso, 2014).

37. Greg Lindsay, "The New Sustainable Cities," *Departures*, October 2011, accessed April 21, 2015, http://www.departures.com/travel/travel/new-sustainable-cities?iid=sr-link2.

38. Ananya Roy, "Postcolonial Urbanism: Speed, Hysteria, Mass Dreams," in *Worlding Cities: Asian Experiments and the Art of Being Global*, ed. Ananya Roy (Malden, MA: Wiley-Blackwell, 2011): 307–35.

2

Heroes of the Special Zone: Modeling Reform and Its Limits

MARY ANN O'DONNELL

He made vigorous efforts to turn the tide, urging boldness and decisiveness in action, and determination to carry out drastic rectification of the seriously chaotic situation caused by the Cultural Revolution. For the benefit of the party and the people, he gave no thought to his own interests or safety and, despite the risk of being overthrown again, waged a resolute struggle against the Gang of Four. This rectification drive was, in essence, an experiment for the reforms afterwards.
President and general secretary Jiang Zemin in his eulogy for former president and general secretary Deng Xiaoping, 1997

Mao Zedong died on September 9, 1976, eight months after Zhou Enlai had passed. Over the next two years, Deng Xiaoping wrested power from Mao Zedong's appointed successor, Hua Guofeng, and the Gang of Four, emerging as post-Mao China's preeminent leader during the Third Plenary Session of the Eleventh Central Committee of the Chinese Communist Party, which convened in December 1978. Deng's political success notwithstanding, he faced a propaganda conundrum: how to partially discredit Mao Zedong while maintaining the legitimacy of the Communist Party. The politburo member Chen Yun famously declared that "the life or death, existence or extinction of the party" was at stake during this transition.[1] The appointed leaders of the Shenzhen Special Economic Zone (SEZ) would play an important role in that transition by providing Deng and his reformist allies with living examples of new socialist models and nationalist goals that were still being defined. Thus Shenzhen not only provided an experimental space for the "reform" policies of the post-Mao era but also represented a political experiment—a venue where certain reformist leaders and cadres conducted bold political experiments. In the parlance of its actors, the political implications of reform were often glossed as "social" reform, downplaying the fundamental restructuring of the Chinese polity at stake in the transition. The stories of their rise and fall, from the founding of the Shenzhen SEZ to the era after the 1989 crackdown, defined for the nation the potential and the limits of both economic and political reforms by carefully circumscribing the limits to "society."

Party structure and internal loyalties facilitated the complicit—albeit highly improvisational—relationship between Beijing policymakers and Shenzhen leadership. The three most lionized Shenzhen leaders were (1) Shenzhen's first party secretary and mayor, Liang Xiang (1981–86);[2] (2) the first standing vice chairman of China Merchants Hong Kong and the director of China Merchants in Shekou, Yuan Geng (1979–93); and (3) the first president of Shenzhen University, Luo Zhengqi (1983–89). All three were Chinese Communist Party members who had demonstrated a personal commitment to national development. And they were all turned into model leaders in the same way—for being willing to take responsibility for actions they believed to be correct, even if those actions were unapproved outside or within Shenzhen's borders at the time. Given that their actions took place immediately after the Cultural Revolution, a time when leaders, individuals, and their families had been harassed, beaten, arrested, and sometimes killed or driven to suicide for not following the party line, the willingness of Shenzhen's first generation of leaders to engage in or overlook quasi-legal activities made them courageous new "heroes" for the reform era. In the new rhetoric, they "dared to take responsibility," which not-so-subtly implied that they were willing to protect others from political repercussions for unsanctioned actions. In fact, this kind of quasi-illicit activity and willingness to give official approval to previously "illegal" actions infused the Shenzhen "experiment" with its national significance, making it a powerful model for the political reorganization of post-Mao Chinese society.[3]

But this rhetoric of heroism had its limits. When these three heroes became political liabilities, that early tolerance for unsanctioned actions provided justifications for removing Liang and Luo from office, while sidelining Yuan from the political process. Vagueness and contradiction had been the hallmarks of Deng's pronouncements on the Reform and Opening policy, of which Shenzhen was the showpiece. But what particular reforms were in fact desirable, correct, appropriate? Which should be recognized and taken up elsewhere? And which should be regarded as dangerous or even counterrevolutionary? Shenzhen's first generation of leaders answered these questions with both their dramatic successes and their later downfall.

Political Appointments

In the first decade of Shenzhen's establishment, political leadership and heroism were defined by the ability of individuals—usually men—to thrive in an uncertain political environment that had been forged on World War II battlefields and through postwar purges.[4] During a 2012 interview in his architectural firm, for example, former Shenzhen University president Luo Zhengqi explained that as an everyday practice, especially at higher levels of responsibility, reform meant "walking farther and farther" from proscribed limits until one had transformed society. He also joked that sometimes the borders snapped back like a rubber band and one was left unprotected and alone in unsanctioned fields.[5] In fact, his metaphorical reference to borders and rivers had an uncanny specificity in early Shenzhen because this rough-and-tumble frontier town was located just north of Hong Kong along the Shenzhen River, where Red Guards had once waved red books, yelled slogans, and promised to liberate their occupied compatriots on the other side. These geographic facts also meant that in Shenzhen, Deng's metaphor of "feeling rocks to cross the river" could be interpreted literally, describing how local residents deployed cross-border work permits to enter Hong Kong for proscribed jobs and then smuggle back consumer goods that were not yet available in the Mainland on their return. Nevertheless, Luo Zhengqi noted that during the early years, whatever happened in Shenzhen constituted "Reform" (*gaige*), and successful leaders quickly learned to act decisively instead of waiting for political recognition, which might (or might not, he implied) come after the fact. Importantly, Luo Zhengqi emphasized that Shenzhen heroes were men who had "dared to act and dared to assume responsibility."

Continuing with Mao-era practice, political appointments to Shenzhen were pragmatically based on revolutionary credentials and demonstrated loyalty to Deng or his reform lieutenants, Hu Yaobang (CCP general secretary, September 11, 1982–January 15, 1987) and Zhao Ziyang (premier, 1980–87; CCP general secretary, November 1, 1987–June 23, 1989).[6] The confluence of revolutionary credentials and personal loyalty shaped the service trajectories of party leaders because credentials made them eligible for promotion, while personal loyalty ensured that they could be trusted to make decisions that would help their benefactor no matter what the prevailing political climate. For example, Liang Xiang, who would become Shenzhen party secretary and mayor, joined the CCP in 1936 and immediately went to study at the Yan'an Central Party School, where he eventually held the position of general party branch secretary. During the war against Japan and the Civil War, he first held the position of party secretary of the Work Committee of Xi'an County, Jilin Province, and was subsequently appointed county head. With the establishment of the People's Republic of China (PRC) on October 1, 1949, the central government deployed him to Guangzhou, where he rose steadily in the Maoist ranks over the next thirty years—from department head of the Guangzhou Planning Commission to Guangzhou's vice mayor, party secretary of the Guangzhou Department of the Secretary, second party secretary of Guangzhou, standing member of the Guangdong Provincial Politburo, and finally vice governor of Guangdong Province. His support in the party came through Xi Zhongjun (father of current General Secretary Xi Jinping) and Zhao Ziyang. In fact, even in the political upheavals in the immediate aftermath of Tiananmen, Liang Xiang demonstrated unwavering personal loyalty to Zhao Ziyang. When Zhao Ziyang's son, Zhao Erjun, escaped China under the assumed name "Chen Xueyang," Liang Xiang reputedly approved his departure through Hainan Island in the south, having his post and party membership stripped one month later.[7]

Similarly, Yuan Geng, who would become a standing vice chairman of China Merchants Hong Kong and the director of China Merchants Shekou (a work unit whose status and definition will be closely examined later) joined the Party in 1939 and was deployed to the Huiyang Anti-Japanese Guerilla Forces in November of that year, where he would later teach military affairs. Beginning in 1940 and through the 1950s, Yuan Geng participated in military campaigns in Jinan and Huainan, saw combat in Huangtan, and ended his military career in 1950 as an advisor to Ho Chi Minh in Vietnam. In the postwar era, Yuan Geng served in foreign affairs under Zhou Enlai, being posted to Indonesia and participating in the Three Worlds Conference in Jakarta. In April 1968, as part of Kang Sheng's purge of senior party officials who opposed Mao, Yuan Geng was incarcerated in Beijing's maximum security Qincheng Prison until Zhou Enlai secured his release in September 1973. In October 1975, Yuan Geng returned to work in the Department of Transportation as the head of Overseas Affairs, Ministry of Transportation. In addition to his ties to Deng through their common sponsor, Zhou Enlai, Yuan Geng found national support in General Secretary Hu Yaobang, Vice General Secretary Li Xiannian, and Vice President Gu Mu.[8]

The youngest of the three special zone heroes by two decades, Luo Zhengqi would become the president of Shenzhen University in 1983. He was a Tsinghua University intellectual, where he had earned the respect and attention of Hu Yaobang in the early reform period. During the Cultural Revolution, as a young Qinghua University architecture teacher, Luo Zhengqi joined the "414 Faction" that supported Liu Shaoqi. During one of the Qinghua student battles, members of the radical Jinggangshan faction, which advocated demolishing the extant power structure, abducted Luo Zhengqi's younger brother, Luo Zhengfu, beat him, and locked him in a car trunk, where he died from asphyxiation. After the Cultural Revolution, Luo Zhengqi was appointed vice secretary of the Tsing-hua University Party. One of the Red Guards who had been convicted of his brother's death applied to Tsing-hua University. The applicant had already served his sentence and Luo Zhengqi personally issued a directive that "historical grudges" were insufficient reason for blocking a candidate's admission to the university. Luo Zhengqi's fair-minded decision came to the attention of Hu Yaobang, who promoted him to the chair of China's Youth Federation. However, almost immediately after he rose to a national position, Luo Zhengqi offended Chen Yun when Chen's son took advantage of his father's position to obtain an opportunity to study abroad. Luo Zhengqi wrote the elder leader a letter asking Deng's stalwart supporter to have his son renounce his scholarship, arguing that such rewards should have been allocated according to merit rather than nepotism. Subsequently, as an implied punishment, Luo Zhengqi was reassigned to Shenzhen, where he was given the responsibility of building the university.[9]

Daring to Take Responsibility

One characteristic of post-Mao heroism was "daring to take responsibility," which meant to serve the interests of Deng and his reformer allies and reproduce and legitimize party hegemony in the transition from the Mao to the Deng eras. In other words, "daring to take responsibility" functioned politically with respect to individual careers and more generally at the level of party hegemony. Maoism conflated belief in Mao with the party's political legitimacy. In the post-Mao era, Deng and his allies acted to create a new political legitimacy that secured the CCP's monopoly on political power by implementing policies—free-market reforms and the privatization of state-owned enterprises, for example—that were not socialist. It is important to note that "daring to take responsibility" did not function as an abstract ideal but instead was actually embodied and performed by officials and then broadcast through reports of model reformers in the state-controlled media. The following analysis of how Liang Xiang became a hero (and model of the post-Mao cadre) highlights how Beijing policy makers used post facto attribution to both create a space for experimentation and maintain control over what constituted a "hero." Ultimately, these policy makers determined who could and could not represent reform and the extent of political reform.

When the first generation of inland cadres, corps of civil engineering members, and workers arrived in Shenzhen and Shekou in 1979, they encountered a relatively weak county administration that was being opportunistically dismantled by local communes and subordinate work brigades and teams. The contract responsibility system had already spread to Bao'an County's twenty-one communes, where local farmers were eager to sell their surplus produce in Guangdong and to Hong Kong people.[10] There were both permanent and occasional markets throughout Bao'an County, which was threaded with dirt roads. The primary form of transportation was walking, although there were wider roads for truck transport. The Kowloon-Canton (Guangzhou) Railway (KCR) had a station at Luohu, and the neighboring commercial area was known as Dongmen. At the end of 1979, the population of the SEZ with household residence was 312,600, not including migrant workers and the reported 1,500 residents with temporary residence permits.[11]

The Special Zone's low population density, lack of infrastructure, and absence of bureaucratic mechanisms to oversee and regulate reform praxis also meant that politicized aphorisms describing uncertain pragmatism—"seek truth from facts" and "it doesn't matter if the cat is black or white as long as it catches mice"—circulated to describe the experience of simultaneously constructing and administering these new social entities. Indeed, the early aphorisms and even the appellation "Shenzhen Speed" all tended to conflate governance with the simultaneous construction of a government itself. In the first years of the SEZ, political success or failure was evaluated with respect to urbanization of the Dongmen/Luohu area of the SEZ and of Shekou, an independent industrial park at the tip of the Nantou Peninsula in western Shenzhen. Crudely speaking, the earliest task of Shenzhen's leaders was to replace the extant collectivist (rural) government with an urban socialist apparatus. Practically speaking, this meant that Special Zone cadres had to figure out how to electrify the area, lay water lines, and pave roads—all without a fully functioning government in place and no oversight from either the provincial government in Guangzhou or the central government in Beijing. This is a key point: because there was no functioning government in place, the first generation of Shenzhen leaders had leeway with respect to governmental institutions and procedures that they established, including higher levels of subordinate independence and decision-making authority than were institutionalized elsewhere in the country.

Prior to the establishment of an official government or official policies, cadres like Luo Jinxing operated on a case-by-case basis, usually with tacit approval from work unit leaders who emphasized results rather than extant Chinese law. In 1979, for example, the Guangdong Post and Telecommunications Bureau transferred Luo Jinxing to Shenzhen, where he was appointed vice director general of Shenzhen's Housing Authority. Luo Jinxing's directive was to construct housing for employees in Shenzhen work units, including government ministries and employees in city-owned companies. At the time, the central government had budgeted thirty million yuan for the construction of infrastructure, but Luo Jinxing estimated that one billion yuan would be necessary to develop a 4 km^2 area in Luohu. By 1980, it was evident that new sources of funding were needed. Under the auspices of the Housing Authority, Luo Jinxing established and became the chief executive officer of China's first real estate company in the post-Mao era, Shenzhen Special Economic Zone Real Estate Company. The new state-owned company entered joint ventures to lease land, construct buildings, and manage its own investment fund. On January 1, 1980, Luo Jingxing and Liu Tianjiu, the first acknowledged Hong Kong investor in Shenzhen, signed an agreement to codevelop the PRC's first commercial residential estate, Donghu Liyuan.[12] The agreement stipulated that Shenzhen Municipality would receive 85 percent and Liu Tianjiu 15 percent of the profits. Phases one and two of the project earned five million yuan for the SEZ government.[13] In a nation where people had been beaten as "capitalists" for raising ducks and geese just five or six years earlier, this was a daring move—even under a vague pronouncement of "change." Thus the most revealing aspect of the Luo Jinxing deal with Liu Tianjiu was that it occurred months before a full Shenzhen government would convene or before the first party secretary and mayor for Shenzhen, Liang Xiang, was appointed.[14]

Liang Xiang would soon work to create a legal framework and government institutions that would retroactively protect these unconstitutional deals, including the rights and profits of foreign companies. Indeed, in November 1982, more than a year after the Donghu Liyuan project went through, the SEZ government released the *Shenzhen Special Economic Zone Outline Plan for Social Economic Development*, which permitted foreign direct investment and provided post facto legal protection for at least three years of development projects and plans. The outline plan also furnished a political justification for this very unorthodox move: Xiang called it "ant theory," which stated that only after a scout ant had discovered a patch of sweetness would other ants be attracted to the area.[15] In other words, liberal policies were the "sweetness" that attracted capitalist "ants," allowing Liang Xiang to redeploy CCP political control over production into capitalist advantage on Chinese territory. On its own terms, ant theory was successful. Between 1980 and 1985, with or without official approval, Shenzhen cobbled together almost six billion yuan in investment capital to construct basic urban infrastructure, including roads, water lines, and electrical networks, which in turn integrated more than twenty residential, commercial, industrial, and tourist parks and zones in the nascent city.[16] But in Beijing, hard-liners like Chen Yun and Bo Yibo worried that these changes could transform China's polity from "socialist" to "capitalist"—a transformation that challenged the Maoist economic system and, by extension, the party's legitimacy.

This history suggests Liang Xiang's boldness was evidenced not only in creating or appropriating capitalist methods but also in his willingness to bestow political legitimacy on those who were already engaging in illegal economic activity. The dangers of political patronage in the immediate post-Mao era made his and other Shenzhen leaders' actions especially daunting. Indeed, when Liang Xiang proposed the outline plan, no one knew how Beijing leaders would respond to the illegal and unregulated activities that had been taking place in the SEZ. On the contrary, everyone knew that there was deep disagreement about whether the country should adopt these kinds of capitalist measures in order to stimulate the Chinese economy. Furthermore, in the immediate aftermath of the Cultural Revolution, everyone was also viscerally aware that disagreement in Beijing could lead to personal tragedy. Even Deng had been punished multiple times in the 1960s and 1970s for his position on economic growth. However, at the time, the need to fund the actual construction of the city[17] and the proximity to Hong Kong investors meant that the most pragmatic methods and capital came by way of the colonial entrepôt—as indeed seemed the intention when Deng located the Shenzhen SEZ on the very border. Liang Xiang's willingness to trust the "capitalist" activities of his subordinates and to assume political responsibility for their actions epitomized the courageous and vexed politics of the Special Zone's early period, where subordinates "dared" to act outside the Constitution (and Cold War borders) because their leaders "took responsibility" for these actions. Here's the important point: when the results of these methods exceeded expectations, the state-controlled national media retrospectively credited Liang Xiang with heroic leadership skills. In fact, the height of Liang Xiang's model status came when Deng voiced his approval of these reforms during the 1984 Inspection Tour. At that moment, Deng performed at the national level the same kind of political patronage that Liang Xiang had given his subordinates in Shenzhen.

Pushing the Limits of Reform

In contrast to Liang Xiang's program in the SEZ, the reforms introduced by Director Yuan Geng in the newly designated Shekou Industrial Zone explicitly challenged Beijing's monopoly on making appointments and determining the form and content of public opinion, converging with popular demand for political as well as economic change. In Shekou, Yuan Geng promoted institutions that reproduced key features of Hong Kong society, including implementing contract labor practices and allowing for an independent press. The result was the clearest articulation of popular representation in the special zone. Importantly, these changes were glossed as "social" change despite their clear political importance. What's more, although "Shekou Spirit" (*Shekou jingshen*) was distinguished from "Shenzhen Spirit" (*Shenzhen jingshen*) within the SEZ, it was Yuan Geng's popular representation and calls for "social change" that became identified with the reforms that made Shenzhen "special" in the larger national imaginary.[18]

Yuan Geng operated with even more de facto political independence than did Liang Xiang because the Shekou Industrial Zone's political designation was even vaguer than that of the SEZ. Although Shekou lay within the borders of the SEZ, it operated under the auspices of the national Ministry of Transportation, making the industrial zone politically independent of the Shenzhen administrative apparatus that Liang Xiang was building. Shekou's relative independence was an artifact of the Maoist system, which had divided from governance into administration by ministry and then administration by government. Roughly speaking, ministries organized and oversaw functions such as transportation while governments organized and oversaw territory such as the SEZ. Many of the conflicts that have characterized Shenzhen government arose as ministries and governments struggled (and continue to struggle) to control functions that must be placed in a territory.[19]

For Yuan Geng, reforming the socialist system meant challenging the complacency of cadres—another instance in which the transition from Maoism was coded as the creation of a new role model. Yuan Geng believed that the low quality of Chinese products could be directly attributed to the mentalities, or *suzhi*, produced by the Communist system, especially with respect to the training of cadres.[20] The difference between the Shenzhen and Shekou models was subtle but important for political change because, in Shekou, Yuan Geng focused on transforming not only the mentalities of migrant workers but also the mentalities of administrators and, by extension, government functionaries, arguably with an eye toward cultivating an active and social minded middle class. Consequently, Yuan Geng's strongest challenge to both local and national politics came at the level of identifying and promoting models at the level of both job appointments and media and the dissemination of information.

In 1980, for example, Vice Premier Gu Mu approved Yuan Geng's petition to directly recruit employees to the China Merchants Bureau rather than requiring him to apply for appointments through the Ministry of Transportation, which was the overseeing line supervisor of China Merchants Shekou. Under the Chinese system of *danwei*, or work unit appointments, the Ministry of Transportation was assigned workers that it then deployed to its bureaus. These employees were assigned jobs for political as well as professional reasons. Indeed, as a lingering effect of the Cultural Revolution, political inclinations were more important than skill sets when assigning jobs. In contrast, Yuan Geng lobbied to hire employees with necessary skill sets and experience rather than accepting employees who had been selected for their political views—an articulation of a technocratic model of employment that many young Chinese intellectuals supported. Two years later in 1982, Yuan Geng extended these reforms to administrative systems, and employees were recruited through relatively open applications and interviews. Yuan Geng and upper level managers also visited university campuses to recruit workers for China Merchants Shekou. Importantly, the debates over Shekou's status and its political structure had clear implications for possible governmental restructuring throughout China. China Merchants Shekou assumed many governmental functions—such as urban planning, building schools and hospitals, and providing basic infrastructure—without concomitant political subordination to a territorial government simply because there was no territorial government in place. In turn, grassroots associations such as the Management Committee (see below) arose to resolve social problems that China Merchants Shekou as an economic entity was not prepared to or did not want to address institutionally.

Yuan Geng's challenge to political organization occurred in 1983, when the Management Committee of the Shekou Industrial Zone was formed. The Management Committee's directive was to administer Shekou society, including urban planning, education, information, social welfare, and so on. Members of the committee would not be appointed but be determined through elections. All China Merchants Shekou party cadres were eligible to be a candidate for the committee and to vote in Management Committee elections, which would be held every two years. Full-time employees of the Shekou Industrial Zone were not eligible to be a Management Committee candidate but were eligible to vote in elections. The criteria for being a candidate ostensibly remained being a vetted party member, coding the elections as an internal party reform. However, Yuan Geng had suspended cadre promotions based on political criteria and instead made cadre promotions based on expertise and contributions to China Merchants' success. This meant that full-time employees could be promoted to cadre and become eligible for election to the Management Committee based on their work in Shekou. This system of work unit internal management effectively democratized Shekou society because almost every worker in Shekou was considered a full-time employee of China Merchants, including factory workers. Yuan Geng had hoped to make elections for political positions universal in Shenzhen, thereby extending ministerial reforms to territorial structures, but Beijing reformers balked at this measure. Instead, in a fascinating and instructive compromise, China Merchants Shekou was incorporated as a limited-holding company wherein the state-controlled China Merchants Bureau selected four members of the new board of directors while employees of China Merchants Shekou elected the remaining seven board members. Importantly, this compromise recoded more far-reaching political reform as an institutional reform within a private company, effectively reproducing the neoliberal business practices of colonial Hong Kong, where Great Britain appointed political leaders but locally owned businesses held board elections.

His organization of the Shekou Ministry of Information most clearly manifest Yuan Geng's efforts to actively transform the employee-residents of Shekou into voting citizen-workers. The first head of the Shekou Information Office was Zhou Weimin, a popular hero of the 1976 Tiananmen movement that had commemorated the death of Zhou Enlai. Zhou Weimin also participated in the Democracy Wall Movement, which led to his incarceration as a political prisoner. After his release from prison in 1978, Zhou Weimin was rehabilitated but nevertheless remained outside the state apparatus without a job and unable to transfer his wife's *hukou* to Beijing, where he lived. Yuan Geng claimed to have based his decision to appoint Zhou Weimin on his skills and experience as a journalist, which were determined through an interview. Nevertheless, Yuan Geng needed to secure the approval of Hu Yaobang in order to finalize the transfer because the Shekou Ministry of Information remained directly under the supervision of the national ministry. Hu Yaobang's patronage of Yuan Geng's reforms importantly underscored the tensions inherent in the management and governance of Shekou.

The contradictions inherent in Shekou's nebulous status first came to a head in 1985, when the *Shekou Bulletin News* received and published an anonymous letter to the editor that criticized Yuan Geng's reforms. The letter targeted business practices in Shekou, including the increasing bureaucratization of establishing businesses, the "low quality" of the foreign companies in Shekou (most were polluting and/or labor intensive manufacturing firms), salaries that reflected neither efficiency nor contribution, and inadequate urban planning in the face of an exploding population. In this letter, the author reworded Yuan Geng's slogan, "Time is money, efficiency is life," to claim instead, "Efficiency is life, and efficiency comes from management (*guanli*). If industrial zone management continues to drag, then it will lose its vitality."[21] This letter directly criticized the leadership of both Yuan Geng and the board of directors by invoking the Management Committee model of democratic organization. The editorial board of the *Shekou Bulletin News* brought the letter to Yuan Geng three times in order to secure his approval to publish the letter. Yuan Geng not only approved the letter but indicated that in the future, any article that did not criticize the party or its political line could be published without being vetted by Shekou leaders, even if it criticized his own leadership.

This unexpected openness led to the bulletin's popular "News Salon," an open forum that met weekly to debate issues of democracy and of changing the legal system in Shekou, which was technically a division of a state-owned industry functioning as an independent society.

As we have seen, Beijing policy makers relied on post facto recognition and withheld it in order to check the actions of Special Zone heroes. By reorganizing the Ministry of Information, Yuan Geng initiated a series of reforms that gave this power to a (relatively) free press. Here then was the challenge that Shekou (and, by extension, Shenzhen and the rest of the Special Zone) posed to the national system of post facto recognition of models and national value: Was the Special Zone a site from which social and political reforms could not only be launched but also valorized locally? After all, from the perspective of the governed, this was precisely the question that the editorial board of the *Shekou Bulletin* had asked Yuan Geng—that is, would leaders continue to protect subordinates even when there was high-level political disagreement (in this case, between the de facto head of government and his minister of information)? And if so, could those disagreements be debated publicly? Even though the *Shekou Bulletin* was a low-level newspaper that explicitly (and technically only) served Shekou and its residents, by the end of the 1980s, it had become one of China's most important experiments in loosening restrictions on journalism, and many of its reports were syndicated in newspapers at every level of government.

The "Dear Deng" Letter

The ideological battle to determine who could engage in post facto recognition of "heroic" actions came in October 1986 when the *Shenzhen Youth Herald* published Shenzhen University professor Qian Chaoying's controversial opinion piece entitled "I Support Comrade Xiaoping's Decision to Retire," which was nicknamed the "Dear Deng" letter because although it was an opinion piece, it was rhetorically presented as a letter to Deng.

In the early years of reform, the *Shenzhen Youth Herald* was, along with Shanghai's *World Economic Herald*, one of the two most independent newspapers in China. Consequently, despite being a small newspaper, the *Youth Herald* had a national subscription base, providing intellectuals throughout the country with a platform for debating progressive ideas and evaluating ongoing experiments in reforming Chinese society. However, as the steps taken before publishing the critique of Yuan Geng in the *Shekou Bulletin* indicated, the *Youth Herald*'s decision to publish the "Dear Deng" letter was politically charged, no matter what steps the editorial board took to downplay political implications through the rhetoric of "social" reform. In the manner of traditional intellectuals, for example, Qian Chaoying's writing style was sincere and humble, but the content was unmistakably radical. Moreover, the piece drew directly on Shenzhen's experience, asking, "Why must the people show our sincere and deep feelings for Deng by sacrificing further reform of the political system?" Qian's meaning was that Deng's voluntary and early retirement would allow China to reflect on and establish a more just political system, a system that was more in keeping with the needs of reform, rather than a return to cult politics, which had characterized the Cultural Revolution's glorification of Mao Zedong.

In April 2012, the youth editor who had published the "Dear Deng" letter, Cao Changqing, published exiled democracy activist and former head of the politics department at the Chinese Academy of Social Sciences Yan Jiaqi's recollections of how party elder Bo Yibo had responded to the letter. Apparently, Bo Yibo not only was furious about the "Dear Deng" letter but also had taken it as an attack on the power of older and already retired leaders because the letter implied that party elders had illegitimately ruled too long. During a closed meeting on political reform, Bo Yibo apparently overheard a conversation between Zhao Ziyang and Peng Chong about the "Dear Deng" letter. Upon overhearing the conversation, Bo Yibo became livid and is reported to have screamed at the younger leaders, "You are already fifty, sixty, and seventy years old. We won't die and you won't rise!" He was implying that the power structure would continue to be headed by the older generation, their retirements notwithstanding. Junior member of the Standing Committee Hu Qili was apparently so frightened that he immediately showed his support for the elders, wishing that the old leaders of the proletarian revolution would live to a healthy old age. Importantly, at that closed meeting, Bo Yibo also called for the party to investigate the situation in order to determine on whose behalf Qian Chaoying had written.[22] In the contentious world of Beijing political reforms, Bo Yibo assumed that neither Qian Chaoying nor the *Shenzhen Youth Herald* was acting as an independent voice, but rather were acting on behalf of one of the young reformers in Beijing seeking to undermine the elders' power, most likely Hu Yaobang.

The opinion piece was published at a critical time in Chinese politics. General Party Secretary Hu Yaobang and Premier Zhao Ziyang, Deng's "right and left hands," were pushing for further political liberalization. Less than two months after the letter was published, students organized public protests in more than a dozen cities to support political and economic liberalization. The famous Chinese astrophysicist Fang Lizhi led the protests, calling for the introduction of political reforms that would ultimately end the single-party system and the continuing use of government as an instrument of party policy. Two other intellectuals, Wang Ruowang and Liu Binyan, also led protests. It was rumored that Deng disliked Fang, Wang, and Liu, and that he directed Hu to dismiss them from the party, but Hu refused. In the fallout, Hu was forced into retirement because it was said he had been too lenient with student protestors.

The *Shenzhen Youth Herald* was also one of the victims of the 1987 crackdown. The newspaper was closed and Cao Changqing was banned for life from working in journalism. At the same time, Hu Yaobang was forced into retirement. Cao Changqing's inability to protect either himself or Qian Chaoying reminds us that the central government could and did take action within the SEZ. It also suggests that in years following Hu's dismissal during the 1987 crackdown, the stakes of daring to take political responsibility in Shenzhen escalated. Politically appropriate ways of figuring the state were being codified and ranked even as at the grassroots-level calls for more extensive political reforms were growing.

Building the Future at Shenzhen University

Both geographically and politically, the newly founded Shenzhen University was situated just between the foreign-investment-financed construction of the new city of Shenzhen and the social experiments going on in Shekou. Moreover, unlike Shenzhen and Shekou, Shenzhen University was explicitly a social institution, a designation that, in the political environment of Shenzhen in the 1980s, provided a protective umbrella for various social and political experimentations. Important also for the establishment of Shenzhen University as a social institution, Shenzhen University President Luo Zhengqi's political patron was General Secretary Hu Yaobang, who had explicitly called for a more liberal political environment. In his opening address for the 1987–88 academic year, for example, Shenzhen University President Luo Zhengqi provided a definition of the "Shenzhen Spirit" that glossed participation in the construction of buildings as the expression of self-sacrifice, patriotism, and the highest form of intellectual life: "Students, where have you come? This is a Special Zone ... Now, you have come here as reinforcements for the brigade building the Special Zone, to supplement our Special Zone Construction team. From this day forward, you are the builders of the Special Zone."[23]

Moreover, the 1987 address clearly articulated the idea that there was more at stake in the construction of Shenzhen than simply economic adjustment. Instead, Luo Zhengqi went on to explain to those in the audience that the purpose of the special zone was to reinvigorate socialism by creating a new kind of intellectual citizen: "Every year when new students arrive, the University offers each one of you four concepts, those concepts are: seek facts, become independent, tolerate diversity, and innovate . . . I just bumped into four students from the senior class of undergraduates and vocational students. They [told me] would rather go to a relatively poor work unit, in order to strive to work in a job that is truly challenging and risky. They'd rather go to a place with a low salary, instead of a high salary. I believe they have ideals. I feel that all Shenzhen University students should be like this."

The graduating students were starting their fifth year at Shenzhen University because there was no campus when they had been accepted to the university in 1983. Instead, there was a plot of land and a promise from Liang Xiang that Shenzhen City would finance the construction, pay staff salaries, and pay student tuitions for the new university—a total bill that, at the time, was more than Shenzhen's entire budget. An architect by training, Luo Zhengqi led a brigade of student workers to embark on a five-year course of study that included designing and building their own campus in addition to attending classes. Moreover, to promote new social-political values such as competition and responsibility, Luo Zhengqi organized students into work teams that developed plans for particular buildings and then presented their proposals to a jury made up of the university president and staff from the university's new school of architecture. The university body then worked together to realize the winning plan. In other words, the construction of Shenzhen University was a metaphor for rebuilding the Chinese nation through new values that would support a more technocratic and liberal government.

Deng endorsed the construction of Shenzhen University as a model for rebuilding the Chinese nation during his first Inspection Tour of Shenzhen. In 1984, when Deng's entourage passed the campus location (on the way from Shenzhen to Shekou), at first they did not believe that Shenzhen University had opened because, from the roadway, it was impossible to see the makeshift construction site, temporary housing, and plastic stools that constituted the classrooms within the lush lychee orchard. However, upon seeing the construction site campus, which had appeared over the course of one academic year, Deng reportedly remarked, "This is Shenzhen speed!" Indeed, upon graduation in June 1988, the students who entered the university in 1983 had actually designed and constructed a functioning campus complete with an office building, a classroom building, and a library. They had also designed and built staff housing, student dormitories, canteens, a student center, and a hotel, which was situated next to a lake and park area. Significantly, the campus layout reflected Luo Zhengqi's ideals. Unlike traditional Chinese campuses, Shenzhen University was not enclosed by a wall but open to everyone in

Shenzhen who could enter the campus without being checked in and, for example, use the library or enjoy a lakeside walk.

The construction of Shenzhen University by its own students, faculty, and president highlighted an important aspect of special zone heroism and new models for participating in national construction. The Shenzhen University model of intellectual citizens, like the Shenzhen and Shekou models of development, arose out of a symbiotic relationship between leader-heroes and their subordinates in the pursuit of practical goals—in this case, building a university. In turn, these goals were not only glossed in the political rhetoric of the Reform and Opening policy but also held up as evidence that the policy was successful. The case of the students who built their own university highlights two critical moments in the creation of special zone heroism: creative agency and post facto recognition by Beijing leaders. In other words, participation in the construction of the special zone was necessary but insufficient for the creation of heroes. It was also necessary that this participation received public commendation in the state-controlled media, which was (even as Luo Zhengqi spoke) being challenged by both the *Shenzhen Youth Herald* and the *Shekou Bulletin News*.

The experience of the students who built their own university also illuminates Shenzhen's symbolic place in ongoing political debates that were being waged in Beijing. Importantly, the students themselves—like construction company workers and, ultimately, Shenzhen leaders—did not determine the meaning and value of their work themselves. Instead, leaders in Beijing reserved the right to determine what did and did not constitute reform. And this is precisely why popular heroism could be defined as "daring to take responsibility"; it entailed making a grassroots reality politically viable.

David Graeber reminds us that society is an active project (or set of projects) in which "value is the way actions become meaningful to the actors by being placed in some larger social whole, real or imaginary."[24] In Shenzhen, the national Reform and Opening policy took the form of practical and measurable results, which ranged from productivity bonuses to campuses and rising GDP. These results became valuable (in Graeber's sense of the term) only to the extent that Beijing reformers ritually acknowledged them within the state-controlled media. In this system, one of the highest accolades was designation as a model—the goal to which all others would then strive to conform. "Reform" in Shenzhen depended upon post facto recognition by Beijing politicians, which made the actions of its leaders vulnerable to changing national political situations. Once-lauded "results," such as a building or accepting foreign investment, were always at risk of being turned around and reinterpreted as "spiritual pollution" or "bourgeois liberalization," as they had been in the antireform campaigns of 1983 and 1987.

Unexpected Consequences

By 1987, the forced retirement of progressive reformer Hu Yaobang and the campaign against bourgeois liberalization indicated that national leaders disagreed to what extent Chinese society should be opened to Western ideas and, by extension, to what extent the party's rule could be challenged. Nevertheless, Shenzhen leaders continued to demonstrate their support for their subordinates' socially progressive actions. Until the 1989 crackdown, both progressives and conservatives used events in the special zone to make their cases for or against further kinds of political reform, respectively. Notably, in the aftermath of the 1989 crackdown, all three of the first-generation SEZ heroes tacitly sided with protestors in both Beijing and Shenzhen, where the university was a hub of prodemocracy activity. Consequently, their fall and subsequent punishments reiterated at a higher level the same process that had left Cao Changchun banned from Chinese journalism and Qian Chaoying a downplayed member of the Shenzhen University staff—their leader-patrons could not or refused to take responsibility for their actions.

Events leading to the 1989 democracy movement began in January 1988, when Li Yangjie, Qu Xiao, and Peng Qingyi of the National Youth Thought Education Research Center conducted an inspection tour of Shenzhen to understand the practical and ideological effects of the Reform and Opening policy on young people. As part of their tour of Shekou, the local branch of the Chinese Communist Party invited the three representatives to speak with a group of young workers. In most inspection tours at the time, exchanges between visiting dignitaries and local leaders and workers would have been tightly scripted and choreographed, with both leaders and workers giving rote answers to expected questions. Indeed, the event began with the leaders giving their opinions on "the party's correct decisions on the policy of Reform and Opening."

Instead of progressing as a ritualized performance of politically correct ideas, however, the exchange quickly escalated into a debate over the meaning of "prospectors" (*taojinzhe*), an explicit condemnation of people whose actions were motivated by the desire for individual self-interest rather than by patriotism or collective interests. The Shekou youths challenged the visitors' claim that an individual's desire for material goods was counterrevolutionary and would not help China modernize, making three interrelated claims. First, they claimed that Shekou had not been built by patriots but by individuals who worked for wages. Second, the youths claimed that their experience had brought home the truth that the concrete and individualized practice of earning money was itself an expression of patriotism because their cumulative efforts improved society. Third, the youths also drew attention to the class inequalities that had already become apparent in Shekou. They emphasized that outside of monetary compensation, there was no reason to work in Shekou because living conditions in construction site shanties and factory dormitories were often lower than in work units in other cities.[25]

After the delegation returned to Beijing, one of the representatives, Guo Haiyan, wrote a document that analyzed the meeting in terms of "pedagogical norms," casting the Beijing representatives as teachers and the Shekou youth as disrespectful students. The *Shekou Bulletin* published a counteranalysis, arguing that pedagogical conversations between leaders and subordinates were outdated. To them, the country needed more debate between equals. Subsequently, this "Shekou Storm" became a national issue when an editor at *The People's Daily*, Zeng Xianbin, interviewed participants from the meeting. He published a transcript of those interviews that highlighted the diversity of opinions about the direction of reform. Importantly, the piece advocated the Shekou youths' second claim—that by working for wages and market incentives, individuals contributed to society in the aggregate.

Significantly, however, the article downplayed the emerging class issues that the Shekou youth had raised. This rhetorical move highlights the extent of and limits to reform in the SEZ. With respect to political reforms, Zeng Xianbin's emphasis on the political implications of the confrontation was consistent with Beijing's ongoing interpretation of events in Shenzhen as a manifestation of the Chinese polity. Moreover, events were interpreted as the effects of policy decisions. Nevertheless, Zeng Xianbin's article emphasized the kinds of social experimentation that had been taking place in Shekou, where the political field had been opened to include a wider public. In contrast to Beijing leaders' emphasis on obedience and order, Zeng Xianbin's article suggested that the rights, responsibilities, and limitations of leaders were at stake in the Shekou Storm—a controversial model of political reform. With respect to the limits of reform, however, by downplaying emergent forms of economic inequality, the article had removed capitalist exploitation from discussion within the political sphere; labor was represented as a choice, shifting responsibility for working conditions from the state or its representative work units to the individual. In this neoliberal model of work, individuals are exploited by their employers because they have chosen to take a job.

In his statement for the Zeng Xianbin article, China Merchants Shekou

Director Yuan Geng distanced himself from the other interviewees by ignoring the content of the meeting altogether. Instead, he addressed the structure of the meeting, a clear indication that he not only saw the political implications of the Shekou Storm but also understood his role as the one taking responsibility for the event and its fallout. First, Yuan Geng confirmed that in Shekou, where democratic discussion was becoming common, pedagogy as a model of public discourse had no truck with the youth. Second, he stressed that Shekou did not enforce "language crimes," further arguing that even if the youths had said incorrect things, the Chinese Constitution gave citizens the right to hold different opinions. As a leader in Shekou, he stated, it was his responsibility to provide such an environment rather than to enforce a correct line of thought. An editor at the *People's Daily* opined, "The so-called 'Shekou Storm' was not just an incident which happened on that tiny plot of land [Shekou]; it is actually related to or reflects an inevitable and nation-wide debate in the ideological sphere. This debate would have come about sooner or later anyway."[26]

The debate moved to Tiananmen Square in Beijing on April 5, 1989, when students gathered to mourn the passing of Hu Yaobang, who had supported all three Shenzhen heroes in their reforms of key areas of the Maoist system. Liang Xiang had worked to reform the Chinese state apparatus, Yuan Geng had challenged that system via its ministerial apparatus, and Luo Zhengqi had advocated a new model of intellectual citizen. Indeed, the youthful protagonists of the Shekou storm arguably embodied the new intellectual citizen that Luo Zhengqi saw as the product of a reformed pedagogy. Consequently, when Beijing students gathered in Tiananmen to mourn the passing of Hu Yaobang and stayed to protest corruption and call for democratic reforms, they did so against and within experimental political models that were being tested in Shenzhen. A month later, when Mikhail Gorbachev visited Beijing, the students were still protesting. The Soviet Union's leader increased the tension in the standoff between Beijing's leaders and students. Gorbachev's morning visit with Deng resulted in the announcement that relations between the Soviet Union and China were normal. His afternoon visit with Zhao Ziyang revealed the inner power structure in Beijing; although Zhao Ziyang was the general secretary of the CCP, Deng was still the head of state. Indeed, for his opposition to using the military to resolve the standoff between the students and central government, Zhao Ziyang was dismissed and placed under house arrest. On May 20, 1989, party authorities declared martial law and deployed more than 300,000 troops into Beijing.[27]

During the democracy protests and after the June Fourth crackdown, the three Special Zone heroes acted to protect their subordinates against the interests of their Beijing patrons. At the height of the democracy movement, for example, Shenzhen University President Luo Zhengqi wrote a letter to Deng, calling on him to end the governmental impasse and to allow the politburo to exercise their constitutional power. He also demanded for an investigation of Premier Li Peng's abuses of power.[28] Liang Xiang had already transferred from Shenzhen to Hainan in 1986, and in the aftermath of the June fourth massacre, he allowed many of the protest leaders to leave the country by way of Hainan, including Zhao Ziyang's son, Zhao Erjun. Moreover, in Shekou, although Yuan Geng did not actively support the student movement, in the immediate postcrackdown years, China Merchants Shekou accepted the transfer of many intellectuals who—like Zhou Weimin a decade earlier— had been sent home to reflect on their mistakes and write self-criticisms. As a result of their actions, all three men were punished. Both Liang Xiang and Luo Zhengqi were expelled from the party and removed from their official positions. In Shekou, Yuan Geng's de facto position as the head of the Shekou government was revoked when the industrial zone was explicitly incorporated into the Shenzhen Municipality as a street office precinct of Nanshan District. Yuan Geng remained in his position as director and CEO of China Merchants Shekou but had no place in the newly established government apparatus.

Party Secretaries Instead of Heroes

In the years immediately following the 1989 crackdown, Shenzhen again provided Deng with a platform from which to secure his power base in Beijing. During his 1992 Inspection Tour, Deng visited Shenzhen's Guomao Building, which he declared had gone up in "Shenzhen Speed," with one story of the skyscraper built every three days. Moreover, in the revolving restaurant at the top of the building, he sat at a window that overlooked the booming city and reminded the country of the need to adhere to the line, principles, and policies instituted since the Third Plenary Session of the Eleventh Central Committee of the Communist Party of China in December 1978.[29]

The key principle of the 1978 Plenary Session was "one center and two basic points," which referred to the central task of economic construction and the subordinate tasks of adhering to the Four Basic Principles and implementing the Reform and Opening policy.[30] The policies on economic growth reversed the Maoist position of "class struggle," advocating economic liberalization, a political move that had been presaged in the downplaying of class inequality during the Shekou Storm. The Four Basic Principles had given a clear indication that at stake in reform was not only making China rich and powerful but also doing so under the leadership of the CCP.[31] More important, post-1989

Shenzhen was no longer a special zone but rather a space of exceptional economic policies that were understood as permitted by the central government, specifically Deng himself. Thus, in the post-1989 era, the (sometimes) progressive laissez-faire governance of the three Special Zone "heroes" was coded in the state-controlled media as an explicit endorsement for economic development without concomitant political reform.

In the words of an early migrant to Shenzhen, the difference between the government during the 1980s and the government after Deng's 1992 Southern Tour could be described as the difference between "heroes" and party secretaries. On this long-term Shenzhen resident's reading, in contrast to the generation of Special Zone heroes who had taken political responsibility for the wellbeing of their subordinates and, by extension, worked for the wellbeing of the people, the new generation of party secretaries identified with their political patrons, taking responsibility for the political well-being of the party and their own bureaucratic careers. After Liang Xiang left Shenzhen, for example, Li Hao was appointed party secretary of Shenzhen, a position he held from May 1986 through April 1993. Li Hao's most pressing task was to create a more comprehensive municipal apparatus, including the integration of Shekou into the territorial administration of the Shenzhen municipal apparatus.[32] Li Hao laid the groundwork to transform the Shenzhen municipal apparatus from a two-tiered organization of weak links into a three-tiered system with a centralized municipal authority. His successor, Li Youwei, was responsible for further rationalizing the Shenzhen municipal apparatus based on this new structure, which included bringing rural organizations into the urban state. In turn, the state-controlled media began promoting the Shenzhen Municipality as the agent of economic reform, and the Shenzhen "Special Zone" no longer entered national discourse as a space and incubator of recognizable heroes who dared to work for political reform (through the rhetoric of social reform) but rather as an exemplar of successful party policy. Over the next decade, Shenzhen leaders acted to normalize party leadership of Shenzhen, strengthen a bureaucratic apparatus that could achieve this goal, and garner public legitimacy for Jiang Zemin, who had replaced Zhao Ziyang as the general secretary of the CCP.

In late 2012, more than twenty years after the "special decade" ended, Special Zone heroes and their accomplishments reappeared in national politics when the newly elected secretary general of the Chinese Communist Party, Xi Jinping, appropriated and redeployed Shenzhen Special Zone history to push forward anticorruption reforms. In his first speech as secretary general, Xi Jinping cited Yuan Geng, who had once said that "empty talk harms the nation, active work helps everyone." Subsequently, Xi Jinping's first trip as secretary general was to Shenzhen, where his itinerary not only retraced key moments in Deng's historic 1984 and 1992 special zone inspection tours but also included visits to two key symbols of the new Shenzhen—the Qianhai Cooperative Zone and Tencent, the software and digital telecommunications company responsible for hugely successful Chinese social media platforms such as QQ and We Chat. Indeed, it is worth noting that although a Beijing leader had once again come to Shenzhen to promote the city as a national model, this time he recognized corporations and government ministries as "heroes" rather than actual cadres. One of the accomplishments of the early SEZ, for example, had been to redeploy the Cold War Sino-British border— the Bamboo Curtain—to allow for gradual integration of the Chinese national and the global capitalist economies. The new general secretary's tour of the Qianhai Cooperative Zone foregrounded the extent to which Shenzhen and Hong Kong had (at least at the level of social infrastructure) integrated their economies, creating a new form of southern Chinese society. Similarly, the tour of Tencent emphasized how special zone policies had flourished and transformed the economic landscape. Twenty years previously, Deng had celebrated compensatory trade practices. In explicit contrast, the visit to Tencent highlighted the emergence of entrepreneurial high-tech enterprises in Shenzhen. In fact, the visit to Tencent also reminded the country that although Shenzhen was home to many of China's largest privately held corporations, the central government still controlled technologies of social recognition and the production of national models.

During Xi Jinping's inspection tour of 2012, many of Shenzhen's residents privately remembered the other, earlier Shenzhen model. They remembered, for example, that during the 1984 tour, Deng had designated Shenzhen University and its discredited former president as the model of Shenzhen speed. This model emphasized the speedy construction of a new kind of society. In contrast, Deng's 1992 designation of Guomao as an exemplar of the effects of correct policy merely celebrated the rapid development of real estate. Xi Jinping's itinerary validated the 1992 Shenzhen model rather than its earlier form. Ironically (for some), by defying Beijing in the aftermath of the 1989 crackdown, Shenzhen's first generation of leaders transitioned from being state-sponsored heroes to popular local heroes—leaders who had embodied the potential of the state apparatus to truly represent the people. Thus, in Shenzhen, Xi Jinping's endorsement of the Shenzhen model prompted another flurry of calls for and speculation about the rehabilitation of both Liang Xiang and Luo Zhengqi in Shenzhen blogs and social media, as well as a concomitant return to the political values of the first Shenzhen model, when leaders dared to take responsibility and had the political clout to support the independent decisions of their subordinates. Thus reconstituted in and against the current Shenzhen model, this latent, re-remembered Shenzhen model of early heroes and post-Mao heroism defined proper government in representative terms, emphasizing both economic opportunity and local self-determination.

Notes

1. This quotation is from Tang Tsou, "The Historic Change in Direction and Continuity with the Past," *China Quarterly* 98 (1984): 384. Chen Yun was one of the most influential leaders during the 1980s and 1990s. He contributed to the development of Deng's economic strategy and was more directly involved in its implementation than was Deng. For accounts of competing visions for post-Mao China, see David Bachman, "Differing Visions of China's Post-Mao Economy: The Ideas of Chen Yun, Deng, and Zhao Ziyang," *Asian Survey* 26, no. 3 (March 1986): 292–321. For an overview of Deng's role in the Reform and Opening policy, see especially Ezra Vogal, *Deng Xiaoping and the Transformation of China* (Cambridge, MA: Harvard University Press, 2011). For an account the Deng years by his successor, read the English version of "Jiang Zemin's Report at the 15th National Congress of the Communist Party of China," which has been reprinted online on the website of the Federation of American Scientists (http://www.fas.org/news/china/1997/970912-prc.htm).

2. In 1985, Liang Xiang resigned as Shenzhen mayor but he remained its party secretary until 1986.

3. My understanding of the "heroism" of the first generation of Shenzhen leaders is based on three sources of information: (1) my interview notes and research in Shenzhen, which has been ongoing since 1995; (2) published newspaper and online accounts of the history of early Shenzhen; and (3) recent biographies of Liang Xiang and Yuan Geng, which attribute the two leaders with foresight and daring. See especially Zhu Chongshan and Chen Rongguang, *Shenzhen Shizhang* [Shenzhen Mayor Liangxiang] (Guangzhou: Huacheng Publishers, 2011); and Tu Qiao, *Yuan Geng Zhua: Gaige Xianchang, 1978–1984* [The Yuan Geng Story: The Site of Reform, 1978–1984] (Beijing: Zuojia Publishers, 2008). In 2014, a group of civic-minded "old Shekou" residents established the Seminar on Yuan Geng's Ideals (Yuan Geng Lixiang Yanjiuhui). The group has organized several lectures and a birthday memorial for Yuan Geng. More importantly, it has served as an incubator for new social organizations in Shekou, including a community-funded trust that aims to promote community welfare and culture outside the direct purview of either a government or business.

4. See Mary Ann O'Donnell, "Path Breaking: Constructing Gendered Nationalism in the Shenzhen Special Economic Zone," *Positions: East Asian Culture(s) Critique* 7, no. 2 (Fall 1999): 343–75, for an account of the symbolic transformation of People's Liberation Army officers and foot soldiers into architects, engineers, and construction workers of reform. For an account of the gender of labor, see Ngai Pun, *Made in China: Women Factory Workers in a Global Workplace* (Durham, NC: Duke University Press, 2005); and Ching Kwan Lee, *Gender and the South China Miracle: Two Worlds of Factory Women* (Berkeley: University of California Press, 1998). For more on the gender of reform in this volume, see Wong, chapter 9; Florence, chapter 4; and Dong and Cheng, chapter 8.

5. Luo Zhengqi, interview with the author, November 3, 2012.

6. See Susan L. Shirk, *China—Fragile Superpower: How China's Internal Politics Could Derail Its Peaceful Rise* (Oxford: Oxford University Press, 2007), for a nuanced discussion of the role that inter-party politics have played in official policy. See also Tang Tsou, *The Cultural Revolution and Post-Mao Reforms: A Historical Perspective* (Chicago: University of Chicago Press, 1986), for his Cold War–inflected analysis of interparty politics as a winner-take-all situation.

7. Fang Yuan, "The 'Huangque Xingdong' zeng Yingjiu Zhao Ziyang zhi Zi Zhao Erjun yi Jia" [The "Operation Sparrow" Saved Zhao Zeyang's Son, Zhao Erjun and His Family], *Da Ji Yuan* [Epoch Times], posted June 5, 2009, accessed February 27, 2016, http://www.epochtimes.com/b5/9/6/6/n2549694.htm.

8. Chen Yushan and Chen Shaojing, *Yuan Geng zhi Mi* [The Riddle of Yuan Geng] (Guangzhou: Huacheng Publishers, 2005), chapter 4, 108–54.

9. Wan Runnan, "*Da Dao Qing Tian: Luo Zhengqi de liang feng xin* [The High Moral Road: Luo Zhengqi's Two Letters]," essay uploaded to the Shenzhen University intranet, written in honor of Luo Zhengqi's seventy-eighth birthday, March 20, 2012, accessed May, 2012.

10. Promulgated in 1979, the contract responsibility system reduced the role of the state in production decisions and increased the scope of free-market forces and self-management in rural areas. Under the system, peasants were contracted to produce and sell certain quantities of commodities at low official prices to the state. They were then free to sell anything over the contracted amount on the free market. For the history of rural reform in north China, see the articles in Edward Friedman, Paul Pickowicz, and Mark Seldon, *Revolution, Resistance, and Reform in Village China* (New Haven, CT: Yale University Press, 2005). To understand the story in Guangdong, see Helen Siu, *Agents and Victims in South China: Accomplices in Rural Revolution* (New Haven, CT: Yale University Press, 1989); and Anita Chan, *Chen Village: Revolution to Globalization*, 3rd ed. (Berkeley: University of California Press, 2009 [1984]). In 1981, a campaign was launched to extend the contract responsibility system to medium and large industrial enterprises. For discussions of the contract responsibility system, see Andrew G. Walder, *Zouping in Transition: The Process of Reform in Rural North China* (Cambridge, MA: Harvard University Press, 1998); Derong Chen, *Chinese Firms between Hierarchy and Market: The Contract Management Responsibility System in China* (New York: St. Martin's Press, 1995); and Anthony Y. Koo, "The Contract Responsibility System: Transition from a Planned to a Market Economy," *Economic Development and Cultural Change* 38, no. 4 (1990): 797–820.

11. All Shenzhen statistics are available through the Shenzhen Government online website, http://www.sz.gov.cn/tjj/tjj/. Annual reports at http://www.sz.gov.cn/cn/xxgk/.

12. Liu Tianjiu was the CEO of Millie's Holdings, a company that produced women's shoes. His first investment in Shenzhen was formalized on June 16, 1979, when he signed a contract with the Shenzhen Food and Beverages Commission to develop the Bamboo Garden Hotel, which catered Hong Kong visitors who came seeking investment opportunities and gustatory pleasure. Early special zone food streets specialized in dog hot pot and wild game dishes that were unavailable across the border.

13. Li Jing, "Auction Fires First Shot for Land Reform," *Shenzhen Daily*, August 11, 2008, accessed January 2013, http://szdaily.sznews.com/html/2008-08/11/content_294973.htm.

14. From January 23, 1979, through June 17, 1980, Zhang Xunpun served as Shenzhen's party secretary and Gu Hua as mayor. Then, from June 17, 1980, through March 6, 1981, Wu Nansheng was appointed both party secretary and mayor. Nevertheless, Liang Xiang was recognized as the leader who decisively shaped early Shenzhen development and ethos.

15. "Ant theory" was Liang Xiang's most famous metaphor for the new economy, and he

was affectionately called the "ant mayor." On April 4, 2012, for example, the Shenzhen newspaper *Jingbao* published "Shenzhen Qinghuai: Tamen Zou le dan xuehan he linghun hai liu zai zheli [Shenzhen Sentiment: They've Left, but Their Blood, Sweat, and Spirit Remain Here]" by Jia Wenfeng, an article that combined nostalgia for Liang Xiang and an account of ant theory. See http://news.sznews.com/content/2012-04/04/content_6625284.htm.

16. *The General Planning of the Shenzhen Special Economic Zone*, Shenzhen Municipal Government, 1986, p. 1.

17. See Weiwen Huang, chapter 3 in this volume.

18. Xu Luo discusses the importance of the Shekou model to post-Mao China in "The 'Shekou Storm': Changes in the Mentality of Chinese Youth Prior to Tiananmen," *China Quarterly*, no. 142 (June 1995): 541–72.

19. See Mason, chapter 10 in this volume.

20. Yuan Geng, "Shekou de Gaige Changshi Tansuo [Reform Attempts and Explorations in Shekou]," in Chen Yushan and Chen Shaojing, *Yuan Geng zhi Mi* [The Riddle of Yuan Geng] (Guangzhou: Huacheng Publishers, 2005), 289–94.

21. Ju Tianxiang, *Zhiyi yu Qishi: Yuan Geng zai Shekou Jishi* [Righteousness and Enlightenment: Records of Yuan Geng in Shekou] (Beijing: Zhongguo Qingnian Publishers, 1998), 223–24.

22. Yan Jiaqi, "Bo Yibo Zhongnanhai Fabiao Qinli Ji [Personal Recollections of Bo Yibo's Violence in Zhongnanhai]," published on Cao Changqing's News Blog, April 4, 2012, http://caochangqing.com/gb/newsdisp.php?News_ID=2782.

23. Luo Zhengqi, "1987 Opening Ceremony Address to Incoming Class of Shenzhen University," transcript uploaded to Shenzhen University intranet on October 19, 2008, by kevinlee, http://mcs.szu.edu.cn/Forum/49742.

24. David Graeber, *Toward an Anthropological Theory of Value: The False Coin of Our Own Dreams* (London: Palgrave Macmillan, 2001), 254.

25. Xu, "'Shekou Storm,'" 543–44.

26. Xu, "'Shekou Storm,'" 545.

27. The Tiananmen crackdown, or June Fourth incident, is considered one of the pivotal moments in post-Mao Chinese society. Roderick Macfarquhar provides lucid analyses in "The Road to Tiananmen: Chinese Politics in the 1980s," in *The Politics of China: Sixty Years of the People's Republic of China* (Oxford: Cambridge University Press, 2011) chapter 5, 337–467. Richard Gordon and Carma Hinton's movie *The Gate of Heavenly Peace* (1995) and associated website (http://www.tsquare.tv) provide more direct access to the youthful leaders of the 1989 democracy movement.

28. Wan Runnan, "*Da Dao Qing Tian: Luo Zhengqi de liang feng xin* [The High Moral Road: Luo Zhengqi's Two Letters]," essay uploaded to the Shenzhen University intranet, written in honor of Luo Zhengqi's seventy-eighth birthday, March 20, 2012, accessed May 2012.

29. This was the meeting in which Deng first took explicit control over the party and government apparatus, promoting the need for economic and social liberalization.

30. Shenzhen Central Ministry of Information, *Deng Xiaoping and Shenzhen: Spring 1992 (Deng Xiaoping yu Shenzhen: Yi jiu jiu er Chun)* (Shenzhen: Haitian Publishing House, 1992).

31. Fang Lizhi, "The Real Deng," *New York Review of Books*, November 10, 2011, accessed electronically January 2013.

32. The inclusion of Shekou in the Guangdong self-governing commercial zone at the end of 2014 has raised both new questions and speculation about the continuing role of China

Merchants in social and political reform. Through its inclusion in the self-governing commercial zone, Shekou has been politically elevated to a position at least equal to that of the Shenzhen Municipality. Moreover, with its Shekou and Qianhai holdings, China Merchants is now the largest property holder in the self-governing commercial zone. At the time of this writing, no one I have spoken with knows what any of this will mean with respect to the practical politics of social reform.

3

The Tripartite Origins of Shenzhen: Beijing, Hong Kong, and Bao'an

WEIWEN HUANG

Shenzhen Speed

Located just north of Hong Kong, Bao'an County was elevated to the status of Shenzhen Municipality in 1979. In 1980, the Special Economic Zone (SEZ) was established as a window and laboratory for China's experiments with the national Reform and Opening policy. Over the course of the next thirty years, Shenzhen exploded, averaging a 12 percent annual population growth rate. In 1980, for example, the population of Shenzhen was 340,000, and by 2010, the official census population reached 10.36 million, while the administrative population (*guanli renkou*) had already risen to 14 million. In the same period, Shenzhen's GDP grew at an average rate of 30 percent, from US$42.9 million to US$152 billion (in 2012 dollars). More precisely, from 1980 through 1995, Shenzhen's GDP grew at an average annual rate of 47 percent and then maintained an average annual growth of 19 percent between 1996 and 2010.[1] But it was the rate of building construction that was the most emblematic of Shenzhen's growth rate. The historical expression "Shenzhen Speed" referred to the construction of the Shenzhen International Trade Building (*Guomao*), for which workers put up one floor every three days, completing the fifty-three-story skyscraper in thirty-seven months.[2] In a word, Shenzhen "exploded" with the release of the economic and social productive forces that the planned economy and extreme leftist policies had suppressed for years.[3]

The concept of "space-time compression" neatly summarizes the general processes characterizing the Shenzhen boom. Space-time compression refers to social processes such as industrialization, mechanization, and technical advances that have reduced the time necessary to produce goods, travel from one place to another, and transmit information.[4] In fact, Shenzhen Speed was achieved by intensifying elements—rural-urban migration, manufacturing for export, and the emplacement of telecommunications infrastructure, for example—that have defined industrial-capitalist urban booms elsewhere in the world; the Shenzhen difference has been the scale at which this transformation occurred and was taken up in other parts of China. Concomitantly, the rapid transformation of Shenzhen's infrastructure constituted an important competitive advantage in the global economy that emerged after 1980, with its global epicenter shifting from the United States to East Asia.[5] The speed with which Shenzhen was built meant that the city could—and did—reorient its economy with each comprehensive urban plan, adjusting construction with respect to changing circumstances. In 1982, the urban plan proposed relatively independent clusters of development that were connected but functionally separate; in 1986, it proposed the building of an integrated, modern, special zone city with a focus on industrial manufacturing; in 1996, it elaborated those earlier goals in order to transform all of Shenzhen into a modern SEZ and global city; and in 2010, Shenzhen aspired to claim a higher percentage of value-added production in IT, finance, and life sciences. Each iteration of the master plan represented the ongoing consolidation of Shenzhen as an entity that could be directed and deployed in order to achieve social and economic goals. Indeed, the plan boldly asserted the importance of urban planning in cities and their concomitant urbanisms.

Interestingly, despite the speed and scale of its urban development, Shenzhen has not experienced housing shortages, serious failures to provide basic infrastructure for most residents, or any other typical symptom of megacity urban blight. In contrast, over the past thirty years, cities as diverse as Mumbai, Rio de Janeiro, and Port-au-Prince have produced urban environments characterized by enclaves of extreme exclusion from the urban grid even as older industrial cities have decayed. The paradoxical appearance of urban ghettos around the world where residents do not have reliable access to potable water, sanitation services, and healthy neighborhoods—even as their fellow citizens inhabit gated enclaves of material wealth—contextualizes how China and the world have evaluated Shenzhen as a developing city enjoying the benefits of economic modernization without any of its stereotypical problems. Domestically, Shenzhen has received national recognition as a national garden city and as a model city, in addition to numerous other awards. Internationally, in 1992, the United Nations awarded the Shenzhen Ministry of Housing an honorable mention in its Habitat Scroll of Honor Awards, and in 1999, the International Union of Architects (UIA) nominated Shenzhen for an honorable mention in the Abercrombie Urban Planning Award. Thus Shenzhen's experience has not only influenced the scale and pace of urbanization throughout China but also been promoted as a model for other developing cities to emulate.

This chapter asks the question, "What were the mechanisms and logic that enabled Shenzhen's high-speed and generally successful urban development?" This chapter argues that the answer to this question can be found through an investigation of the heterogeneous forces—political, economic, and local—that came into play in the early history of the SEZ. It is worth mentioning that in both local urban practice and academic (especially economic geography) debates, these forces have been represented and imagined as originating in particular places on a shifting map. "Beijing" is used as a metonym to stand in for national power, "Hong Kong" is used to stand in for international capital and ideas that came into Shenzhen both from and by way of its southern neighbor, and "Bao'an" is used to stand in for the residents and collective system that was in place before the establishment of Shenzhen. In other words, the urban construction of Shenzhen remains a complicated process of unmaking Bao'an, planning Shenzhen, and responding to changes in the world system, which, as we will see, begs the question of just how portable the so-called Shenzhen model may actually be.

Political Restructuring

Within the regional context, it is useful to think of Shenzhen as a site where colonialism and collectivism were redeployed to achieve both political and economic goals. From the perspective of Beijing, for example, the transformation of Bao'an County into Shenzhen Municipality entailed the dismantling of the socialist planned economy and its restructuring into a more efficient, productive apparatus to meet national goals and to secure Communist party power in the immediate post-Mao era.[6] From the perspective of Hong Kong, the opening of Shenzhen to foreign investment and international trade reactivated a dense regional economy that socialism had disrupted, allowing for the reemergence of the Pearl River Delta as an important node in the global economy. Placing Beijing in rhetorical opposition to Hong Kong makes salient the multiplicity of stakeholders and competing claims to Shenzhen.

In 1898, Qing and British officials established the Shenzhen River as the border between Shenzhen's historical precursor, Xin'an County (which was renamed Bao'an County in 1914) and the British colony of Hong Kong, which implemented a free-trade and market economy (figure 3). The establishment of Hong Kong as a colonial entrepôt both intensified regional marketing and production networks that had historically focused on Guangzhou and allowed the British to extract the surpluses that this system generated. With the establishment of the People's Republic of China (PRC) in 1949, the economic situation on the banks of the Shenzhen River became radically different. On one side of the Shenzhen River was the British colony of Hong Kong. On the other side was Bao'an County, which implemented collective agriculture and fishing under the Chinese planned economy. While Hong Kong developed within the Cold War expansion of the global economy, the PRC implemented the socialist centrally planned economy, including an attempt to make a "great leap" over the early stages of Communism that Marx had described. Nevertheless, at the level of infrastructure development, the two areas were not as separate as the barbed wire of the Bamboo Curtain made them appear. Central policy makers, for example, began integrating Shenzhen and Hong Kong infrastructure through cross-border projects such as the East River Shenzhen Waterworks Project, which began in 1959 under the personal direction of Zhou Enlai.

Nevertheless, from the point of view of Bao'an residents, the pull of Hong Kong society was difficult to ignore, because at one point, Hong Kong residents earned as much as one hundred times the income of Bao'an residents. This difference created a thirty-year economic pull, during which time more than one million southern Chinese left for Hong Kong illegally—risking death by drowning, being shot, or incarceration—all in order to enjoy the colony's relative prosperity. Local origin stories about the rise of Shenzhen emphasize that before the establishment of the SEZ, only women and children remained in decrepit villages because all the young, able-bodied men had escaped to Hong Kong in order to make better lives for themselves. In fact, several Bao'an villages, including Maliao, Jingdu, and Gaoling, became "ghost villages" because all their residents opted to flee to Hong Kong.

Upon his third rise to power, Deng Xiaoping opined that the problem of economic refugees fleeing from Bao'an County to Hong Kong was a result of poor policies, and therefore it was useless to rely on border patrols and military measures to solve it. A pragmatic politician, he also realized that the economic gap between China and the developed world had created a situation of increasing political risk. In 1978, Deng Xiaoping thus implemented the Reform and Opening policy. Bao'an County was selected to be the "window" for China's first post-Mao experiments in opening to the outside, bringing in Hong Kong capital, technology, and management know-how. In addition, Shenzhen would be the test site for transforming the planned economy elsewhere in the country. Consequently, the administrative stutter steps that preceded this transformation occurred almost immediately after Deng Xiaoping's rise. In 1979, Bao'an County was elevated to Shenzhen Municipality. In 1980, an 84.5 kilometer administrative boundary, known as the Second Line (*erxian*), divided the municipality into the SEZ, which occupied 327.5 km^2 of

THE TRIPARTITE ORIGINS OF SHENZHEN   69

the municipality and the rest. New Bao'an County was formally reinstated in 1981. Importantly, the SEZ was not the first effort to reform Bao'an; the China Merchants Shekou Industrial Park was first by several months. The timeline below points to the scale and type of rural-urban contradictions that constituted Shenzhen Municipality and its subsequent redistricting as the Shenzhen SEZ and New Bao'an County during the years 1978–82, culminating in the promulgation of the 1982 plan for the Shenzhen SEZ. Shenzhen celebrates its birthday on June 26, 1980, the day the Regulations on SEZs in Guangdong Province were approved by the Standing Committee of the National People's Congress.

Timeline: The Establishment of Shenzhen Municipality and Its Subsequent Division into the Special Economic Zone and New Bao'an County, 1978–82

March 1979: Bao'an County is elevated to Shenzhen Municipality.
July 1979: Shekou Industrial Park is established under the direction of the China Merchants Bureau, a state-owned enterprise of the national Ministry of Transportation.
August 1980: Shenzhen SEZ is established (327.5 km^2), consisting of Shenzhen Market, Shatoujiao Market, and four communes—Fucheng, Futian, Nantou, and Shekou.
October 1981: Bao'an County is restored as an administrative division of Shenzhen Municipality and is formally renamed New Bao'an County (1,625 km^2). It comprises seventeen communes: Dapeng, Kuiyong, Pingshan, Longgang, Pingdi, Henggang, Pinghu, Buji, Guanlan, Longhua, Shiyan, Xixiang, Shajing, Fuyong, Songgang, Gongming, and Guangming Dairy.

The 1979 elevation of Bao'an County to Shenzhen Municipality and its subsequent restructuring in 1981 as New Bao'an County and the Shenzhen SEZ have had critical consequences for urbanization in Shenzhen. First, although political restructuring facilitated increasing forms of decentralization, it did not recognize the political autonomy of local communities, such as historic villages or emergent neighborhoods around construction sites and industrial parks. Instead, the organization of Shenzhen Municipality into the SEZ and New Bao'an County reproduced the Maoist division between urban work units and rural collectives as the basic structures of social administration and regulation. Second, the central government mobilized national and provincial ministries to invest in the SEZ while ignoring New Bao'an. In turn, these state-owned enterprises enjoyed privileged access to national investment capital, technical know-how, and natural resources, reproducing

Beijing's hegemony in situ, legitimating reforms within the borders of the SEZ, and effectively delegitimizing subsequent urbanization in New Bao'an County. Third, in the absence of or in addition to domestic investment, both New Bao'an and the SEZ sought international capital and technical investment in order to jumpstart their enterprises. During these early years, the majority of external investment came from or through Hong Kong. Consequently, the so called Shenzhen speed was not simply the result of step-by-step political "loosening of restrictions" on different sectors of the economy and the concomitant releasing of productive forces, in this case allowing industrial manufacturing (formerly an urban activity) to be located in a rural area. Rather, and more precisely, the restraints governing the Cold War separation of Beijing and Hong Kong were destroyed, and then differently placed rural and urban actors and enterprises achieved new integrations as New Bao'an and the SEZ, respectively.

Within the context of Shenzhen urban planning, the division of Shenzhen Municipality into New Bao'an County and the SEZ was an unintended experiment on how to reform the *hukou*, or household registration system. Beginning in the 1950s, the *hukou* system allowed the central government in Beijing to plan and regulate the nation by restricting population movement between rural and urban areas and relocating redundant urban residents to rural areas. The government achieved this goal through policies that fed, employed, educated, and provided health care to individual Chinese citizens based on their legal address. For example, an individual with a Bao'an County household registration could only have a house in Bao'an, be allocated grain and other staples such as cooking oil and soy sauce in Bao'an, be employed by a local commune or work brigade, send his or her children to school in Bao'an, and receive medical care in a Bao'an clinic or hospital. Importantly, the quality of housing, food, employment, education, and medical care in cities was significantly higher than in rural areas such as Bao'an County; the *hukou* system not only allowed for national economic planning but also institutionalized inequalities between rural and urban China. In other words, the structural division of Shenzhen Municipality into the SEZ and New Bao'an County reformulated the Maoist contradiction between rural and urban China. Importantly, although Shenzhen now has some of the most progressive *hukou* laws in the country and former Bao'an County residents have Shenzhen household registration, nevertheless, the Maoist segregation of rural and urban societies continues to inform urban planning (and by extension, governance) in Shenzhen, where urbanized villages have been both a condition for rapid urbanization and a symptom of the inequalities inherent to this process.

Beijing: Delineation and the Enclave Development Model

In 1992, the song "The Story of Spring" celebrated Deng Xiaoping's second visit to Shenzhen and his support for establishing SEZs with the lyric, "An old man painted a circle at the edge of the Southern Seas." This allegory for Beijing's power to determine the life conditions of Chinese people referred to a 126 kilometer wire cordon with seven checkpoints, which segregated Shenzhen from the rest of China.[7] The common name for the border was "the Second Line," as opposed to the "First Line" that separated Shenzhen from Hong Kong. In fact, according to the memory of the director of the Shekou Industrial Zone, Yuan Geng, there was an actual incident of a Beijing leader drawing a line on a map to establish the borders of an area. In November 1978, Yuan Geng approached Vice General Secretary Li Xiannian in Beijing and asked him to approve the Shekou area (in Shenzhen) for the China Merchants to industrialize. As Yuan Geng recounted the story, he showed Li Xiannian a map of Hong Kong and pointed to the edge of the map, at Shekou, which was a harbor located on the southern tip of the Nantou Peninsula. Li Xiannian took out a pen and drew a line, saying, "Take this peninsula!" The area was more than 80 km^2, and Yuan Geng was reluctant to take that much responsibility. Instead, drawing a smaller circle, he assumed responsibility for Shekou, an area that originally was 2.14 km^2 and surrounded the harbor. Nevertheless, that map remains in the archive of the China Merchants in Shekou today, the two circles drawn in Beijing still apparent, delineating China's first experimental field or experimental management area, the China Merchants Shekou Industrial Zone (figure 4).[8]

The song "Story of Spring" and its version of the story of Shekou's political origins illustrate how the state-controlled media represented political restructuring in Shenzhen. Here, the key point is that the paramount leader Deng Xiaoping was credited with the delineation of a "circle" or enclave drawn by his subordinates as an ad hoc planning strategy, and then the work of local leaders and urban planners became filling in the circles that Beijing and its local representatives drew and redrew on the changing map of Shenzhen. This model of delineating circles and then allocating the corresponding tracts of land to ministries, national enterprises, and even commercial real estate developers was a deliberate policy aimed at attracting capital and producing quick results. In some cases, enterprises and commercial developers were entrusted with the management of social institutions and services, functions that elsewhere in China were clearly the responsibility of government. In retrospect, as Shenzhen's economy boomed, post facto recognition by the state-controlled media recast drawing circles on a map as an intentional and therefore legitimate policy in action. Thus Deng Xiaoping came to be revered as Shenzhen's master planner.

In addition to the larger circle created by the Second Line and the smaller circle of the Shekou Industrial Zone, circles of various sizes and densities came to fill Shenzhen. Consider, for example, the developmental history of the Nantou Peninsula, which is located in western Shenzhen and extends into the Pearl River. Shekou is located at the southernmost tip of the Nantou Peninsula. In 1979, the China Merchants planned its territory at a human scale with an integrated network of industrial, residential, and consumption areas. In 1983, Shenzhen University was allocated 2.9 km^2 at the northern approach from the mainland onto the peninsula. In 1984, in the area between Shekou and Shenzhen University, roughly 23 km^2 was allocated to the Southern China Petroleum Company (Nanyou) for joint ventures. Development in Nanyou was chaotic and lacked a surrounding-connection traffic network. In 1989, the independence of enclaves on the Nantou Peninsula was reinforced when the Shenzhen University campus was segregated from the surrounding area with a wall and guarded gates.

As we move across the rest of the municipal territory, we discover that the enclave logic illustrated by the development of the Nantou Peninsula was repeated throughout Shenzhen. From the northern gate of Shenzhen University, we step off the Nantou Peninsula onto the mainland. In 1985, the 10 km^2 located along the eastern border of Shenzhen University was designated the Shenzhen Hi-Tech Park. This area was planned to support research and development of high-technology products and was designed with a low population density, which would be dispersed across a mono-functional space of broad streets and wide greenbelts. The same year that the technology park was established, its eastern neighbor, the 12 km^2 Shahe Farm, was subdivided into the Overseas Chinese Town (OCT; 5 km^2) and Baishizhou (7 km^2). Baishizhou became one of Shenzhen's largest urban villages, and the OCT developed commercial real estate. Its development strategy reproduced enclave delineation, as both the OCT and Baishizhou were further partitioned into smaller areas for theme parks, bedroom communities, and limited industrialization.

Other Shenzhen delineated circles include the Shangbu Industrial Park, which was developed in 1980 by national ministries of electronics, aviation, tools, and textiles, and the Bagualing Industrial Park, which was developed by local engineering companies in the same year. In addition, the national government financed the development of large-scale residential circles, including Yuanling (1983), Bashaling (1987), and Binhe (1989). Since the first land auctions were held in Shenzhen in 1987, real estate development companies have created even more tightly enclosed enclaves and restricted circles in the form of commercial property. In 1989, the Tian'an Industrial Park was developed with a Hong Kong joint venture. The 1.4 km² Futian Bonded Area was a self-delineated circle started in 1991. The national government continued to invest in Shenzhen after the 1992 Southern Tour, delineating the first and second Lianhua Villages (1992), North Lianhua Village (1993), Yitian Village (1996), and Meilin Village (1998).

The planning goals, functions, size, and morphology of each of Shenzhen's constitutive enclaves were diverse, resulting from their independent planning and development. These concentrations of diverse resources and development strategies stimulated rapid development. However, over time it became apparent that the relative independence of all these enclaves negated the effectiveness of an overall urban plan for the city. For example, the enclaves had different traffic networks and some did not allow for easy passage between and through enclaves, creating a situation that further isolated those living even in adjacent enclaves. One result was increasing reliance on motor vehicles and highways to travel from circle to circle even as inhabitants increasingly sought ways to circumvent intermediary enclaves on their way to a destination. These areas have since become subject to integration projects that override their initial independence. Shenzhen Municipality, for example, planned and implemented a citywide transportation network called the "seven horizontal and thirteen vertical" freeways and overpasses. The project goal was to further facilitate motor vehicle movement between enclaves. Ironically, however, given the internal layouts of the enclaves, the highway and overpass system had the effect of cutting the city into even smaller enclaves. The most glaring example is how the Binhai expressway has cut the city off from its coastline.

The household registration system both depended on and reinforced this strategy of delineating enclaves. Shenzhen's official population was based on the *hukou* system; ostensibly, the city only planned for residents with Shenzhen *hukou*. Consequently, in official planning, population estimates were based on the extant and projected *hukou* population figures despite the fact that in everyday practice, Shenzhen officials were aware that the number of unregistered residents in the SEZ (let alone in New Bao'an County) far exceeded the number of residents with Shenzhen *hukou*. The discrepancy between *hukou* and actual population figures highlights the dysfunctional role that abstract, proscriptive planning from Beijing has often played in Shenzhen's urbanization. Consider, for example, the massive rural-urban migration that provided the labor for constructing the city and manning the factories. This large and undocumented population has remained one of Shenzhen's most pressing social problems—how does one simultaneously provide for and ignore a surplus population of several million people?

Moreover, successive iterations of the Shenzhen Comprehensive Urban Plan ignored both local rural areas (New Bao'an County and the new villages) and rural migrants without Shenzhen *hukou*. In fact, the 1982 and 1986 comprehensive plans only addressed the territory within the SEZ and excluded the village settlements (including land holdings) within it because these areas were designated as "rural" (figure 5). This discrepancy between proscriptive planning and the actual population settlement and use of space repeatedly made it difficult (if not impossible) to design and implement a plan for the entire city, because the data only represented a small percentage of the population and total area of the city. This strategy of ignoring rural areas continued even after New Bao'an County was rezoned as two urban districts—Bao'an and Longgang—within Shenzhen Municipality. During the first and second decades of the Shenzhen experiment, for example, the discrepancy between the projected *hukou* population and actual population was vast. In 1985, it was estimated that the 2000 *hukou* population of the SEZ would be 1.1 million, while in 1996 it was projected that in 2010, the SEZ *hukou* population would have grown to 4.3 million. In 1996, it was also estimated that in 2000, the *hukou* population of Shenzhen Municipality would be 4 million. However, according to 2000 census data, the population of Shenzhen Municipality was 2.56 million within the SEZ and another 7.01 million in Bao'an and Longgang Districts (formerly New Bao'an County). Here is the important social conundrum that enclaves and *hukou* have created for Shenzhen: Increasing areas of land have been segregated for use by an increasingly smaller percentage of the population. In turn, more and more migrants have had to make do in ever-smaller enclaves composed of residual rural holdings.

Shenzhen's bounded communities derived from the legacy of central planning for residential modernism and communist work unit compounds. Importantly, both modern residential areas and communist work units assumed single functions, homogenous populations, and relative privacy. There was insufficient densification, diversity, and accessible street networks to allow for microcirculation and pedestrian street life. In 1984, during the creation of Shenzhen's first master plan, planners from the China Academy of Urban Planning and Design in Beijing advocated the clusters model for the urban structure of the Shenzhen SEZ. In fact, this model was another form of circle delineation. The roots of the urban clusters structure can be traced back to 1958, when the master plan for Beijing included the "separate clusters model," in which large greenbelts were used to separate different development groups.

These greenbelts were meant to provide the rapidly expanding city with fresh air and natural spaces as well as preempt certain forms of urban blight. In 2005, all the green circles in Shenzhen became one big one—the control line for ecologic preservation, covering half of its 1,954 km². Delineating this particular circle required courage and foresight to be successfully implemented, because it aimed to integrate areas that had been administered by competing levels of government or state-owned enterprises and were thus considered bureau resources rather than municipal resources. Moreover, with respect to Shenzhen's complete development and the rapid exhaustion of land resources, the value of this circle was immense. The exception to this kind of development was the spontaneous emergence of new villages, which will be discussed in part 3. For now, it is important to note that this top-down model of delineating circles for independent planning and development was at odds with Shenzhen's actual population and infrastructure needs.

Hong Kong: Selectively Loosening Constraints

When we track urban development in Shenzhen, we see that connections with Hong Kong have remained one of the most decisive geographic factors for Shenzhen's development. In fact, anyone who arrived in Shenzhen during the 1980s and 1990s felt the pervasive influence of Hong Kong on the everyday life of Shenzhen residents. Migrants who did not meet the criteria for a travel pass to Hong Kong could apply at the local police station for a special pass to enter Zhong-Ying Street (Sino-British Street) in Shatoujiao, in the eastern section of Shenzhen. Zhong-Ying Street was, as its name suggests, a road that divided China from the British colony of Hong Kong. On Zhong-Ying Street, it was possible to purchase goods available in Hong Kong but not in China, as well as to experience something of Hong Kong life, such as watching R-rated movies. In the 1980s context of the relatively backward and isolated PRC, the "Hong Kongification" of everyday life in Shenzhen was simultaneously stimulating and alarming. It was also the product of allowing common Hong Kong business practices such as joint ventures, shareholding ownership, stock exchanging, public bidding for engineering projects, piecework incentives, paid use of land, and open job recruitment to be openly practiced in Shenzhen.

With the establishment of Shekou and the SEZ, economic and government protocols and institutions that were common in Hong Kong and elsewhere became the object of large-scale social experimentation in Shenzhen. Five major sites of political loosening legitimized these capitalist practices in Shenzhen. First, under the household responsibility system, farmers were permitted to farm their land and sell produce in markets. This loosening had the concomitant result of freeing up labor that had been restricted to agricultural production. Second, factories implemented a reward system for meeting targets and did not cap the amount of workers' remunerations. Third, entrepreneurs were permitted to organize production and outlays in order to meet market demands. Fourth, restrictions on market activities were eased, allowing for the relative free trade and pricing of production and consumption goods, including land. Fifth, restrictions on capital were loosened and investment was no longer characterized as leading to exploitation. As each political constrained was loosened, the suppressed economic and social forces could be directed toward production. At the same time, capital and technology that had accumulated in Hong Kong became the means through which China's vast reserves of cheap labor and land resources were sutured to the international economy. In this sense, China's suppressed productive forces found their first globalizing "blow out" in Shenzhen.

During the process of relaxing constraints, for example, the SEZ introduced economic and government tools such as the 3+C model. The 3+C model referred to Shenzhen's intermediary role in global processing chains and dispersed investment strategies. The C referred to compensation trade, in which incoming investments would be repaid from the revenues generated by that investment. The 3 referred to three aspects of the processing trade and assembly manufacturing, a business activity of importing all or part of the required raw and auxiliary materials, parts and components, and accessories and packaging materials from abroad in bulk and reexporting the finished products after processing or assembly. In terms of global production, China imported intermediary goods from surrounding Asian countries (by way of Shenzhen) and then reexported finished goods to North America and the European Union (by way of Hong Kong). The 3+C model represented the introduction of "flying geese formation" capital accumulation strategies into Chinese territory.[9] Within Japanese discourse of capitalist expansion, the flying geese formation hierarchically integrated and subordinated East Asian economies to Japanese investment (the first goose) and then ranked investments from Korean, Taiwanese, Singaporean, and Hong Kong firms. Flying geese formation practices in East Asia enabled Shenzhen to borrow production methods that had already been tested elsewhere, significantly lowering the inherent costs of trial and error. However, unlike Japan, which needed to externalize its investment strategy because it had limited territory, China deployed Shenzhen to both join global production chains and establish similar production chains domestically, assuming the position of the first goose with respect to the Chinese interior. Within China's vernacular discourse of capitalist expansion during the early reform era, however, Shenzhen's function was metaphorically described as being a "window" or a "laboratory," which rhetorically elided the hierarchy expressed in the Japanese model by emphasizing instead the state's agency in "opening" a window and "conducting" experiments.

Three historic points highlight the importance of the 3+C model of manufacturing in Shekou and the SEZ with respect to the emergence of Township and Village Enterprises (TVEs) first in Shenzhen (especially New Bao'an County) and then throughout China. First, processing trade and assembly manufacturing were relatively simple and therefore easily copied systems of industrial production. During the 1980s, even as state-owned enterprises were building 3+C industrial parks in Shekou and the SEZ, neighboring villages (still technically considered rural work teams, or production brigades) used their land to attract investment capital to build 3+C industrial parks. Second, after Deng Xiaoping's 1992 Southern Tour, when economic reforms were introduced throughout the country, 3+C manufacturing was the easiest "reform" to export because it required relatively small levels of capital investment, a disciplined but relatively uneducated workforce, and access to container ports that were located in Shenzhen but most importantly in Hong Kong. TVEs were built both along and in anticipation of the roads and railways that integrated (and would connect) Shenzhen to the port of Hong Kong. In fact, one of the largest production areas in early Shenzhen was located in Buji Township in New Bao'an County. Located directly north of the Shenzhen–Hong Kong border at Luohu, Buji Township and its constituent villages had both railway and road access to the port of Hong Kong. Third, as Shenzhen upgraded its economic base to secure higher value-added production in financial, IT, and biotech sectors of the economy, much of this manufacturing moved further north to Dongguan, where TVEs flourished along the region's expanding transportation network and Shenzhen's expanding port system, which has increasingly competed with Hong Kong to control the logistical flow of manufactured goods from the region to the rest of the world.

The answer to the ongoing question of whether the "Shenzhen Miracle" can be reproduced elsewhere is no. Shenzhen was the exceptional product of an exceptional time, place, and demographic; it is impossible to meet all these conditions in either contemporary China or other places in the world. The temporal predicates of the Shenzhen Miracle were the contemporaneous redistribution of global production and China's Reform and Opening policy. Cold War isolation had produced the extreme discrepancy between the social economic systems in Bao'an County and neighboring Hong Kong. What's more, although the economic differences between the SEZ and

New Bao'an County—and subsequently between Shenzhen and Dongguan municipalities—created relative production advantages, nevertheless, these discrepancies have never achieved the level that existed between Bao'an County and Hong Kong at the end of the Cold War. In addition, the human context of the Shenzhen Miracle was importantly diverse. In the late 1970s, China had a group of leaders who had experienced the Cultural Revolution but remained idealistic about social transformation. Locally, Bao'an farmers supported the expropriation of land by city government by transferring their land at cheap prices. Moreover, the country also had large reserves of manual labor and well-trained technical workers, including engineering corps, for example, who were eager to work in Shenzhen because, however "cheap" they were on the international market, their wages and salaries were the highest levels of compensation available in China at the time. At the level of micro-implementation within Shenzhen, a variety of decentralized (and usually unregulated) experiments could be located within enclaves that had been allocated to a state-owned enterprise, a governmental industrial park, or a township or village enterprise area without direct governmental regulation.

The political decision to systematically open Shenzhen via Hong Kong to global exchanges of resources, commodities, information, and ideas not only sent shockwaves through the Chinese system but also amplified these effects, creating an urban space based on a market economy as well as structuring new functions and development. The resulting city, Shenzhen, should be regarded as a dissipative system with open, nonequilibrium fluctuation. Importantly, as early research on SEZ developmental possibilities has emphasized, Hong Kong could influence Shenzhen because the two cities shared kinship and trade networks as well as culture and history. The colonial division of what had been one geopolitical entity into two distinct regions and subsequent segregation of these two regions during the Cold War resulted in the increasing inequality between these two areas, including population loss as Bao'an locals fled to neighboring Hong Kong. However, the decision to allow for selective integration of the two societies immediately reversed this process. Elements of Hong Kong society that were directly used in Shenzhen ranged from surplus capital to the relocation of manufacturing from the New Territories to the SEZ. In addition, the SEZ also adopted Hong Kong protocols for purchase ordering, technical and management experience, as well as government policies. With the increasing influx of these Hong Kong resources, the persistent flow of people into Hong Kong that had characterized the Mao era was immediately transformed into a structure that countered the dissipation of resources into one that could sustain rapid evolution and sustained growth.[10]

The SEZ was a rare case of a city with both an international and a domestic border—the First Line separated it from Hong Kong and the Second Line separated it from the rest of China. How these two borders differently regulated diverse "flows" of people, materials, information, and capital were the conditions of possibility for urbanization in Shenzhen. In a 2007 paper on the Shenzhen–Hong Kong border, I used the figure of a hydraulic power plant to metaphorically describe the flows across the borders, because "the larger the drop height of a river, the more efficiently it generates power to convert into electricity."[11] This thought experiment allows us to think of the SEZ as functioning as a link in a series of locks in a shipping canal. At the respective locks along the First and Second Lines, the flows of people, objects, and water—indeed every flow between uneven landscapes—could be linked and exchanged. If we make sections along the First and Second Lines, we see clearly the flows across these two borders. There were seventeen checkpoints for crossing the First Line. In one year, there were 167 million human crossings and 15.37 million vehicular crossings, including semitrucks and cargo ships. There were 400 million human crossings and 125 million vehicular crossings across the Second Line. As the largest land port for shipping people and goods in China, the Shenzhen–Hong Kong border at Huanggang is open twenty-four hours every day, like a river that never stops flowing. At the same time, the First Line created its own commercial opportunities, including customs, agents, parking, gas stations, and a large service industry.

When we examine the layout of Shenzhen from the perspective of cultural geography, we can clearly see traces of the entropy flows and energy dissipation between Shenzhen and Hong Kong. People cross the border from Shenzhen into Hong Kong for tourism and business, to study abroad, purchase designer goods, give birth, watch censored movies, make international flights, and work in the sex industry.[12] From Hong Kong to Shenzhen, people cross to invest; work; practice Chinese opera; retire; purchase pirated DVDs and other pirated goods, books, and vegetables; as well as enjoy nightlife or their "second family" life. Moreover, Shenzhen–Hong Kong traffic, resources, and energy distribution determined the core area and hotspots of each era of development in Shenzhen. For example, during the early 1980s, the primary areas of development were Luohu, which is near the local station of the Guangzhou-Kowloon railway; Shekou, which is next to the extant water ports that connect settlements in Shenzhen and Hong Kong; and Shatoujiao, which is located along Zhong-Ying Street. Since the 1990s, this development pattern has continued. Key sites include Futian, which is next to Huanggang Port, China's largest land border crossing and the subway/railway connection between the two cities; urban plans for peripheral Longhua, which is a stop on the Guangzhou-Shenzhen-Hong Kong high-speed railway; Houhai, which is near the Western Corridor Bridge over Shenzhen Bay and its concomitant checkpoint; and Qianhai, which is the site of the planned Shenzhen-Hong Kong airports connection and the Shen Kong Cooperation Zone. These sites of integration with Hong Kong transportation and logistics networks stretched between Mainland China and Hong Kong, informing the development of each stage in Shenzhen's development as well as the conditions through which land use changed.

New Bao'an: Local Self-Organization

Shenzhen Municipality used urban planning, land appropriation, and repatriation in order to implement its policy of delineation and enclave development. The goal of this practice was to secure land that could then be given to government-approved developers. Nevertheless, this model of heteroorganization did not result in all local land transferring from indigenous settlements to the municipality. Leftover spaces included roughly two thousand village settlements that were scattered over both the SEZ and New Bao'an County, new development that had been erected outside the plan, and land that had not yet been expropriated from the villages. These spaces existed outside the city's master plan in the cracks between designated circles. Local villagers, some agricultural workers and laborers from Bao'an communes, and even guest workers began to develop this space outside the plan. This self-organization gave rise to another kind of circle within Shenzhen—the urbanized village.[13]

The defining characteristics of Shenzhen's urbanized villages have been as follows: The household is the developer and the area of development was a 100 m^2 area on which a two- or three-story family villa (during the 1980s) or a six- to eight-story (sometimes even higher) commercial rental property (during the 1990s) was built. The distance between each building ranged from one to eight meters, creating dense concentrations of freestanding buildings. At the same time, village collectives developed industrial parks and commercial areas and collectively held rental properties. The high-density fabric of urban villages has stood in contrast to the modernist or superscale space in city-planned enclaves. The urbanized village enclaves have developed independently without municipal administration such as planning, design and building approval, quality control, property registration, or any other regulatory procedure. Compared to official circles, property rentals were cheaper in the urbanized villages, attracting low-income families, migrant workers, low-cost business, entertainment, and recreation service industries. As part of this transformation, village collectives reorganized as stock-holding corporations that were responsible for building and maintaining infrastructure; the provision of water, gas, and electricity; and public security. In short, the villages operated to a large degree autonomously from the municipality.

Shenzhen's urbanized villages are another manifestation of the "loosening restrictions." The critical difference, however, is that the eruption of urbanized villages did not originate with the government or policy. Instead, it resulted from local villagers who acted according to their own interests, proactively participating in and benefiting from urbanization in Shenzhen. The Chinese land system had two defining characteristics: (1) collective land that has not been expropriated by the government cannot be legally developed until it has been converted to state-held land and (2) land that was restored to villagers for their houses could only be developed within fixed parameters for specified uses. For example, according to Shenzhen law, irrespective of the location of a village settlement or adjacent development, at first the total area of their buildings was restricted to 240 m^2, which was increased to 480 m^2 in 1999. Nevertheless, once local villagers realized the discrepancy between the compensation price for expropriated land and the value that land had on the newly instituted property market, they decided to make the most of the land that remained in their hands and develop with respect to market prices. If any remaining policies continued to restrict their development, they simply ignored them, proceeding to develop their land according to market demands, which in turn led to increasing density and economic vitality. Some villages, such as Caiwuwei or Gangxia, are today located next to some of the highest skyscrapers and most expensive property in the city.

Shenzhen Municipality originally had 320 administrative villages, with a population of roughly 330,000. The total area of their private homes was less than 10 percent (93 km^2) of the total building area in Shenzhen. Nevertheless, according to 2007 statistics, more than half of Shenzhen's 13 million inhabitants lived in that area, for a population density of 70,000 per square kilometer. In fact, Shenzhen's urban villages have functioned as low-income housing for the city. Rent in the villages is cheap because capital outlays for village construction were low, because they paid neither for the land nor any associated registration fees. More important, the distance between villages was walking distance. The traditional agricultural principal of walking to work in the morning and returning at night became the layout for Shenzhen's "low-income housing" because there was a village located next to any government-sanctioned circle within the city. Every commercial area, industrial park, or new center abutted a village where workers could live cheaply and commute easily to work. Moreover, the location of the villages supplemented what the urban plans had clearly overlooked: sufficient housing for low-income workers and recent arrivals. The existence of the villages ameliorated the effects of Shenzhen's rapid development in two ways. First, they ensured that there was enough housing for the influx of migrants. Second, the location of the villages meant that migrants lived within walking distance of their place of employment, lessening the need for massive investment in traffic infrastructure.

The self-organizing development of the urbanized villages supplemented blind spots in the urban plan as well as insufficient infrastructure to manage the burgeoning population. In turn, this allowed for the improvement of urban land functions and efficiency, provided a self-sufficient system of low-income housing, enriched urban services for the public, and lowered the cost of services and starting businesses. Shenzhen's urban villages could be seen as a self-regulating mechanism that contributed to the efficient running of the city or simply as one type of basic urban infrastructure. Unfortunately, the important function of urbanized villages in Shenzhen's development was not completely or objectively understood. Instead, planners focused on the problems of high density, sanitation, and aesthetics. These complaints, combined with capitalism's ongoing need for profitable land, resulted in the complete razing of several urban villages and their redevelopment as expensive real estate enclaves. Village companies have usually received satisfactory compensation when their residual territory was razed, and indeed many individuals became billionaires. Nevertheless, from the perspective of the rights of low-income people to affordable and convenient housing, as well as the creation of a heterogeneous urban life, history, culture, space, and community, the decision to raze the urbanized villages has overlooked their fundamental and irreplaceable social function within Shenzhen society.

The contribution that inhabitants of Futian, Shangsha, and Xiasha villages made to the flourishing businesses and manufacturing in Huaqiangbei and Chegongmiao may not have been immediately apparent to the swarms of people walking along the street or crossing pedestrian overpasses. City planners and developers often overlooked the important social fact that village self-organization was itself an urban response to the city's need for inexpensive housing and convenient access to consumer goods. In addition, the villages contributed to the sustainability of urban networks, the heterogeneity of its social ecology, and the creation of cheap public transportation within and between Shenzhen districts. The interests of developers and government dovetailed in the search for profits and the decision to rebrand the city, constituting a strong force to raze the villages and renew these areas. More and more urbanized villages have been bulldozed. The result has been intensive gentrification, and the naturally low-carbon "layout of low- and middle-income housing" is being diminished and dismantled. As a class, migrant workers have been forced to live farther and farther from the city center. They are moving to Meilin, Buji, and other former checkpoints along the former Second Line. In turn, rapid unplanned population growth in these areas has resulted in environmental degradation and a natural diminishment in workers' quality of life. Even though the city has responded by increasing investment in public transportation, nevertheless, these new measures cannot substitute for the urban functions that the villages have historically provided. In fact, when urbanist John Friedmann visited Shenzhen, he opined that when compared with the sterile and lifeless places that were controlled by the city's master plan, the robust life in the urbanized villages was, in fact, the best representative of Shenzhen's unique urban character.

Conclusion

Shenzhen's rapid development had heterogeneous origins: Beijing's policies, Hong Kong's extant integration into global capitalist networks, and the pre-existing Bao'an village network. These origins provided a perspective on and a structure for researching Shenzhen's development within and against global economic restructuring, and with respect to the future, these three factors will continue to shape Shenzhen and its possible urbanisms.

In 1980, the Shenzhen SEZ was created through exceptional policies (tax exemptions) and concomitant delegation of power (to establish laws, for example). Thirty years later, these exceptions were made contiguous with the city's borders. The Qianhai Cooperative Zone, for example, was honored by being designated a national reform experimental site, demonstrating Beijing's continued support of exceptional policies in the city. Over this same period of time, Shenzhen pursued, and in some senses overtook, Hong Kong, which nevertheless continued to influence both the economic development and the concomitant layout of the city, especially through enthusiasm to cooperate in the Qianhai and Lok Ma Chau Loop areas. In contrast, the fate of the villages has been less sanguine. Either the number of villages will continue to shrink through gentrification projects or another more collaborative model of development needs be introduced. In either case, as of this writing, how the city decides to handle the remaining urban villages will profoundly alter its urban fabric and civic identity. Consequently, it needs to carefully research and evaluate the role of urban villages over the previous thirty years of successful high-speed development. Indeed, if the city's urban planners remain unable to learn from, integrate with, and tolerate self-generating spaces, the city's prospects for a flexible, heterogeneous, and livable urbanity are bleak.

Notes

1. Statistics are available through the Shenzhen Government online website, http://www.sz.gov.cn/tjj/tjj/.

2. See O'Donnell, chapter 2 this volume, for another version of the origin of the expression "Shenzhen Speed."

3. As part of its policy decision to be more transparent, the Shenzhen government (and each of its constituent districts) has uploaded statistical and planning documents to their respective websites. For statistics, visit the website of the Shenzhen Bureau of Statistics (http://www.sztj.gov.cn). For urban planning maps, visit the website of the Municipal Urban Planning Department (http://www.sz.gov.cn/cn/xxgk/jqgh). For an ongoing analysis of urban planning in Shenzhen, visit the Urban Planning and Design Institute of Shenzhen (http://www.upr.cn). Shenzhen studies have been central to Chinese thinking about economy, society, and cultural change since Deng Xiaoping's 1984 Southern Tour confirmed the value of transforming the planned economy to a market economy. In addition to national- and provincial-level campaigns to study the Shenzhen experience, Shenzhen University established the Institute for the Study of Special Zones, Hong Kong and Macau, in order to learn from and disseminate information about Shenzhen. Early studies of Shenzhen are simultaneously replete with information about and enthusiasm for the experiment. Representative texts include Lu Zhenhua, *Shenzhen* (Beijing: Zhongguo Haiyang Publishers, 1985); Chen Weshan, *Shenzhen, xin tizhi yanjiu: Tizhi gaige diaocha yanjiu baogao* [Shenzhen, New System Studies: Research on the Reform of the State Apparatus] (Shenzhen: Shenzhen government publication, 1988); and Ni Yuanlu, Peng Lixun, and Shen Yuanzhang, eds., *Shenzhen: Maixiang shehui zhuyi shichang jingji* [Shenzhen: Striding toward a Socialist Market Economy] (Beijing: Renmin Publishers, 1999).

4. For the now classic discussion on space-time compression, see David Harvey, *The Condition of Postmodernity: An Enquiry into the Origins of Cultural Change* (Cambridge, MA: Blackwell, 1990). For an earlier treatment of the social production of space, see Henri Lefebvre, *The Production of Space* (Oxford: Blackwell, 1991); and Mark Gottdiener, *The Social Production of Urban Space* (1985; repr., Austin: University of Texas Press, 1994).

5. For the argument that the epicenter of global capitalism is shifting east, see Giovanni Arrighi, *The Rise of East Asia in World Historical Perspective* (Binghamton, NY: Fernand Braudel Center for the Study of Economies, Historical Systems, and Civilizations, 1996). See also Arif Dirlik, *After the Revolution: Waking to Global Capitalism* (Hanover, NH: Wesleyan University Press, 1994); and Dirlik's edited volume, *What Is in a Rim? Critical Perspectives on the Pacific Region Idea* (Lanham, MD: Rowman and Littlefield, 1998). See Jonathan Bach, chapter 1 of this volume, for an account of how zones of exception, like Shenzhen, have contributed to the current global restructuring.

6. See O'Donnell, chapter 2 of this volume, for an account of how elite politicians in Beijing used Shenzhen to forward controversial changes to the state apparatus.

7. See Ma and Blackwell, chapter 6 in this volume, for a detailed discussion of the establishment and spatial effects of the Second Line.

8. See Tu Qiao's *Yuan Geng Zhuan: Gaige xianchang, 1978–1984* [The Yuan Geng Story: The Site of Reform, 1978–1984] (Beijing: Zuojia Publishers, 2008) for a detailed history of the Shekou model of reform.

9. Martin Hart-Landsberg and Paul Burkett have been particularly critical of China's capitalist turn and the restructuring of East Asia. See, in particular, "Contradictions of Capitalist Industrialization in East Asia: A Critique of 'Flying Geese' Theories of Development," *Economic Geography* 74, no. 2 (1998): 87–110.

10. For an early Hong Kong perspective on what was and was not shared across the border, see Wenhong Chen and Shuji Zeng, *Shenzhen Pouxi* [Shenzhen Analysis] (Hong Kong: Cero Consultants, 1985).

11. Weiwen Huang, "On Borders (Shuo Jie)," *Chengshi Zhongguo* [Urban China] 24 (2007), http://www.china-up.com/newsdisplay.php?id=1441954&sib=1&unam=.

12. See Hirsh, chapter 11 in this volume.

13. In chapter 5 of this volume, O'Donnell details Shenzhen's vernacular geography of urban villages.

4

How to Be a Shenzhener: Representations of Migrant Labor in Shenzhen's Second Decade

ERIC FLORENCE

Shenzhen is the resting post of our youth, we should therefore extend our youth in this resting post, since we have walked on the road to *dagong*, so it is a youth without regret and a shameless destiny.
SHENZHEN SPECIAL ZONE DAILY, March 25, 1994

They did not do anything extraordinary, they just offered their youth silently to the Chinese people . . . Yesterday, they awoke Shenzhen with the sound of their feet; they have created the Chinese miracle. Today, they have stepped into the new century and have initiated a more beautiful and more resplendent tomorrow.
SHENZHEN SPECIAL ZONE DAILY, March 29, 1998

Shenzhen Spirit: New Models of Reform

During the 1980s, the task of modeling post-Mao reforms had fallen to the first generation of party leaders in Shenzhen. However, in the decade after Deng Xiaoping's 1992 Southern Tour, the task of modeling the new socialist citizen fell to migrant workers from China's hinterlands. This chapter examines how migrant workers in Shenzhen were enlisted in this political discourse through encouraging their participation in the public sphere of newspapers and magazine publications. Based on a close reading of the Shenzhen press in the years 1994, 1998, and 1999, this chapter examines shifting definitions of the model rural migrant to Shenzhen within and against the hegemonic construction of "Shenzhen Spirit."

In 1994, the official model of the Shenzhen migrant worker exalted self-reliance and initiative in contrast to the Maoist valuation of self-sacrifice.[1] However, as Shenzhen was increasingly integrated into the world economy during the late 1990s, the Special Economic Zone (SEZ) became closely associated with the notion of youth. Looking at the period following the Fifteenth Party Congress (1997), which restructured state-owned enterprises and solidified President and Secretary General Jiang Zemin's emphasis on economic development, this chapter tracks how the Shenzhen press in the period from

1998 to 1999 idealized migrant workers with respect to the challenges and opportunities of the SEZ's competitive environment. This modified model called upon migrant workers to liberate and nurture their individualized and individualizing abilities, ideals, and aspirations.

As with the 1980s construction of post-Mao socialist heroes, the 1990s construction of model workers encapsulated a set of values that linked Mao-era identities to the Reform and Opening policy identities. With respect to labor, the 1990s Shenzhen model linked Chinese Communist Party ethics of "self-sacrifice" and "contribution" to new values such as "autonomy" and "competition." The rhetorical vehicle for this social construction was "Shenzhen Spirit," which in turn was a local articulation of national policy—the simultaneous construction of material and spiritual civilization.

The Shenzhen Spirit was formulated over the late 1980s but officially concretized in 1990. A "bull clearing the wilderness" (*tuo huang niu*) and "opening up" had long been the two metaphors most closely associated with the early construction of the Shenzhen SEZ. At a 1987 meeting of the Shenzhen City Working Group on Thought and Political Work, the values of that early construction period were expanded to include "opening up" (*kaituo*), "creating" (*chuangxin*), and "devoting one's whole life" (*xianshen*).[2] "Clearing the wilderness" and "opening up" entailed the idea of doing something for the SEZ, while "sacrificing oneself," or "devoting one's whole life," to the nation reprised Mao-era ethics, when individual interests were to be subsumed under the collective interest of constructing socialism. In Shenzhen, the new "collective interest" took the form of the economic development and prosperity of the SEZ, and this required a new kind of ideal subject, "a person able to transform her- or himself and the socialist world."[3]

The notion that the ideal Shenzhen subject would be able to create something new drew from the novelty of the SEZ itself and was framed in contrast to the economic failure of Maoism. In the construction of an identity for Shenzhen, socialist values such as self-sacrifice and collectivism were downplayed in favor of rhetoric of pioneering and newness.[4] In 1990, the City Party Standing Committee added "unity" (*tuanjie*) to the list of official values, and this new synthesis was then officially called the "Shenzhen Spirit" and was approved by Jiang Zemin.[5] So defined, Shenzhen Spirit celebrated the values of "deciding for oneself, strengthening oneself, autonomy, competition, taking risks and facing danger, equity, effectiveness, and legality [*zizhu, ziqiang, jingzheng, ganmao fengxian de gainian, pingdeng, xiaolü gainian, yiji fazhi gainian*]." The Shenzhen Spirit was seen as central to the construction of post-Mao society.

Building "spiritual civilization" in Shenzhen opened a new space for the shaping of a new post-Mao era socialist subject along with the values, norms, and attitudes this subject ought to embody and exemplify.[6] The ideological link between "material and spiritual civilization" and the Shenzhen Spirit solidified during Deng Xiaoping's 1992 Southern Tour. Spiritual civilization referred to cultural, scientific, and ideological dimensions of society in contrast to the economic dimensions of material civilization. In fact, both a 1995 volume edited by Shenzhen's then mayor, Li Youwei, as well as a 2000 Shenzhen-level officially sponsored volume reviewing twenty years of development of Shenzhen emphasized that during his 1992 Southern Tour, Deng Xiaoping distinguished "Socialism with Chinese Characteristics" from other political systems because, in China, both material and spiritual civilization had yet to develop.[7] The relationship between spiritual and material civilization was often referred to as "seizing with both hands, [and] both hands need to be firm." The goal of material civilization was clear—building China's economy. However, the goal of spiritual civilization work was less clear because it entailed modeling a new kind of worker, "the four-haves person"—that is, "a new person with ideals, culture, ethics, and discipline."[8] In fact, the model of the four-halves person was not adopted until the Third Plenum of the Fourteenth Party Congress in 1994, well into the second decade of Shenzhen reforms.

This model illustrates the extent to which productive forces were linked with building a market economy that rests on a neo-liberal conception of the market which "is not taken to be a natural formation, but [as] both a system and a subjectivity that has to be actively produced and facilitated."[9] In other words, rural migrants to Shenzhen had to be taught how to be workers. In turn, as they learned to be Shenzheners, they also became a model for the rest of the country. Ironically, while rural migrant workers became increasingly central to Shenzhen's economic development—and indeed that of the entire nation—their exploitation challenged a ruling party whose founding narratives continued to reject capitalist exploitation.

Chinese Migrant Workers in Western Theory

The experiences of rural migrants to and in Shenzhen factories not only influenced the construction of post-Mao models of labor and laborers in China but also shaped Western models of emergent labor regimes in postsocialist China. Through reforms initiated in Shenzhen, China's rural labor became the core element of a "labor-squeezing strategy of development."[10] Simultaneously, former workers and employees of state-owned enterprises lost their lifelong employment and their subsidized access to social welfare. Together, these processes unleashed what Ching Kwan Lee separated into "three patterns" of the working-class transition in urban China: "the making of the global peasant worker," "the remaking of the socialist worker," and "the unmaking of the redundant worker."[11]

At the same time, Shenzhen catalyzed the political and economic restructuring of the Pearl River Delta, where local governments produced no longer in accordance to a national plan but in competition with one another to provide investors with land, infrastructure, and labor.[12] The regional relationships between cities changed from integration to competition, and this spurred maximum labor flexibility and kept wages relatively low, even in comparison with other Southeast Asian countries.[13] In Shenzhen, this was accomplished through the combination of labor control mechanisms and population control alongside intensification and concealment of exploitation.[14] By "externalizing" migrant workers and exerting everyday "routine repression,"[15] the household registration system (*hukou*) and the several certificates and permits required for employment and residence in the city further enabled the implementation of a highly flexible production regime, or what Robin Cohen has called a "labor repressive system."[16]

Modeling Labor: How to Be a Shenzhener

> In the field of experimentation of socialism with Chinese characteristics, how should we shape the appearance of the Shenzhen person?
> SHENZHEN EVENING DAILY, March 17, 1994

Throughout the late 1980s and early 1990s, Chinese newspaper accounts of migrant workers presented a generally homogenizing and threatening picture of them.[17] Rural people migrating to or staying in cities were described chiefly as masses flowing into Chinese cities. A very simplified narrative structure of such press accounts usually explained that poverty compelled migrants to leave the countryside, forcing them to "pour blindly" into the cities, which disturbed the urban social order. These press accounts would then call for strong measures by urban authorities to control or expel the migrant workers from the cities. In these articles, rural migrants were seldom given a personality of their own or described as individuals with any personal will. Instead, they were portrayed as masses entirely motivated by the search for profit, a drive that would possibly lead them to commit crimes.[18] On the whole, rural migrants were not asked by journalists to express their personal or collective experiences.

Shenzhen's newspapers also depicted migrant workers as homogenized types; however, articles dedicated to rural workers were not as numerous—and the tone was not as passionate—as those found in Beijing newspapers. More important, at the end of the 1980s and early 1990s, several Shenzhen magazines appeared that focused on migrant workers' experiences, including *Dapengwan, Dagongmei,* and *Wailaigong*. More broadly, in Shenzhen newspapers, the narration of migrant work (*dagong*) in articles written by both professional journalists and migrant workers themselves became increasingly popular.[19] These dedicated magazines encouraged migrant workers to write about their experiences through articles, poems, short stories, or novels.[20]

Both in these magazines and within the Shenzhen official press, the depictions of voiceless masses of people that were flooding into Guangdong Province gave way to descriptions of smaller groups of people often interviewed by journalists. This change was also reflected in the use of pictures of smaller groups of often smiling people rather than of indiscriminate crowds. While this change could eventually be seen nationwide by the 2000s, it was apparent as early as 1994 in the newspapers of Shenzhen and the Pearl River Delta—for instance, in Guangzhou's *Yangcheng Wanbao* and *Nanfang Ribao* and *Shenzhen Tequbao* and *Wanbao*. This change can be explained by the important place that migrant workers occupied in Shenzhen and other SEZs. Moreover, as the inflow of foreign capital increased during the 1990s, attracting cheap and young migrant labor to Shenzhen actually became necessary for local authorities. In the delta, earlier than elsewhere, migrant workers were soon identified as a component of economic reforms the state could not do without. It became also more and more important to present an attractive image of Shenzhen and an image of a hardworking and disciplined labor force.[21] Such shifts in representation are hence evidence of a gradual change in the official conceptions of rural-to-urban migration.[22]

The "How to Be a Shenzhen Person" debate was launched in 1994 by party authorities. In fact, this campaign was part of the spiritual civilization propaganda work that was aimed at "mobilizing the participation and consciousness of moral construction among Shenzhen people."[23] It was initially planned to last for four months, but because of its unexpected success, was extended five more months. It took place through the media in newspapers chiefly via letters to the editors, radio and television programs, as well as within workplaces. Local officials and white collars were the main participants in the debate. Very few rural migrants participated in the discussions, and the few workers who did participate expressed frustration about their second-class status in the city. Clark put it this way: "What emerged from the debate was less a collective Shenzhen identity hoped for by authorities, than a realization of 'communities' proliferating throughout the city, marked by class, education, native place and goals."[24]

Based on a combination of the "How to Be a Shenzhen Person" campaign, Deng Xiaoping's thought, and the party's basic guidelines, the Shenzhen Party Committee published a policy entitled "Norms of Shenzhen Inhabitant Behavior" in 1994, which was addressed to all its inhabitants. These norms included such phrases as "love one's country," "build Shenzhen," "open up and create," "unite and offer contributions," "do all one can at work," "serve the public," "respect discipline and the law," "fair competition," "be civilized and polite," and "love the environment."[25]

Many of Shenzhen's defining values are gathered in the text that launched the debate in March 1994 and were used recurrently to describe rural migrant workers. In this text, the reference to the sacrifices one ought to make for the SEZ actually pointed to early 1980s narratives of Shenzhen as a desolate place to which people came with the ideal of constructing the SEZ. Stressing the disinterested nature of these early "builders" allows them to be distinguished from the "gold diggers."[26] The next paragraph, however, pointed to the highly recurring narrative of "going forward and grasping opportunities" as well as "the need to adapt to competition," which would emerge as a core theme in the second half of the 1990s. The rhetoric of "being able to grasp opportunities" in order to avoid elimination echoed the post-Mao era ethos of economic reform and was a major mode of justification for social stratification.[27]

The World of Dagong

In January 1994, the leading newspaper in the SEZ, the *Shenzhen Special Zone Daily* (*Shenzhen Tequbao*) began to dedicate a full page to migrant workers. This page was called "World of *Dagong*" and included articles, essays, poems, and images of migrant workers. From 1994 to 1999, "World of *Dagong*" presented migrant workers as a crucial component of the Shenzhen identity, allowing for the articulation of the defining values of the "Shenzhen Spirit" detailed above. The term *dagong* was a Cantonese term from 1960s Hong Kong denoting mobile, commodified labor.[28] However, via "World of *Dagong*," commodifying one's labor would emerge as a new—and valorized—social identity. The editorial of the very first issue of the "World of *Dagong*" special page, published on January 7, 1994, announced, "Millions of migrant workers have created the myth of Shenzhen, and we eventually possess a world that is ours . . . *Dagongren* has been the noun for manual laborers: but today, its connotation has gone far beyond this meaning. Blue collars are *dagong* people, white collars are also *dagong* people, all the workers who are laboring industriously are *dagong* people. Today, for this growing community of *dagong* people, we have solemnly created this special page called 'World of *Dagong*.' In this world, you will see a 'happy nation' that silently makes sacrifices for the construction of Shenzhen."[29]

This editorial also hinted at the extension of the meaning of *dagong* to larger categories of workers. Although this 1994 statement may be read as a euphemization and a concealment of the class-relation dimension of the very condition of rural migrant workers, it may also be conceived as an annunciator of the upheavals that were going to characterize the labor condition from the second half of the 1990s.[30]

Throughout its five years of publication, "World of *Dagong*" celebrated labor and repeated notions and values common within the Shenzhen press at the time, including the idea that *dagong* should include all forms of laborers, regardless of their *hukou* status. One of the most recurrent images it provided for migrant workers was their contribution to the economic achievements of the SEZ.[31] In these depictions, migrant workers were often labeled the "builders of the Special Zone."[32] Migrant workers' role in the city's success were also strongly expressed in phrases such as "Without these migrant workers Shenzhen would be an empty city" and "Migrant workers are the ones who have built the modern city we live in."[33]

In these accounts and descriptions, "World of *Dagong*" drew a causal link between migrant workers' efforts and sacrifices and the rewards they received in the form of successful employment mobility or the improvement of their own lives. Sometimes this reward also took the form of having a "sense of belonging" and "feeling at home" in Shenzhen. Although articles usually stated that migrant workers could not become legal permanent residents of Shenzhen, the articles stressed that migrant workers could "feel" that Shenzhen belonged to them via their contribution to the development of the SEZ and through the sense of pride that this contribution provided. Migrant workers' merit and efforts, exemplified by their "sweat," "tears," and "blood," were said to be embodied in Shenzhen's buildings, which eventually should provide them with a sense of pride and belonging and enable them overcome the pain and suffering they endured in their work.

It should be observed here that the stress in these articles on the notion of "feeling at home" was related to an official concern expressed in Shenzhen pronouncements relating to the construction of the "Shenzhen Spiritual Civilization." In 2000, for example, a municipal publication explained that there were vast differences between the "highly qualified permanent population" and the "low quality workers from outside." At the time, Shenzhen had in an official population of four million people, among which approximately two-thirds held temporary registration permits. It was significant that these temporary workers were encouraged to "psychologically feeling at home" because they were expected to leave when their job was done.[34] "World of *Dagong*" picked up this theme, exhorting workers to use and nurture their potential skills, because "one may not have a Shenzhen 'green card,'[35] but what one can surely not fail to possess is to have ideals and aspirations, knowledge and competences, and dignity."[36]

Closely associated with the notion of merit and contribution was the conspicuous use of the historically loaded expression, "To pay with one's blood and sweat" (*fuchu xuehan*). This expression had been used during the Mao era in reference to the sent-down urbanites who were to learn from and, in return, educate peasants. These "educated youth" (*zhiqing*) had left their cities wholeheartedly, "offering their best years of youth respectfully to the country." Significantly, a collective justification that had been used to rusticate urban youth during the Mao era was redeployed in Shenzhen to legitimate the new labor regime. In a culminating tale in 1998, for example, the productivist exaltation embodied in the sacrifice of the migrant workers' youth assumed poetic and ideological form:

> The reason why the production line is so beautiful is because it is dressed up in youthfulness? These rows of youngsters sitting there are like green grass and flowers sitting along the water, they are contending vigorously [*zhengqi douyan*] with life . . . ?[37] The value of youth is flowing away smoothly along the production line. I myself am immersed in the production line, neither impulsive nor weak, my vigor is getting stronger . . . When my hands and eyes move, embracing the production line's rhythm, my heart feels so good . . . What makes me even more relieved is that, in my struggle towards the "zero fault" goal, every day, "red stars" shine on my attendance record while the red flag flitters in the wind.[38]

In this passage, specific terms related to the socialist era were reworked in support of the Reform and Opening policy. Importantly, both periods subordinated rural people to urban goals. The differences between these eras, however, illustrate how Shenzhen mediated the transformation of the Maoist labor regime. Under Mao, rural workers were socially and geographically bound to the land and had little choice but to work for the rural collectives (the socialist cooperatives and later the collectively organized popular communes), as the whole economy was planned centrally by the party-state. Moreover, individual workers had to submit to the collective goals of the party, which was said to express the common will of the people. In contrast, during the era of the Reform and Opening policy, rural workers became mobile—migrating from hometowns to Shenzhen. In turn, through their labor, migrant workers modeled how individual goals contributed to a state-lead project of integration into global capitalism.

A Place for Dagong: This Warm Earth That Is Shenzhen

The image of Shenzhen as an environment in which a worker's potential could be realized became more pronounced by the end of the 1990s. In the local media, there was a recurrent association of the SEZ with self-determination, autonomy, and adaptability on the part of its migrant workers. These descriptions repeatedly contrasted Shenzhen and the south (referring to the PRD) from the countryside and the interior of the country.[39] The expression, "This warm earth that is Shenzhen" was widely repeated in the Shenzhen press and in *dagong* literature in order to describe the SEZ's dynamism, competition, challenges, and opportunities. The descriptions of Shenzhen were utopian, and its explicit foil was the backward and lackadaisical countryside.[40] By the 2000s, Shenzhen had become a place—at least in the public discourse—where migrant workers could nurture their dreams and aspirations and make the most of their potential.[41]

This idea of nurturing aspirations and dreams in Shenzhen was also tightly connected to an intense notion of youth.[42] For instance, in an earlier piece from February 1998, Shenzhen's skyscrapers and avenues are associated with "youthfulness, struggling and a pioneer spirit."[43] In a March 1998 *Shenzhen Special Zone Daily* text, another migrant worker nurtured the dream of being successful in Shenzhen by imagining herself and her future "as more and more magnificent," and as having to "fly conscientiously higher and further in the Shenzhen sky."[44] In short, Shenzhen was meant to attract individuals who were dissatisfied with their hometown situation. Young migrants willingly left the countryside for Shenzhen to polish their determination. Importantly, although the rhetorical link between individual effort and success similarly permeated the workplace in many factories of the Pearl River Delta,[45] in Shenzhen, writers assumed that the desire to transform oneself through migration to Shenzhen was shared across all social classes. Thus the Shenzhen press described typical migrants not only as workers from the countryside but as people with urban backgrounds, such as doctors, teachers, and older people with more education.[46]

Turning the Outside In: Lowering the Expectations among Urban Migrants

In addition to stressing the need for all migrant workers to continuously learn from experience, the rhetoric of Shenzhen as a place where people improved themselves conflated the experiences of migrants from the countryside and those from other cities, modeling appropriate attitudes for success in the SEZ. During its last year of publication, "World of *Dagong*" invited laid-off workers from state-owned industries, unemployed Shenzheners, and those facing the obstacles of "not getting highly qualified jobs and not being willing to do lower jobs" to write to the new column, "My Experience Searching Again for Work."[47] The introductory text for the column stated that workers faced issues such as "how to raise their own quality [*suzhi*] and how to change their conceptions on looking for work."[48] In these press accounts, "the interior" with which Shenzhen was contrasted no longer referred to the countryside but rather to bankrupt state-owned enterprises where workers had become redundant.

In this column, laid-off workers were encouraged to nurture an attitude that would allow them to reevaluate their position in the socioprofessional hierarchy. These texts emphasized that independent study (instead of formal education) and "learning from one's experiences" would be possible when laid-off workers and the unemployed acknowledged their deficiencies and lowered their expectations for jobs and salaries. The result of nurturing an attitude of acceptance would be "replenishment" (*chongshi* or *huibao*), which was, in fact, a new job.[49] This rhetoric suggested that at least some unemployed laid-off workers were not quite fit for Shenzhen nor were they ready to do the kinds of tasks they had previously scorned—hence the need to change one's outlook when searching for reemployment. This rhetoric portrayed laid-off workers as too narrow minded, too choosy, and not daring enough for the Shenzhen labor market. In order to successfully find a job in the SEZ, these workers needed to lower their demands and "start from scratch."[50]

The column also modeled how recent college graduates could succeed through humility and hard work. In one text, for example, a university graduate who downplayed his university degree in order to get a job as a waiter—a position clearly beneath what his education had prepared him to do—said, "Half a month later, I went to a hotel to work as waiter ... I no longer dared to show my university diploma that had resulted from my hard work ... During that period, as I was holding the bucket to the toilets and as I was offering subservient services to others, I held back my tears and did not think that I was a university graduate. I was just considering myself as an apprentice who did not understand anything."[51]

His new attitude paid off. The graduate was later informed that he would be appointed a group leader precisely because he had put down his university student airs and he was willing "to become a primary student again."[52] In this text, "starting from scratch" meant that a highly educated person had to reevaluate his employment expectations and be ready to start at a low rung, such as that of a waiter. In another instance, a female migrant worked in a hospital laundry without being paid a salary. She worked so fast and steadily that she was eventually rewarded with a paid job. The moral of this tale is expressed in the very last sentence of the text: "The person who can help you best is yourself."[53]

The repetition of such exemplary tales exhorted migrants to not give up searching for work, to remain self-confident, and to be ready to accept any job. The key words that emerged in the 1998–99 "World of *Dagong*" pages emphasized attitude as central to employment. These words included "to replenish oneself," "to charge one's batteries," "to know oneself again," "to find one's direction," "to look for oneself again," and "to search for one's value." Several of the 1999 texts were thinly veiled warnings addressed to state-owned enterprise workers or laid-off workers. The warning sent was clear: those who did not actively look for work and were unwilling to independently acquire new knowledge and learn new techniques would most likely be laid off or unable to find employment on the Shenzhen labor market.

The following fragments nicely illustrate the tone of these warnings:

> When you feel at a loss, don't forget: life does not believe in tears, only those strongly determined talents will manage to reach the shore of victory.[54]

> During all this period, [she] had not stopped going forward, through independent study she earned a specialized technical degree. This because she realizes that in a society of tough competition, only those who endlessly keep replenishing themselves may continuously make progress, otherwise they will be eliminated.[55]

The message sent to former state-owned enterprise workers, although not always explicitly expressed, was that everyone should be ready to lower his or her work expectations in order to adapt to the Shenzhen labor market.

The Political Work of Modeling Migrant Labor

During Shenzhen's second decade, migrant workers embodied the values of the Shenzhen Spirit, constructing an identity for the SEZ. Shenzhen's young migrant workers embodied a series of values and attitudes belonging chiefly to two major rhetorics. The first rhetoric cited the ethics of the Mao era, such as "devoting one's whole life" and "offering a contribution" for the prosperity and economic development of the SEZ and of the country. The second rhetoric was associated with the development of commodified labor, including values such as "adaptation to competition," "self-confidence," and "autonomy."

Processes of state-making and subject-making through the figures of Chinese migrant workers share a number of commonalities with similar processes in the history of population mobility and the formation of nation-states in other parts of the world. First, Chinese internal migrants share with international migrants in other countries "a form of institutionalized discrimination," which produces a layering of statuses and citizenship, with different categories of migrants enjoying differentiated rights and duties, as well as being the object of various degrees of social control.[56]

Second, the experiences of Chinese migrant workers in Shenzhen overlap with those of (im)migrants elsewhere in the world. Practices of legal, bureaucratic, and narrative categorization have occupied a central role within the never fully completed processes of state-formation and reinvention. For instance, in nineteenth- and early twentieth-century France, the very notion of the "national citizen" was shaped in opposition to a number of "interior others" constructed as "the vagrant," "the refugee," "the destitute," or "the homeless." Concomitantly, the state defined the very attributes of what constituted a "legitimate citizen," or the "national community," in contradistinction to various categories of interior and alien others and the various forms of social ills they were associated with.[57] As empty signifiers, migrants—be they labeled "peasant workers," "illegal migrants," "undocumented workers," or "refugees"—can be loaded with specific meanings and values, enabling the sovereign and diacritic power of the state to define who belongs and who does not, who is a legitimate citizen and who is not, and what forms of labor are more or less valued. They can also enable the state to determine which forms of labor should be concealed, thereby delineating areas of state intervention.

Third, migrants may be turned into useful figures at the level of rhetoric and for their economic roles, especially in times of changes in the relationship among the party-state, capital, and other social groups. The very features that render Chinese rural migrant workers, laid-off workers, and international migrant workers elsewhere in the world attractive—their "invisibility, marginality and vulnerability"—are the same characteristics that make them "hard to control and legalize."[58] As Bach has argued in chapter 1 of this volume, in their management of rural communities, the Shenzhen authorities combine a kind of laissez-faire with periodic attempts at restoring social order, a management pattern that "enables the very migration that is simultaneously necessary and 'illegal.'" Within the political economy of both Shenzhen and many Western countries, the ambiguous and often concealed links between the "formal" and "informal," or "legal" and "illegal," ends of the economy and the roles played by various categories of migrant workers offer specific ways for the state to manage the politics of labor, making visible its sporadic sovereign affirmation of authority.[59] In Shenzhen as elsewhere, the "power of the national state sometimes seems more visible and encroaching and sometimes less effective and less relevant."[60]

Throughout the 1990s, rural migrant labor occupied an increasingly important position in the economic growth of the Shenzhen SEZ. Moreover, through the conflation of rural hometowns and state-owned enterprises, the figure of the rural migrant worker enabled Shenzhen to model lowered job expectations for urban migrants, who were represented as "starting from scratch." On the whole, the inclusion of migrant workers in the party-state rhetoric of "the Shenzhen miracle" concealed the precariousness and liminality that characterized and continues to define migrant workers' conditions and the politico-institutional arrangements that have enabled the exploitative regimes of production. Indeed, the kinds of narratives documented above constructed an image of Shenzhen that diluted the class antagonism that characterizes labor relations in Shenzhen. As Pun put it, "The language of class is subsumed so as to clear the way for a neoliberal economic discourse that emphasizes individualism, professionalism, equal opportunities, and the open market."[61] In other words, the intense cultural construction of migrant workers and of *dagong* demonstrates the party-state's ability to adapt its system of signs and symbols to the conditions of global capitalism. To do so, it must reconcile exploitative forms of labor with the party-state founding discourse and identity, as well as shape legitimate forms of relationship with its subject-categories.

During the 1990s, Shenzhen was culturally constructed as a space in which rural and urban youth might realize aspirations for social mobility. Importantly, the relative debasement of rural migrants was mobilized within the narrative of Shenzhen Spirit to both encourage rural migrants to improve their situation and discipline urban migrants for having excessive expectations. Nevertheless, many of the values related to self-reliance sponsored in 1994 in the Shenzhen official press have turned into core societal values defining social mobility in China today because the Chinese state is requiring that people from all social categories become self-reliant and find ways to provide for their own well-being—to adopt an attitude of acceptance.

In the 1990s, rural migrant workers in Shenzhen were the ideal subjects to mobilize to model these values for all migrants to the SEZ. Rural migrant workers could be transformed into positive models of self-reliance precisely because, under the *hukou* system, they were more debased than their urban counterparts; their lives did materially improve through commodified labor because the party-state had not guaranteed food, employment, or welfare to rural people in the way it had for urbanites. In contrast, urban migrants had to be taught the precariousness of their condition. In this sense, the figure of the migrant worker as it was constructed in the Shenzhen official press should be thought of as a precursory icon crystallizing a major socioeconomic and political transformation being carried out by the post-Mao Chinese leadership—the vast commodification of labor that would to touch almost all categories of the population. Therefore, the subject of this chapter is actually the very heart of the party-state and its search for new modes of legitimization of the social hierarchy.

Notes

1. In this chapter, I do not study the ways in which these officially sponsored values and norms are reworked and negotiated by migrant workers in their everyday lives. For studies that deal with this, see Tamara Jacka, *Rural Women in Urban China: Gender, Migration and Social Change* (New York: M. E. Sharp, 2006); Wanning Sun, *Subaltern China. Rural Migrants, Media, and Cultural Practices* (Lanham, MD: Rowman and Littlefield, 2014); and Eric Florence, "Migrant Workers in the Pearl River Delta: Discourse and Narratives about Work as Sites of Struggle," *Critical Asian Studies* 39, no. 1 (2006): 121–51.

2. Youwei Li, *Shenzhen jingji tequ de tansuo zhi lu* (Shenzhen: Guangdong Renmin Chubanshe, 1995), 232–34.

3. Mary Ann O'Donnell, "The Ambiguous Possibilities of Social and Self-Transformation in Late Socialist Worlds," *Drama Review* 50, no. 4 (2006): 97.

4. Georges T. Crane, "Special Things in Special Ways: National Economic Identity and China's Special Economic Zones," *Australian Journal of Chinese Affairs* 37 (1994): 76, 83, 89.

5. Youwei Li, *Shenzhen jingji* (Shenzhen: Shenzhen Tequ Chubanshe, 1995), 231–32.

6. In this chapter, for reasons of space constraints, the analysis is limited to the exploration of positive of values that are fostered by the party within Shenzhen Spiritual Civilization work in the 1980s and 1990s, which aim at rearing Shenzhen's "the new person." But spiritual civilization also implies defining this new person or the legitimate migrant worker in opposition to the kinds of behaviors and people that are to be rejected. In Shenzhen's case, the people whose presence is undesired and unsightly are mainly the "three withouts" people.

7. This focus on spiritual civilization was actually a major way in which Deng Xiaoping, Hu Yaobang, and other CCP leaders legitimized the economic reforms in the early 1980s and 1990s. The development in the material (economic) sphere had to be checked by a continuous emphasis on "spiritual civilization," which referred to moral and social order as well as to party-state-sponsored normalizing and disciplining of the various groups of Chinese society. See Borge Bakken, *The Exemplary Society: Human Improvement, Social Control, and the Dangers of Modernity* (Oxford: Oxford University Press, 2000).

8. Yuange Ni, Lijun Peng, and Yuanzhang Shen, *Shenzhen: Maixiang shehuizhuyi jingji* (Beijing: Renmin Chubanshe, 1994), 234.

9. Hairong Yan, "Neo-Liberal Governmentality and Neo-Humanism: Organizing *Suzhi/* Value Flow through Labor Recruitment Networks," *Cultural Anthropology* 18, no. 4 (2003): 492.

10. Eli Friedman and Ching Kwan Lee, "Remaking the World of Chinese Labour: A 30-Year Retrospective," *British Journal of Industrial Relations* 48, no. 3 (2010): 507–33.

11. Ching Kwan Lee, "Three Patterns of Working-Class Transition in China," in *Politics in China: Moving Frontiers*, ed. Jean-Louis Rocca and Francoise Mengin (New York: Palgrave Macmillan, 2002), 62–92.

12. Yuen-Fong Woon, "Circulatory Mobility in Post-Mao China: Temporary Migrants in Kaiping County, Pearl River Delta Region," *International Migration Review* 27, no. 3 (1993): 578–604; Shen Tan, "The Relationship between Foreign Enterprises, Local Governments, and Women Migrant Workers in the Pearl River Delta," in *Rural Labor Flows in China*, ed. A. West Loraine and Yaohui Zhao (Berkeley: University of California Press, 2000), 292–309. Beginning in the 1980s, with its mix of pro-investment measures and deregulation of labor, the Pearl River Delta witnessed an unprecedented parallel increase of foreign investments and rural migrant workers. From 1987 on, the population holding Shenzhen temporary household registration certificates outnumbered the permanent population by 51.8 percent, and by 1994, this proportion of temporary population had reached 72 percent. In 2004, out of a total population for Shenzhen of 5,975,000, 4,324,200 were holding a temporary registration. *Shenzhen Statistical Yearbook* (Beijing: China Statistics Press, 2005).

13. Chris-Chi Chan, *The Challenge of Labour in China: Strikes and the Changing Labour Regime in Global Factories* (Abingdon: Routledge, 2010). Concurrently, in the face of increasingly conflicting labor relations, the party-state has developed a whole body of legislations and regulations such as the 1994 Labor Law, the much-debated 2008 Contract Law, and the Conciliation and Arbitration Law, which was also promulgated in January 2008.

14. From 1988 on, for each worker they wanted to hire, enterprises had to comply with a series of complex procedures at the District Labor Bureau in order to hire workers temporarily. The permits obtained had to be renewed on a yearly basis. Once these procedures were accomplished, the enterprises had to apply to the Public Security Bureau in order to obtain a certificate of temporary residence registration. Thereafter, the enterprises had to apply for a temporary household registration (*hukou*). Eventually they had to apply to the District Public Security Bureau for a one-year temporary residence certificate. Pun Ngai, *Made in China: Women Factory Workers in a Global Workplace* (Durham, NC: Duke University Press, 2005).

15. By "routine repression," Scott referred not to outright use of force—or even the memory of massacres or stormy repressions that peasants would still retain fresh memory of—but to "the steady pressure of everyday repression backed by occasional arrests, warnings, diligent police work, [and] legal restrictions." In most Chinese cities, each category of rural migrants has been subjected to differentiated degrees of state intervention, both formal and informal, in spheres of residence, employment, reproductive practices, and so on. James C. Scott, *Weapons of the Weak* (New Haven, CT: Yale University Press, 1985), 274.

16. Robin Cohen, *The New Heliots: Migrants in the International Division of Labour* (Gower: Aldershot, 1988), 20.

17. The newspapers under scrutiny in this chapter are all linked institutionally to Shenzhen authorities and are all under the supervision of the party propaganda department. The *Shenzhen Special Zone Daily* (*Shenzhen Tequbao*; hereafter STQB) is under the direct control of the Shenzhen municipal authorities; the *Evening of Shenzhen* (*Shenzhen Wanbao*; hereafter SWB) is linked to the latter paper but targets a more popular audience, while the *Shenzhen Legality Daily* (*Shenzhen Fazhibao*; hereafter SFB) is linked to and intended for Shenzhen administrations of

justice and public security. All articles relating to migrant workers in these newspapers for the years 1989, 1990, 1994, and 1998–99 have been systematically analyzed (January, February, March and August for each year).

18. Delia Devin, *Internal Migration in Contemporary China* (New York: St. Martin's Press, 1999); Eric Florence, "Migrant Workers in the Pearl River Delta: Discourse and Narratives about Work as Sites of Struggle," *Critical Asian Studies* 39, no. 1 (2007): 120–50.

19. From the middle of the 1990s on, there has been an increase in the number of such magazines in China. Some of these magazines may be linked to mainstream newspapers or to government or party authorities while others may not have such straightforward institutional links. One should also distinguish between magazines that are officially registered and those that are not. This distinction may be of importance, since it may imply a different editing process—that is, officially registered magazines are likely to entail a more constraining ideological editing process.

20. The first of these writers is Anzi. In 1987, she started writing in the *Special Zone Culture* magazine. Her first book, *The Resting Post of Youth: The True Story of a Shenzhen Female Workers*, was published in 1991. In 1992 she started to work for the Shenzhen propaganda department.

21. See O'Donnell, "The Ambiguous Possibilities," 96–119. The need for the preservation of the quality of the environment for investors was actually stressed very recurrently in Shenzhen's official newspapers' accounts of the cleansing campaigns aimed at the unwanted rural migrants, the "three withouts." This also enabled the local party-state to present an image of itself as capable of preserving such an environment. Ann Anagnost makes such an argument in her work on "the quality of the population." For her, the whole teleological discourse on "civility" (*wenming*) and on the low quality of the rural population, which is located on a line from backwardness to civility, provides the Party with a justification for its role as the entity that can help the rural masses move toward civility and raise their quality. Ann Anagnost, *National Past-Times: Narrative, Representation, and Power in Modern China* (Durham, NC: Duke University Press, 1997).

22. Dorothy Solinger, *Contesting Citizenship in Urban China: Peasant Migrants, the State, and the Logic of the Market* (Berkeley: University of California Press, 1999).

23. Constance Clark, "The Politics of Place Making in Shenzhen, China," *Berkeley Planning Journal* 4 (1998): 104.

24. Clark, "Politics of Place Making," 103–25. See Bach, chapter 7 of this volume, for a similar characterization of a complex and "contradictory, even cacophonous" identity for Shenzhen, which is actually both urban and rural.

25. Clark, "Politics of Place Making," 238–39.

26. See O'Donnell, chapter 2 of this volume.

27. We may consider that in post-Mao China, in the background of social comments on the social hierarchy, there stands a strong criticism of Maoist society "as a means of radical disengagement from Maoist socialism." Lisa Rofel argues, "Economic reform is also and most significantly a space of imagination." Lisa Rofel, *Other Modernities: Gendered Yearnings in China after Socialism* (Berkeley: University of California Press, 1999), 29, 98.

28. Ching Kwan Lee, *Against the Law* (Berkeley: University of California Press, 2007), 204–6. *Dagong* means the kind of work performed by rural migrant workers, which often entails highly intense regimes of production and various degrees of exploitation. The term has slowly been used to include wider sectors of laborers and employees.

29. *Shenzhen Special Zone Daily*, January 7, 1994, 6.

30. It is worth noting that most of the migrant workers I interviewed from 1999 to 2010 in the Pearl River Delta clearly expressed that they wanted to be distinguished from the "white

collars" and that their work was characterized by its painful and unstable character. Some of them actually said they ought to be called the "black collars." Fieldwork notes, 1999, 2001, 2003, 2006, 2008, 2010.

31. See, for example, *Shenzhen Special Zone Daily*, January 1, 1994, 6; *Shenzhen Special Zone Daily*, January 14, 1994, 6; *Shenzhen Special Zone Daily*, January 28, 1994, 6; *Shenzhen Evening Daily*, February 13, 1994, 3; *Shenzhen Evening Daily*, February 21, 1994, 1; and *Guangzhou Ribao*, February 4, 1994, 1.

32. The term "builder of the zone" was used to oppose the term "gold diggers" in the debates on the values of Shenzhen society that ensued within the "Shekou storm." O'Donnell highlights that Shenzhen University's former president Luo Zhengqi viewed participation in the construction of buildings as expressing patriotism and self-sacrifice and said that it was "the highest expression of intellectual life." See O'Donnell, chapter 2 of this volume.

33. See *Shenzhen Evening Daily*, February 2, 1994, 3; and *Shenzhen Evening Daily*, March 29, 1998, 6 for other illustrations.

34. Wu Zhong et al., *Zouxiang xiandaihua: Shenzhen 20 nian tansuo* (Shenzhen: Haitian chubanshe, 2000). Furthermore, in the same publication, it is added that considering how to solve the problems of education and management of this rural migrant population is another important issue for Shenzhen municipality cadres: As "so many of the workers from outside have a rather low quality, in an environment of fierce competition, they often are under heavy psychological pressure. Therefore, what most of them need, it is explained, is spiritual consolation." If they were to lack such consolation or provision of comfort, it is stated, they "could easily try to get support from religious or even all kinds of small informal organizations."

35. The use of this expression actually refers to the possibility for those who could afford it to buy a Shenzhen household registration, referred to in the 1990s as "blue *hukou*." The use of the term "green card" in this article is awkward but may be related to Shenzhen's fascination with the United States in the early years as an "immigrant city."

36. *Shenzhen Special Zone Daily*, March 8, 1998, 9.

37. This phrasing may be derived from the well-known expression "Let a hundred flowers blossom and a hundred school of thought contend" (*baihua qifang, baijia zhengming*), or "Let a hundred flowers blossom, weed through the old to bring the new" (*baihua qifang, tuichen chuxin*), used by Mao during the One Hundred Flowers campaign (1957), which itself came from the Warring Kingdoms period (403–221 BC).

38. *Shenzhen Special Zone Daily*, March 8, 1998, 6.

39. See Bach, chapter 7 of this volume, on the narrative construction of Shenzhen as based on the opposition of urban and rural.

40. Yan makes the argument that as the countryside is discursively constructed as a space of lower value for economic development, it is often depicted as space of death by rural people. Yan Hairong, *New Masters, New Servants: Migration, Development, and Women Workers in China* (Durham: Duke University Press, 2008). It is actually interesting to note that as migration to the cities increases, the countryside becomes closely associated with the sheer impossibility of achieving one's plan of personal emancipation.

41. See Winnie Wong, chapter 9 in this volume.

42. See Xia Guang, *Shenzhen yekong bu jimo* (Shenzhen: Haitian chubanshe, 1999), 1–47, for illustrations of this association of Shenzhen with youth. This volume is a selection of transcripts of a radio program in which a female journalist replies to the queries of people who are living in Shenzhen.

43. *Shenzhen Special Zone Daily*, February 15, 1998, 6.

44. *Shenzhen Special Zone Daily*, March 8, 1998, 6.

45. I draw this insight from Pun Ngai, who wrote that the female migrant workers were the object of the triple oppression of "global capitalism, state socialism, and familial patriarchy." Pun, *Made in China*, 4. According to the evidence provided by a Foxconn 2010 study carried out in several Foxconn plants, slogans such as "Run towards the magnificent dream," "To suffer is the basis for richness," and "You will extend your dream from here to your future" were placed on the walls of the plant. Foxconn, "Liang an gao xiao diaocha yanjiu baogao," unpublished report, 2010. I develop this argument further through an analysis of migrant workers' rationale of why they left their village in Eric Florence, *Struggling around Dagong: Discourses about and by Rural Migrant Workers in the Pearl River Delta* (PhD dissertation, University of Liege, 2008), http://hdl.handle.net/2268/109931.

46. See Mason, chapter 10 in this volume.

47. On how laid-off workers are represented in Beijing tabloids, see Zhao Yuezhi, "The Rich, the Laid-off and the Criminal in Tabloid Tales: Read All about It!," in *Unofficial China: Popular Culture and Thought in the People's Republic*, ed. Perry Link and Richard Madsen (Boulder, CO: Westview Press, 2002), 111–35.

48. *Shenzhen Special Zone Daily*, March 15, 1998, 6.

49. *Shenzhen Special Zone Daily*, February 15, 1998, 6; *Shenzhen Special Zone Daily*, March 8, 1998, 6; *Shenzhen Special Zone Daily*, January 17, 1999, 6; March 28, 1999, 6.

50. *Shenzhen Special Zone Daily*, January 31, 1999, 6.

51. *Shenzhen Special Zone Daily*, March 28, 1999, 6.

52. *Shenzhen Special Zone Daily*, March 28, 1999, 6.

53. *Shenzhen Special Zone Daily*, March 28, 1999, 6.

54. *Shenzhen Special Zone Daily*, January 31, 1999, 6.

55. *Shenzhen Special Zone Daily*, March 28, 1999, 6. On this March 28 page, three articles conveyed a very similar kind of warning.

56. Solinger, *Contesting Citizenship*, 7.

57. Gerard Noiriel, *Etat, nation et immigration: Vers une histoire du pouvoir* (Paris: Belin, 2001), 201.

58. Kitty Calavita, "Italy and New Immigration," in *Controlling Immigration*, ed. Wayne Cornelius, Philip Martin, and James Hollifield (Stanford, CA: Stanford University Press, 2004), 319.

59. It is a well-known fact, for instance, that an important informal sector developed in Shenzhen within the segmented labor market at the end of the 1980s and the beginning of the 1990s. Many of the people who were part of the "illegal" segment of this workforce (the "three without people") were part of this informal sector. According to Liu, for the Township and Village Enterprises of the two major industrial districts of Bao'an and Longgang, only 50 percent of the migrant workers in factories had applied for temporary registration, most of whom represented a major component of the "economic prosperity of the city." Liu Kaiming, *Bianyuan ren: Migrant Labor in South China* (Beijing: Xinhua Chubanshe, 2003), 62–63.

60. Michael R. Trouillot, "The Anthropology of the State in the Age of Globalization," *Current Anthropology* 42, no. 1 (2001): 126.

61. Pun, *Made in China*, 11.

PART II

Exceptions (1992–2004)

5

Laying Siege to the Villages: The Vernacular Geography of Shenzhen

MARY ANN O'DONNELL

In 2004, Shenzhen became the first Chinese city without villages. Full stop. This fact bears repeating: legally, there are no villages in Shenzhen. As of 2016, Shenzhen Municipality had a five-tiered bureaucracy consisting of the municipality (*shi*), districts (*shiqu*), new districts (*xinqu*), subdistricts, or streets (*jiedao*), and communities (*shequ*). This chapter provides an overview of how "urban villages," or "villages in the city" (*chengzhongcun*), emerged within Shenzhen's shifting landscape. The goal of this chapter is to suggest the historic sources of administrative power that have enabled the rise of urban villages as unofficial yet legitimate social forms (figure 6).[1]

Under Mao, rural areas were China's revolutionary heart, and "villages surrounded the city" was an explicit political, economic, and social strategy for revolutionary change by cutting off cities from both rural supply chains and other cities.[2] The Mandarin expression "surrounds" (*weirao*) can also be translated as "lays siege to," highlighting the rural basis of the Chinese Revolution. Early Chinese Communists had followed the Russian example and entered cities to organize workers. However, when nationalist forces led by Generalissimo Chiang Kai-shek violently suppressed Communist organizations in Chinese cities, the Communists retreated to the countryside. Moreover, Communists and local people identified colonial ports such as Shanghai and Guangzhou with the proliferation of traitors, parasitic merchants, and corrupt officials. Consequently, while Marx claimed that modern history was the urbanization of the countryside, the Chinese revolution aimed to reoccupy and purify the cities with rural people and values. Beginning in 1927 until the occupation of Beijing in 1949, the Communists organized rural resistance to both Japanese invaders and Nationalist hegemony, literally surrounding the cities with an estimated five million rural soldiers.

Historically, there were legally constituted villages in Shenzhen. The present ambiguity over the status of villages and villagers is a result of contradictions between Maoist economic planning and post-Mao liberalization policies and on-the-ground negotiations over the current status of "villages" as legal subjects of the state. Locally, this process has been called "cities surround the countryside," which not only resonates ironically in post-Mao China but also identifies urban poverty with rural status. The establishment of Shenzhen signaled the beginning of a new era in Chinese history—"cities surround the villages" (*chengshi weirao nongcun*).[3] Under Mao, the country was segregated into rural and urban areas. In rural areas, villages were designated production teams and organized into work brigades that were administered by communes. Communes had to meet agricultural production quotas that financed industrial urbanization and socialist welfare policies in cities, which were tellingly defined as "nonagrarian" (*feinong*). Importantly, the *hukou*, or household registration policy, literally kept people in place—the allocation of food, housing, jobs, and social welfare took place through *hukou* status. Food and grain coupons were city specific, for example, and a Shanghai meat coupon could not be legally exchanged in a neighboring city, let alone in Beijing. In rural areas, however, communes and production brigades provided neither food coupons nor housing to members. Instead, brigade members produced their own food (usually what was left over after production quotas had been met) and built their own homes, or rural dormitories, as they were known in the Maoist system.[4]

The elevation of Bao'an County to Shenzhen Municipality created an anomalous situation within socialist China because the administrative division of Shenzhen into the SEZ and New Bao'an County only legalized new economic measures; it did not transfer traditional land rights from brigades and teams to the new municipal government. Instead, the first task of urban work units that came to the SEZ was to negotiate the equitable transfer of land rights from the collectives to the urban state apparatus. The goal was to ensure that rural workers would continue to have space for housing and enough land to ensure agricultural livelihoods. And this is where historical village identities reasserted themselves. In theory, the urban work units negotiated with brigade and team leaders to transfer the administration of land from the rural to the urban sector of the state apparatus. In turn, the brigades and teams would continue to produce food for the new urban settlements. In practice, however, brigade and team leaders acted on behalf of their natal villages and co-villagers, asserting a prerevolutionary social identity rather than straightforwardly adhering to the Communist administrative hierarchy or "system (*tizhi*)," as it is glossed in the vernacular.

There were four key dates in the enfranchisement of Shenzhen's (former) villages. In 1979, when the Guangdong provincial government elevated Bao'an County to Shenzhen Municipality, the area was legally rural, and the majority of its estimated 300,000 residents had household registration in one of 21 communes, which were further organized into 207 production brigades. However, *hukou* status notwithstanding, the integration of brigades and teams had not been complete and members continued to identify with traditional village identities.

In 1980, the central government further liberalized economic policy in Shenzhen by establishing the area that bordered Hong Kong as a Special Economic Zone (SEZ). This internal border was known as the "Second Line" in contrast to the Sino-British border at Hong Kong, or the "First Line."[5] The redesignation legalized industrial manufacturing and foreign investment (primarily from Hong Kong) in the new SEZ. Outside the Second Line, Shenzhen Municipality established New Bao'an County, which was still legally rural and administered through collective institutions. Today in Shenzhen, the areas within the Second Line are known as the "inner districts" (*guannei*) and the areas of New Bao'an County are known as the "outer districts" (*guanwai*).

In 1982, the amended Constitution formally outlined the different property rights under rural and urban government. According to Article 8 of the Chinese Constitution:

> Rural people's communes, agricultural producers' co-operatives, and other forms of co-operative economy such as producers' supply and marketing, credit and consumers co-operatives, belong to the sector of socialist economy under collective ownership by the working people. Working people who are members of rural economic collectives have the right, within the limits prescribed by law, to farm private plots of cropland and hilly land, engage in household sideline production and raise privately owned livestock. The various forms of co-operative economy in the cities and towns, such as those in the handicraft, industrial, building, transport, commercial and service trades, all belong to the sector of socialist economy under collective ownership by the working people. The state protects the lawful rights and interests of the urban and rural economic collectives and encourages, guides and helps the growth of the collective economy.[6]

In contrast, according to Article 10, land in cities is owned by the state: "Land in the rural and suburban areas is owned by collectives except for those portions which belong to the state in accordance with the law; house sites and private plots of cropland and hilly land are also owned by collectives. The state may in the public interest take over land for its use in accordance with the law. No organization or individual may appropriate, buy, sell or lease land, or unlawfully transfer land in other ways. All organizations and individuals who use land must make rational use of the land."[7] The legal slippage between collective identity within China's rural state apparatus and collective identity through membership in a traditional village arose because, although the Constitution and subsequent Land Law of 1986 stated that rural farmland belonged to the collective, neither document went so far as to define what a collective actually was in law.

In October 1986—at least six years after its official establishment—Shenzhen Municipality redesignated its Maoist districts and *xiang* as market towns and villages, respectively. In practice, this meant that the city's "villages" in fact comprised village settlements that were adjacent to but traditionally independent of each other. An administrative village would comprise three to possibly five or six traditional villages, which (over time) would allow for significant restructuring from the ground up.

The difference between rural and urban property rights has been the foundation for post-Mao reforms, first in Shenzhen and then throughout the country. Moreover, the contradiction between the fact that villages no longer have legal status in Shenzhen and their traditional claims to land rights and social status—both of which are recognized by Shenzhen officials and residents—has constituted a serious political challenge for Shenzhen officials, who have viewed the villages as impediments to "normal" (*zhengchang*) urbanization. Officials have defined "normal" urbanization with respect to the Shenzhen's Comprehensive Urban Plan, which has already gone through four editions (1982, 1986, 1996, and 2010; see figure 7). In other words, "normal" urbanization has referred to either formal or informal urbanization that has secured legal recognition. In contrast, Shenzhen's urban villages emerged informally as local residents not only built rental properties to house the city's booming migrant population but also developed corporate industrial parks, commercial recreational and entertainment centers, and shopping streets. As of January 2015, for example, it was estimated that half of Shenzhen's population lived in the villages. At the time, the official population was roughly eleven million, but in January 2016, the city's new party secretary Ma Xinrui confirmed that Shenzhen's administrative population was twenty million.[8]

In Shenzhen, urban villages have been the architectural form through which migrants and low-status citizens have claimed rights to the city.[9] These densely inhabited settlements also provided the physical infrastructure that has sustained the city's extensive grey economy, including piecework manufacturing, spas and massage parlors, and cheap consumer goods. Importantly, informal urbanization in the villages has occurred both in dialogue with and through disregard for formal plans. On the one hand, informal urbanization in Shenzhen urban villages has ameliorated many of the more serious manifestations of urban blight that plague other boomtowns. Unlike Brazilian favelas, for example, Shenzhen urban villages are not located at the edge of the city but are distributed throughout the entire city, and many urban villages occupy prime real estate. Consequently, Shenzhen's urban villages have been integrated into the city's infrastructure grid and receive water and electricity and also have access to cheap and convenient public transportation. Moreover, as Shenzhen has liberalized its *hukou* laws, urban villages have been the form through which migrants have had access to social services, including schools and medical clinics. Thus Shenzhen's urban villages have provided informal solutions to boomtown conditions. On the other hand, the lack of a formal legal status for urban villages and, by extension, the residents of urban villages, has allowed the municipality to ignore residents' rights to the city via the convenience of centrally located low-income neighborhoods. In fact, the ambiguous status of urban villages became even more vexing in 2007 when the Shenzhen government initiated a plan to redevelop urban villages. It has been widely assumed that the government promulgated the new plan in order to benefit from the real estate value of urban village settlements. Critically, the municipality's plans for urban renovation have compensated original villagers while ignoring the resettlement needs of migrant residents. Thus the status of at least half of Shenzhen's population suddenly entered into public discourse, as it has become apparent that although the urban villages resulted from informal practices, they have been the basis for the city's boom.

Each of the following sections in this chapter explores the social antagonisms that have emerged through the transformation of Bao'an County into Shenzhen Municipality via informal urbanization in the villages. With respect to recent Chinese history, this level of specificity aims to make salient how Shenzhen enabled national leaders to reform Mao's rural revolution. With respect to contemporary research on megacities, this essay draws attention to the ways in which architectural forms have facilitated contingent urbanisms that exclude the poor from desired futures.[10]

Concentric Occupations: The Nantou Peninsula

Shenzhen's vernacular geography references to at least five historical social institutions: the imperial salt monopoly, Ming and Qing lineage villages, British colonialism, Maoist collectivization, and post-Mao reforms.[11] Analysis of the vernacular geography of the Nantou Peninsula indicates how each of these social institutions was built into the landscape and, in turn, how they shaped and were shaped by subsequent institutions and concomitant spatial transformation. In contemporary Shenzhen, these histories are manifest through not only place names but also settlement patterns. These patterns are especially visible on the Nantou Peninsula, where new settlements have grown at the edges of older sections. In turn, the older settlements have been downgraded and converted into low-income neighborhoods.

Archeological evidence suggests that sea salt was harvested along the eastern banks of the Pearl River for at least two thousand years. The Han officially incorporated local salt production into China's imperial monopoly in 116 BCE when the area was designated the "Eastern Office" (*Dongguan*) of the imperial salt monopoly in Guangzhou Prefecture (present-day Dongguan, Shenzhen, and Hong Kong). The *Dongguan* salt intendant was located on the Nantou Peninsula, which had protected harbors and access to Guangzhou via the Pearl River; the Eastern Office *yamen* was built at the walled city, Nantou. Over centuries, local sea salt production waned as larger, more efficient salt works were built in Tianjin and Sichuan, but it was successive bans on ocean travel (beginning with the Ming ban in 1371 and continuing intermittently until the Kangxi Emperor forced coastal peasants to move inland) that effectively destroyed the salt fields each time when peasants were relocated inland. Nevertheless, by the Ming dynasty, the area was famous for its oyster and pearl production, lychees, and "fragrant wood"—the tree used to produce incense. In addition, the harbors of the Pearl River's eastern coastline are significantly deeper than those on its western coastline, a fact that has shaped both Chinese and foreign access to Guangzhou. Chinese maritime access to the South China Sea traditionally went via Humen and Nantou. Consequently, despite the Ming ban on ocean travel, Guangzhou remained the southern gate to China, and the ports on the eastern coast of the Pearl River became even more coveted by international traders.[12]

During the first year of the reign of the Ming Wanli Emperor (1573), Dongguan was partitioned into Dongguan and Xin'an Counties. Xin'an County comprised present-day Shenzhen and Hong Kong. Nantou remained the county seat of the newly established county (see figure 3). By the late eighteenth century, Guangzhou had become not only an important financial center but also the center of opium trade. The first Opium War ignited when Lin Zexu dumped the opium stocks of British traders into the Pearl River. In turn, the traders successfully pressured the British government to use military force to secure compensation for their losses. China's defeat in the Opium Wars resulted in British colonialization of southern Xin'an County, including Hong Kong Island, the Kowloon Peninsula, and the New Territories. The Sino-British border was drawn along the Shenzhen River and passed just south of Shenzhen Market. The laying of the Kowloon-Canton Railway (KCR) in 1913 further shifted the flow of goods and people toward Hong Kong and away from Nantou. Small-scale trade between settlements on the Pearl River continued, although Nantou no longer played a dominant role in the regional political economy. Instead, Shenzhen Market, the first station on the Chinese side of the KCR became the political and economic center of Xin'an County, which was renamed Bao'an at the start of the Nationalist era.[13] In fact, the establishment of Shenzhen explicitly invoked colonial history, making the return of Hong Kong to Chinese sovereignty one of the key political impulses behind economic liberalization. Maoist modernization of Nantou, for example, included a two-lane road (today known as New South Road, *Nanxin Lu*), which was laid parallel to the ancient South Gate Road (*Nanmen Lu*) and connected the peninsula villages to the national railroad and highway system. In the post Mao-era, however, state investment has aimed to urbanize the area rather than to integrate rural settlements into the state apparatus. Land reclamation along the Pearl River coastline gives the clearest indication of the scale and ambition of these plans—replacing Hong Kong and possibly even Guangzhou in the global organization of South China trade (figure 8).

The reform-era transformation of the Nantou Peninsula illustrates the broad contours and social contradictions that have characterized "cities surround the countryside." During the Ming dynasty, a pounded earth wall enclosed Nantou, but by the time of the first Opium War, the wall had crumbled into disuse, and only the southern and eastern gates still stood. A road stretched from the decrepit Southern Gate and along the coast of the Pearl River to Nanshan Village, which was located at the foot of Nanshan Mountain. Between Nantou Old City and Nanshan Village, six villages—Guankou, Yongxia, Tianxia, Xiangnan, Beitou, and Nanyuan—claimed land that included access to the Pearl River, a portion of South Gate Road that they identified as Village Main Street (*Lao Jie*), and farmlands that extended inland. However, through land reclamation and the emplacement of a grid of four- and six-lane roads such as the Qianhai Thoroughfare, Shenzhen's rural origins have been surrounded and have isolated South Gate Street neighborhoods from the larger city (see figure 9).

Neoliberalizing the Bamboo Curtain: Luohu and Dongmen

Two factors—the economy and politics—motivated the 1953 decision to move the Bao'an County seat from its historical site at Nantou on the Pearl River to Caiwuwei, a village located next to Shenzhen Old Town and the Luohu checkpoint, which was the first station on the Chinese side of the Kowloon-Canton Railway. Economically, the socialist planned economy relied on an extensive railway system to transform the scale of the Chinese economy from a traditional economy of peasants to a modern economy based on mass transfers of goods and people, and the Shenzhen train station connected the area to the national railway system. In addition, the location of the new county seat also facilitated processing of foodstuffs that were sold for hard currency in Hong Kong (see figure 10). Politically, Shenzhen Market was located at the Sino-British border where the Chinese military was stationed after England supported the American action in Korea. This border became metaphorically known as the Bamboo Curtain, a reference to the Cold War Iron Curtain that split Europe into capitalist and Communist blocks. Luohu Bridge was the southern entry point into the People's Republic. However, the Bamboo Curtain was porous. Beginning in 1955, it is estimated that between one million and two-and-a-half million Mainlanders attempted to escape through Bao'an to Hong Kong, with mass exoduses occurring in 1957, 1962, 1972, and 1979.

The establishment of the Shenzhen SEZ in 1980 was also motivated by the desire to take advantage of the proximity of Hong Kong in order to achieve national goals. The earliest plan for the Shenzhen SEZ was to develop the 50 km^2 area that extended east and west from the KCR railway tracks, upgrading extant roads and developing the rice paddies and lychee orchards that surrounded the county headquarters, commercial area of Old Town, and extant villages. One of the most important decisions was to restructure the traffic flow of the area. First, the railroad tracks that traversed county headquarters were removed. Industrial parks were built along the remaining extension line and the northern portion of the railroad. Second, the area's main road, Jiefang (Liberation), was widened west beyond county headquarters and east near Huangbeiling. The stretch of Jiefang Road that traversed Old Town remained intact. The People's Engineering Corps lay a segment of new road that went around the southern border of Old Town to Dongmen, connecting the newly widened sections of Jiefang Road to the area's largest market. This new road was called Shennan Thoroughfare and its layout informed all subsequent urbanization of the area. Subsequent development followed the railroad either north toward Buji Township or west on Shennan Thoroughfare toward Guangzhou.

The collective was the institutional form through which villages took advantage of the first reforms. Village teams immediately adjacent to Luohu Bridge, Wenjing Crossing, county headquarters, and the KCR railroad tracks boomed. In 1980, the village teams had four primary sources of revenue: monetary compensation for land rights transfer from collectives to the state, profits from agricultural produce sold to the immigrants, rental properties, and contraband goods that were smuggled into Shenzhen and sold in either local markets or a stall in Dongmen market. However, very quickly, the village teams also built leisure facilities and commercial areas that targeted Hong Kong day-trippers, who enjoyed services and bought products at prices well below Hong Kong rates. Indeed, by Deng Xiaoping's 1984 tour of the SEZ, the Luohu village teams had become the symbol of "small prosperity" (*xiaokang*), the material quality of their homes, furniture, and income even surpassing that of workers in state-owned industries, let alone in the rest of China.

The most famous Luohu village team was Yumin (Fishermen), which held an important place in both national Chinese and local Shenzhen symbolic geography for three reasons. First, the name "Yumin" indicated the ongoing smoothing of local hierarchy and integration of Dan households first into Bao'an County and then the city. Yumin villagers were ethnically Tanka (literally, "egg households"), the group of South Chinese fishermen who did not have land settlement rights. Historically, local governments did not permit Dan to wear shoes when they came ashore, to use red lanterns at wedding ceremonies, to marry land villagers, or to participate in the imperial examination. Under Mao, the Dan had been given land from Caiwuwei Village teams (location of Bao'an County headquarters) and had moved onshore to build homes.[14]

Second, Yumin was one of the first village teams to take advantage of reforms. In 1979, even before the official establishment of the SEZ, the Yumin Secretary, Deng Zhibiao, organized the purchase of tractors to increase the size of Yumin fish farms by converting all unused land into fisheries, increasing production from several to more than one hundred *mu* (or 666.7 m^2). According to Deng Zhibiao's calculations, at the time, one *mu* of fish produced several thousand yuan. Within a year, the village team had saved enough money to collectively build two- to three-story private homes as well as factories. Yumin Village thus had the distinction of being the first "ten-thousand-yuan village" in the country. When Deng Xiaoping visited Shenzhen in 1984, he was taken to view one of the small two- to three-story houses that the team had built for each household and shown a modern parlor, complete with a television, curtains, and new furniture. In news reports about Deng's 1984 Southern Tour, Yumin was mistaken for Shenzhen's "original settlement" and the myth that Shenzhen was once upon a time a small fishing village embedded itself in future reports about the city.

Third, Yumin's location meant that they were positioned to develop rental properties for the massive influx of Shenzhen migrants. Even as Deng Xiaoping was pushing for reforms in the fourteen coastal cities, by 1986 when Shenzhen's collectives were officially designated "villages" within the municipal apparatus, Yumin villagers were already razing the original private homes and putting up six- to eight-story handshake buildings to take advantage of rental opportunities. After all, Yumin Village was conveniently located next to the train station. Consequently, in 2000, when Luohu District began to negotiate village renovation with Yumin Village, the stakes had been raised significantly. At the end of the process in 2004, Yumin Village had been rebuilt as an upscale residential area under a single village-owned property management company. The New Village consisted of eleven twelve-story buildings and one twenty-story multipurpose building. Each village household was given thirty units within the new complex. And Yumin "Village" was itself an incorporated limited corporation.

Importantly, Yumin was only one of the Luohu-area village teams. Each of the other village teams—Caiwuwei, Hubei, and Xixiang, for example—underwent similar transformations with one important exception. Unlike Yumin, Caiwuwei, Hubei, and Xixiang had histories that stretched into the Ming-Qing dynasties. This meant their land holdings were more extensive than Yumin, and that gave them a stronger bargaining position vis-à-vis the state apparatus. Moreover, since the 2007 decision to make urban villages the focus of urban renewal, the Luohu villages have been the sites of the strongest popular resistance to upgrading for two reasons. First, as of 2013, the villages remained the cheapest and most convenient housing option for the downtown working poor. Second, the older sections of the villages represented the history of Shenzhen, both ancient and contemporary. More than thirty years after the establishment of the SEZ, Luohu has become an object of nostalgia for many early migrants, second-generation Shenzheners, and young professionals. Not unexpectedly, perhaps, the villagers themselves have been willing to sell their housing rights to the highest bidder, while low-income families have viewed the villages as gateways to better living conditions in one of Shenzhen's formal housing estates.

Informal Urbanization in the Outer Districts: National Highway 107

Shenzhen Township and Village Enterprises (TVEs) in the outer districts (formerly New Bao'an County) were quick to take advantage of reform policies.[15] By 1985, village teams, township brigades, and the recently reestablished Bao'an County government had already registered more than fifty industrial parks. Nevertheless, this massive social restructuring occurred outside and despite municipal urban planning. First, the total area of Shenzhen's informal industrial urbanization was more than four times greater than planned urbanization in inner districts (original SEZ). Second, urbanization in the outer districts occurred outside official urban planning. Moreover, the density of industrialization along National Highway 107 was not included in the 1986 Master Plan map of extent development. Indeed, as targets of urban planning, the outer districts did not appear in official maps until the release of the 1996 Shenzhen Comprehensive Plan. Third, the scale of development in the outer districts indicates the high level of informal organization in the villages. Informal urbanization did not arise sui generis but through the redeployment of TVEs, which represented not only the economic interests of the collective but also traditional identities and social constituencies—villages and local lineages.

In addition to the New Bao'an County TVEs, the outer districts also saw the development of private stock companies formed by alliances between individual villagers and investors, often from Hong Kong or an overseas Chinese community. The diversity of ownership notwithstanding, all these enterprises engaged in low-tech, labor-intensive assembly manufacturing. The factories themselves were long, concrete buildings, usually four to six stories high. These buildings usually had electricity and water hook ups and, depending on the complexity of assembly, individual tables for detail work. Outside firms contracted TVEs to assemble a product according to spec. Earliest manufactured goods included textiles, toys, and cheap electronics. These early industrial parks also included four- to six-story dormitories for migrant workers. Dorm rooms were narrow and fitted with four bunk beds. Opposite the door was a small window, while two bunk beds were placed back to back along the remaining walls. There was enough space between the bunk beds for residents to walk to their bed. Consequently, most conversations occurred sitting on a bed or outside on the lawn. Importantly, these factory complexes were built along either National Highway 107 or the railway, along which goods were transported to the port of Hong Kong and then shipped overseas. A 1992 map of the county seat of New Bao'an County illustrates that the planning for territory outside the Special Zone, including the county seat of New Bao'an County, was to be treated as an industrial park with limited social development. The map notes that the county seat was "1 kilometer from Nantou, 9 kilometers from Shekou, and 23 kilometers from Shenzhen" (figure 11). This sense of Shenzhen as distant and indeed separate from the Special Zone has remained part of everyday conversation, where residents of Bao'an, Longgang, and even Nanshan District speak of "going to Shenzhen." In contrast, Luohu and Futian District residents are more geographically precise and will say "going to Dongmen."

Development along National Highway 107 illustrates how *hukou* has not only shaped immigrant possibility but also influenced Shenzhen's urban form.

In the early 1980s, when the Shenzhen TVEs opened, there were no other manufacturing jobs available to rural workers, who were still tied to collectives and mandatory agricultural production quotas through *hukou* status. This meant that the Shenzhen TVEs had labor reserves throughout Guangdong and neighboring provinces, where rural workers were effectively excluded from wage labor. The number of migrants who took advantage of these opportunities constituted the Shenzhen population boom, transforming the landscape in three important ways. First, more people came than there were jobs and, by extension, dormitory space. This created an immediate need for rental properties. Second, the increasing population also needed food and social services, which in turn created more jobs for migrants across sectors, but primarily in construction and service industries. Third, urban construction of this area was largely informal. By the time that Shenzhen Municipality had elevated New Bao'an County to Bao'an and Longang Districts in 1990, local collectives had emerged as the de facto urban planners outside the SEZ. Moreover, when considered in terms of population and territory, urbanization in the outer districts constituted the primary form of urbanization in Shenzhen. By the mid-1990s, many domestic and international companies chose to rent use rights from the collectives and build campuses in the other districts, especially Longgang near the railway. Indeed, two of the world's highest-profile electronics manufacturers, Foxconn and Huawei, for example, built their campuses just beyond the Second Line in Bantian (see figure 5).

The fuzzy nature of ownership rights overdetermined the haphazard direction of urbanization in the outer districts. Indeed, throughout Shenzhen, the foresight of a collective leader and the willingness of members to coordinate development have shaped the quality of life in specific villages. In the post-Mao era, land ownership rights belonged to the collective, while "use rights" were delegated to members of the community. This slippage provided a brief window of opportunity for individual villagers to engage in individual profit-seeking activities; however, the most successful enterprises belonged to the county, townships, and villages that expropriated use rights by exerting their ownership rights. Indeed, conflicts between Shenzhen Municipality and its "urban villages" have also arisen due to the distinction between ownership and use rights. As of 1992 in the inner districts and 2004 in the outer districts, Shenzhen Municipality owned all land within its borders. However, through housing and industrial parks, the collectives continued to exercise use rights. Indeed, since 1992 and 2004, villages and developers have been negotiating compensation for transferring these rights; Shenzhen Municipality has mediated these transfers through its master plans.

Baishizhou: Neighborhoods for the Working Poor

In 2015, Baishizhou was the largest of the so-called urban villages in Shenzhen's inner districts. With respect to the overall layout of Shenzhen, Baishizhou occupied both the southern and northern sides of Shennan Middle Road at the peripheries of both Luohu (moving west) and the Nantou Peninsula (moving north), making it one of the most centrally located transit centers in the inner districts. As of 2013, Baishizhou had a total area of 7.4 km^2 and an estimated population of 140,000 residents, of whom roughly 20,000 held Shenzhen *hukou* and 1,880 were locals. The population density of Baishizhou had breached 189,000 people per square kilometer, more than twenty-five times that of municipal average of 7,500 people per square kilometer, a statistic that in 2012 made Shenzhen the fifth-most densely populated city on the planet. There were 2,340 low- and mid-rise buildings in the area with an estimated 35,000 units. Monthly rents ranged from RMB 700 to RMB 3,000, which were significantly cheaper than in neighboring Overseas Chinese Town (OCT) or nearby housing estates, where a "cheap" apartment could rent for RMB 4,000—the starting salary of many young architects, graphic designers, and other skilled workers of the OCTs Loft Creative Park (see figure 12).

Until January 2014, many Shenzhen garbage collectors for the area lived in Baishizhou's cheapest rentals, rural Mao-era dormitories where it is possible for three workers to share a 30 m^2 dorm room for RMB 200 a head, plus electricity and water. Old Cai, for example, was sixty-five years old when interviewed. He came to Shenzhen after retirement because his monthly pension was RMB 40 per month, while he and his wife needed RMB 20,000 annually—or about RMB 1,700 a month—to meet their expenses. In Baishizhou, he made a living collecting and reselling cardboard boxes and other garbage. He said he could save money because although there was no real profit, he made enough to support himself and to bring a little home for Chinese New Year. After the dormitories were condemned, Old Cai moved to a nearby handshake building, where he shared a smaller space with two other garbage collectors. However, the diversity of Baishizhou residents also includes working families who have lived in the area since migrating to Shenzhen more than twenty years ago and young professionals who are sharing their first flat independent of their families. One family from Sichuan, for example, rented a 60 m^2 two-bedroom apartment for RMB 1,700 a month, which the husband, his wife, her mother-in-law, and their two children shared. During the day, the parents worked at one of the OCT theme parks, while the mother-in-law took care of the children and housework. In addition, many of Shenzhen's young designers and architects who work in the OCT Loft, a renovated factory area for creative industry, live in higher-end handshake buildings, which sometimes include parking space for a car.

In addition to rental properties, the first floor of most Baishizhou buildings is used for commercial purposes, and the area boasts several commercial streets and at least two night markets and entertainment areas in addition to independent vendors and office space for independent carpenters, builders, and handymen. There is an elementary school and three nursery schools. Moreover, in between two of the abandoned factories of the Shahe Industrial Park, enterprising migrants have set up the Baishizhou Pedestrian Street, which mimics the Dongmen Walking Street. There are food stalls, toy vendors, and several juvenile rides.

Clearly, using the term *village* to describe this level of settlement density and diversity is misleading. Baishizhou is a vibrant urban area composed of five neighborhoods—Baishizhou, Shangbaishi, Xiabaishi, Xintang, and Tangtou—five historical villages that, under Mao, had been organized into a state-owned agricultural collective called Shahe Farm. Shahe itself was a branch of the Guangming Overseas Chinese Farm, a state-owned farm that was held by the Overseas Chinese Office of Guangdong Province. In the early 1980s, a 12.5 km^2 area of Shahe Farm was partitioned into two enterprise areas— OCT in the eastern section and Shahe Enterprises in the western section. In the mid-1980s, both OCT and Shahe built factories for assembly manufacturing. However, the management teams and access to investment capital were significantly different. OCT was a state-owned enterprise and its management team included educated professionals from China's major cities. In contrast, the former collective leaders managed Shahe and its development. In the post-Tiananmen era, when Shenzhen's low-tech, low cost manufacturing had ceased to be as profitable as during the 1980s, OCT developed theme parks—Splendid China, Window of the World, and Happy Valley—to stimulate the economy. In turn, this investment also enhanced the rental value of the area and drove the redevelopment of the former industrial park into a 798-like creative area.

Ruralization: The Ideology of Global Inequality

This chapter has aimed to show that Shenzhen's so-called urban villages are in fact urban neighborhoods that grew out of previous rural settlements through Maoist collectivization and rapid industrial urbanization in the post-Mao era. Nevertheless, the designation of "rural" or "village" still clings to these neighborhoods, making them the target of renovation projects and ongoing calls for upgrades. In turn, these calls justify razing neighborhoods and displacing the working poor with upper- and upper-middle-class residential and commercial areas. Recently, Caiwuwei was razed and rebuilt as the KK 100 Mall, while Dachong was razed and rebuilt as Huarun City, a massive development of office spaces, malls, and upscale condominiums. Hubei, the old commercial center in Luohu, has been designated as the next major area to be razed, while in late 2012, the Shenzhen government and Lujing developers announced their intention to raze and rebuild Baishizhou as a centrally located luxury development.

In Shenzhen, ruralization is primarily an ideological practice through which neighborhoods for the working poor and low-income families have been created by denying the urbanity of these neighborhoods and their residents. In an ideological reversal of the "cities surround the countryside" history of development in Shenzhen, the city's rural history is invoked to demonstrate that neighborhoods that grew out of villages are continuations of the village rather than the result of informal urbanization. Indeed, there are few actual remains of Shenzhen's rural past. Instead, the target of official rural renovation projects are, in fact, the informal housing and industrial parks that were built roughly between the mid-1980s through 2004–5, when the municipal government began actively preventing informal construction.

All this to make a very simple point: when we speak of rural urbanization in Shenzhen, we are—to redeploy Maoist language—speaking of the process through which "the wealthy lay siege to poor neighborhoods," or more simply, redevelopment with Chinese characteristics.

Notes

1. For a detailed discussion of how physical borders have been mobilized to manage and arbitrage the difference between rural and urban Shenzhen, see Ma and Blackwell, chapter 6 in this volume. For a discussion of urban villages as an incubator for citizenship in Shenzhen, see Bach, chapter 7 in this volume. For a nuanced discussion of the tension between rural and urban forms of socialism, see Maurice J. Meisner, "Utopian Socialist Themes in Maoism: The Relationship between the Town and Countryside," in *Marxism, Maoism, and Utopianism: Eight Essays* (Madison: University of Wisconsin Press, 1982), 28–75.

2. On the ongoing relevance of peasants and rural revolution to the construction of modern China, see Liu Chang, *Peasants and Revolution in Rural China: Rural Political Change in the North China Plain and the Yangzi Delta, 1850–1949* (New York: Routledge, 2007); James Z. Gao, "From Rural Revolution to Urban Revolutionization: A Case Study of Luzhongnan," *Journal of Contemporary China* 10, no. 27 (2001): 233–52; Edward Friedman et al., *Chinese Village, Socialist State* (New Haven, CT: Yale University Press, 1991); Sulamith Heins Potter, *China's Peasants: The Anthropology of a Revolution* (Cambridge: Cambridge University Press, 1990); Helen F. Siu, *Agents and Victims in South China: Accomplices in Rural Revolution* (New Haven, CT: Yale University Press, 1989).

3. During fieldwork conducted between 1995 and 1998, when Shenzhen first included Bao'an County in its master plan (1996), urban planners and scholars used this phrase to joke about how Shenzhen had reversed rural revolution. The Shenzhen 1996–2010 master plan (*Shenzhenshi chengshi zongti guihua*) can be accessed online at http://www.szpl.gov.cn/main/csgh/ztgh/plan/plan.htm. For a discussion of how the rural has functioned in the construction of Shenzhen, see Weiwen Huang, chapter 3 in this volume.

4. For a concise review of the complicated institution, see Fei-Ling Wang, *China's Evolving Institutional Exclusion: The Hukou System and Its Transformation* (New Zealand: Contemporary China Research Centre, 2009).

5. See Ma and Blackwell, chapter 6 in this volume.

6. Constitution of the People's Republic of China, adopted on December 4, 1982, accessed February 26, 2013, http://english.people.com.cn/constitution/constitution.html.

7. Constitution of the People's Republic of China, adopted on December 4, 1982, accessed February 26, 2013, http://english.people.com.cn/constitution/constitution.html.

8. The discrepancy between Shenzhen's *hukou* population and its official population is so glaring that in January 2015, the municipality decided to raise the number of *hukou* residents from 2.8 to 4 million. However, when I interviewed variously placed residents (those with and without *hukou*) about this policy decision, there was general consensus among them that instead of ameliorating living conditions for Shenzhen migrant workers, the policy merely highlighted the city's *hukou* catch-22: those who are eligible for Shenzhen *hukou* don't actually need it, while those who need Shenzhen *hukou* are not actually eligible for it. A year later, in Shenzhen Party Secretary Ma Xinrui's public call for better management of human resources, he confirmed that the city's administrative population was twenty million (Tencent News Web, January 30, 2016, http://gd.qq.com/a/20160130/009936.htm).

9. For an overview of the situation of rural to urban migrants since Deng Xiaoping's 1992 southern tour, see Li Zhang, *Strangers in the City: Reconfigurations of Space, Power, and Social Networks within China's Floating Population* (Stanford, CA: Stanford University Press, 2001); and Dorothy J. Solinger, *Contesting Citizenship in Urban China: Peasant Migrants, the State, and the Logic of the Market* (Berkeley: University of California Press, 1999). For a discussion of how Shenzhen migrants in particular modeled the transition of peasants into factory workers, see Florence, chapter 4 in this volume. Anita Chan and Jonathan Unger discuss how remittances from migrants to their hometowns reshaped Shenzhen's rural hinterland in *Chen Village: Revolution to Globalization*, 3rd ed. (Berkeley: University of California Press, 2009).

10. To contextualize Shenzhen's experience with respect to globalizing urbanization elsewhere, see Saskia Sassen, *Cities in a World Economy*, 4th ed. (Thousand Oaks, CA: SAGE/Pine Forge, 2019).

11. Peter Y. L. Ng published an English-language introduction to the *Xin'an County Gazetteer* from 1819: *New Peace County: A Chinese Gazetteer of the Hong Kong Region* (Hong Kong: Hong Kong University Press, 1983). For discussions of this history, see David Faure and Helen F. Siu, eds., *Down to Earth: The Territorial Bond in South China* (Stanford, CA: Stanford University Press, 1995). David Faure tracks the cultural consequences of these transformations in *Emperor and Ancestor: State and Lineage in South China* (Stanford, CA: Stanford University Press, 2007).

12. For history of the Ming ban on maritime travel (*haijin*) and its effects on littoral Guangdong, see Geoff Wade, "Engaging the South: China and Southeast Asia in the Fifteenth Century," *Journal of the Economic and Social History of the Orient* 51 (2008): 578–638.

13. I discuss the cultural reworking of this narrative in terms of displaced nostalgia, when early Shenzheners and locals talked about Hong Kong's history as the history they should have had if not for a socialist twist of fate. See "Becoming Hong Kong, Razing Bao'an, Preserving Xin'an: An Ethnographic Account of Urbanization in the Shenzhen Special Economic Zone," *Cultural Studies* 15, no. 3/4 (2001): 419–43.

14. For a discussion of both Tanka culture as well as inquiry into how we might pay attention to the marginalities of the marginalized, see Barbara E. Ward, *Through Other Eyes: Essays in Understanding "Conscious Models"—Mostly in Hong Kong* (Hong Kong: Chinese University Press, 1985).

15. Township and Village Enterprises have simultaneously functioned as both a symptom and success of the Reform and Opening policy. For a discussion of the impact of the Reform and Opening policy in Guangdong, see Ezra F. Vogel, *One Step Ahead in China: Guangdong under Reform* (Cambridge, MA: Harvard University Press, 1989). For case studies of TVEs, see Daniel Buck, *Constructing China's Capitalism: Shanghai and the Nexus of Urban-Rural Industries* (New York: Palgrave Macmillan, 2012); Calvin Chen, *Some Assembly Required: Work, Community, and Politics in China's Rural Enterprises* (Cambridge, MA: Harvard University Asia Center, 2008); and Hong Yi Chen, *The Institutional Transition of China's Township and Village Enterprises: Market Liberalization, Contractural Form Innovation, and Privatization* (Aldershot: Ashgate, 2000).

6

The Political Architecture of the First and Second Lines

EMMA XIN MA AND ADRIAN BLACKWELL

Constructing Differences and Provoking Flows

All borders function to preserve differences between adjacent territories. A border needs to be constructed when valuable political and economic differences are produced that cannot be preserved through cultural inertia or spatial distance. Yet, despite its apparent function as a barrier, every border is a machine that generates movement. The production of differences provokes sequences of flows of capital, labor, and even something as apparently fixed as land. This chapter examines Shenzhen's "Second Line," the border that separated the Special Economic Zone (SEZ) from the remainder of the People's Republic of China (PRC), from its construction in 1982 to its official decommission in 2010, by examining the differences it constructed and preserved and the flows that these differences still provoke (see figure 13).

Contributors to this volume[1] have shown how the early development of Shenzhen was organized through zones of experimentation, which were constituted through the strategic use of a border. During the period of its official existence, 1982–2010, the Second Line continuously generated the production of exceptional spaces, which range from the zones of economic liberalization within the Shenzhen SEZ; to the socioeconomic boundaries between different subdistricts; to the gates of specific urban villages, residential neighborhoods, and labor controls of factory compounds. As O'Donnell and Huang have shown, zones such as the Shekou Industrial Zone and Overseas Chinese Town (OCT) grew through loosely delineated areas, often without clearly defined boundaries. This chapter will examine how these experimental spaces nevertheless depended on a political and spatial organization. The space was defined by two fixed boundaries: (1) the Sino-British Border (the First Line) along the Shenzhen-Bao'an border and (2) the Shenzhen SEZ Administrative Line, colloquially known as the Second Line. Whereas the "First Line" became the boundary of the One Country, Two Systems policy, the "Second Line" became the boundary of a local "One City, Two Policies" debate. Although the impetus for Shenzhen is officially presented as the importation of foreign free trade zone policies in other East Asian cities, the examination of the First and Second Lines in this chapter situates it with the history of the Pearl River Delta.

The Sino-Chinese Border

A long history of land division in the Pearl River Delta sets the stage for the Shenzhen SEZ in its present form. In the late Qing dynasty, the area of present-day Shenzhen and Hong Kong was known as Bao'an County, an area composed of agricultural villages fed by interspersed market towns, including Nantou and Shenzhen.[2] The Opium Wars and resulting Treaty of Nanjing in 1842 severed an island from the administrative control of southern Bao'an County, placing it under British sovereignty. The border that divided the island of Hong Kong from China moved northward in steps as the British negotiated the cessation of Kowloon in 1860 and the New Territories in 1898.[3] In the thirty years following Mao Zedong's Great Leap Forward in 1958, several hundred thousand Chinese fled political turmoil and persecution through Bao'an County by swimming across Shenzhen Bay. Throughout the political upheavals of the state under the Chinese Communist Party (CCP), significant migratory movements took place in 1957, 1962, 1972, and 1979, from or through Bao'an County to Hong Kong,[4] creating ties of familial kinship across Shenzhen Bay (see figure 14).[5]

Some of these illegal migrants across the Sino-British border escaped to the Kowloon Walled City, a 1,000 m^2 site of a former Qing military fortress deep inside the New Territories of Hong Kong, which was never ceded to British control under the 1860 and 1898 treaties. As a result, a settlement space that was especially welcoming to illegal migrants from across the First Line was built up into a self-governed enclave that was estimated to hold the world's highest urban density. It was not until the 1985 Sino-British Joint Declaration, which set out the terms for the handover, or return, of Hong Kong to Chinese sovereignty in 1997, that the British colonial government in Hong Kong (presumably with the tacit agreement of the PRC's government) began the concerted effort of criminalizing, dispossessing, and eventually razing the entire settlement in 1993–94. This occurred even as Hong Kong was being reconstructed as the first Special Administrative Region (SAR)—a space of exceptional policy legitimated by the success of Shenzhen. The Kowloon Walled City was replaced with a public park in 1995 that displays the preserved foundations of the Qing fortress that existed beneath the massive human settlement, with few traces of the forced-evacuated community acknowledged. What the Kowloon Walled City Park reminds us is that exceptional spaces have a long history in this region, and that microhistories of migration and settlement are constantly erased by the redrawing of boundaries.

The creation of Shenzhen in 1979 and the SEZ in 1980 was thus part of that wider readjustment of Sino-British relations in the Kowloon Peninsula and the redrawing of geopolitical boundaries at the waning of the Cold War. However, as with elsewhere during the early reform era, events on the ground anticipated, and in some sense forced, this redrawing. For example, Shenzhen party secretary Wu Nansheng identified two homonymous villages on both sides of the Sino-British border (both named Luofang Village) in which the income of villagers on the Shenzhen side was one tenth that of the villagers on the Hong Kong side. He presented the two Luofang villages almost as if one were a test group in a controlled experiment that legitimated the selective dismantling of the border.[6]

Whereas the Kowloon Walled City and the Hong Kong Luofang Village functioned as exceptional spaces through illicit practices and quasi-recognition, the creation of the SEZ and related spaces within it like the Shekou Industrial Zone[7] legitimized territorial exceptionalism as a strategy of development. The state-imposed territorial separation of the SEZ occurred in Bao'an County precisely because of its geographic adjacency to Hong Kong, the region's cultural distance from the political capital of Beijing, and the province's historical connection to world trade since the Ming dynasty. Its location was designed to make use of the unique agglomeration of foreign capital, technical development, and management expertise next door, while offering up an abundance of labor and land at a fraction of what they cost in the British colony.[8] Nevertheless, one important difference between the two lines is that the First Line was an externally imposed division of land to mark the spoils of British imperialism while the Second Line was a self-imposed apparatus of segregation put in place by the Chinese government to experiment in integration with, and separation from, the global market economy.

With the establishment of the SEZ, a policy was put in place to facilitate movement across the Sino-British border to Shenzhen and back for Hong Kong entrepreneurs. In contrast, few Shenzhen residents could obtain permission to visit Hong Kong. Until the turn of the millennium, the border remained an international boundary that was difficult to pass for Chinese citizens.[9] As Vogel detailed in the preface to this volume, crossings at the outset were done on foot, walking across the Luohu Bridge.

In 1980, only Luohu (Lo Wu) checkpoint existed for passengers and Wenjindu (Man Kam To) checkpoint existed for goods.[10] To meet the increasing demands of transportation in and out of the SEZ, Shatoujiao (Sha Tau Kok) was added in 1985 for passengers and Huanggang (Lok Ma Chau) in 1989 for goods. In 1995, Lo Wu accounted for 85 percent of all passenger crossings between Shenzhen and Hong Kong. To facilitate access and relieve congestion, a new passenger rail, the new Lok Ma Chau Spurline (or Futian checkpoint), was built at Lok Ma Chau in 2003, and the Shenzhen Bay crossing was built in 2007.[11] In a telling inversion of the earlier illicit exceptional spaces on the Hong Kong side (such as Kowloon Walled City and Luofang Village), the establishment of the SEZ catalyzed the creation of exceptional spaces for Hong Kong consumers—this time on the Chinese side. Urban villages near the boundary checkpoints transformed themselves to meet the market demands and commodified desires of Hong Kong day-trippers. At Luohu border, for example, several malls servicing Hong Kong consumers became known for selling counterfeit designer goods, portrait photography, and custom-tailored products like window curtains and clothing. At the Huanggang checkpoint, spas, massage parlors, and the sex trade became particularly prominent.[12]

For Shenzheners who could not go into Hong Kong, these same commodified desires were met on Sino-British Street (Chung Ying Street). Since 1898, Sino-British Street, located in the Hong Kong town of Sha Tau Kok (Shatoujiao), had straddled the Sino-British border itself. On one block, one side of the street was on the Chinese side and the other side of the street was on the British side. The street became a major route for the illicit traffic,[13] but sometime in the 1990s it became possible for Shenzheners to visit and purchase foreign or Hong Kong goods there.

The spaces documented above (whether legitimate or not) illustrate the extent to which territorial exceptionalism was a constitutive feature of the Sino-British border, both before and after the establishment of Shenzhen. Indeed, such spaces of exceptional consumption anticipated the future select privileges extended to mainland Chinese consumers in Hong Kong; for example, the "Disneyland" visas were introduced in 2008, which allowed Shenzhen residents to visit Hong Kong Disneyland (but not the rest of Hong Kong) in tour groups. These spaces provided experiences of radically different and sometimes imaginary economic life. They were spatial manifestations of the adage that Shenzhen was a "Window to the World," and also legitimated and modeled the territorial exceptionalism that would be tested at the Second Line.

The Second Line: Construction and Administration

After Deng Xiaoping's initial establishment of the Second Line in Beijing, the central government sent helicopters over Bao'an County to perform three aerial surveys in order to determine the exact dimensions of the SEZ they had but delineated on a map,[14] fixing its initial area at 327.2 km².[15] It would later be increased to 396 km². The State Council approved the final plans for the Second Line in 1982, and construction began later the same year. The main body of the line and checkpoints were completed in 1983, which allowed some areas to begin the process of personal identification inspection. The supporting infrastructure and utilities were put in place in 1984, and the border became fully functional as an administrative apparatus in 1985. The final construction cost of the Second Line was 1.38 hundred million RMB.

Much of Bao'an's undulating topography and agricultural fields, especially along its coastline, was flattened and razed for the implementation of the urban SEZ. Major inland geographic features were retained to provide the basis for parks and recreation. The Second Line was strategically designed to take advantage of natural topography as a barrier, running along the Wutong and Yangtai mountain ranges. The difficulty of navigating these high elevations dissuaded illegal crossing. Over time, many changes were made to the line. For example, Silver Lake, which was originally situated outside the boundaries of the Second Line, was later incorporated into the SEZ for its value as a tourist attraction.[16]

To prepare for the construction process, the engineering team surveyed the entire length of the Second Line on foot over a period of two and a half days. Considerations of the in-depth examination involved the assembly of the fence and the exact placement of watchtowers on the difficult terrain. The surveyors unexpectedly came across a number of agricultural villages that were located directly in the projected path of the line. The central government's abstract planning on a map and the aerial overview surveys had neglected to take the local people into consideration.[17] It became evident to the residents of the borderline villages that it would be beneficial to them to be included within the SEZ's boundaries, as the delineation would inflate their land prices and make the inhabitants privy to the benefits of the SEZ. However, the government negotiated with the villages on a case-by-case basis, which resulted in the inclusion of some villages in the SEZ and the exclusion of others. On site negotiations continued to change the shape of the physical border and caused variances from its planned position on the map. These border villages would later influence the construction of the Second Line and contribute to the legal and illegal economies and practices of the Shenzhen SEZ.

Despite the improvised processes defining the border's precise path, its architectural construction projected a fixed and unambiguous boundary. Eight checkpoints were distributed over the length of the line (the number would later be increased to fifteen in addition to Shekou Port). Between checkpoints, the Second Line was defined by a wire fence supported on concrete posts. A cobblestone-paved patrol road totaling 90.2 kilometers was guarded by 163 watchtowers. Each watchtower was inhabited by a single guard who kept the particular segment of the line under surveillance and prevented illegal passage.[18]

The administration of the line became the jurisdiction of the Seventh Detachment of the Guangdong Province Border Police, a sector of the People's Liberation Army. As the major checkpoints were the main points of access into the zone, every person (with a few notable exceptions) who passed into the SEZ was subject to inspection and required to display their People's Republic of China Boundary Region Pass in addition to government-issued personal identification. The boundary pass was difficult to obtain and was preceded by months of paperwork, which included the provision of proof of employment and political affiliation, an invitation from within the SEZ, and a monetary deposit. The process of entering the SEZ was akin to that of entering another country.[19]

Checkpoints provided both pedestrian access through the main administrative building as well as vehicular passage in adjacent booths set up across the roadway. Auxiliary facilities such as training and lodging grounds for the guards of the Seventh Detachment were also within the scope of the development of the line. The guards patrolled the fences adjacent to checkpoints closely in order to prevent civilians from approaching the vicinity of the boundary and to discourage illegal crossing.[20]

Auxiliary Means of Crossing the Second Line

The distance between the major checkpoints of the Second Line made life particularly inconvenient for residents of border villages because an impassable wire fence ran through their villages. The long route to a checkpoint and back, compounded by long lines and wait times for processing and inspection, made the situation unviable. In response, the administration placed a series of twenty-four "agricultural gates," better known as "minor checkpoints," along the Second Line for easier access.[21] Resident villagers of these border villages could obtain special agricultural passes from the major checkpoints, after which they would be able to travel across the minor checkpoints. Two or three soldiers would inhabit each minor checkpoint to inspect the agricultural passes. The relative frequency of minor checkpoints alleviated congestion and unnecessary wait time in the daily routine of the indigenous farmers. The minor checkpoints were open twenty-four hours a day, justified by the long hours demanded by agriculture. This is another example of how the rural was the rationale for exceptions within the "urban" policy (see figure 15).[22]

The differences in land value and employment opportunity became daily markers of inequality for residents of Shenzhen. For example, the priority the SEZ placed on economic and industrial growth greatly increased employment opportunities within the zone. However, the workers who were employed in the lower tiers of industry in the SEZ could not afford the high housing costs within the zone. So they necessarily lived outside the SEZ where housing was more affordable. For these workers and other commuters, crossing the Second Line was a daily occurrence. The long travel distances made biking and walking impossible, and the personal ownership of vehicles was not yet commonplace in China. Many lower-middle-class workers depended on public transportation to get to work and services. Throughout the life of the Second Line, drivers and passengers of buses were required to disembark at checkpoints, wait in line to be individually inspected by border guards, then re-embark on the same bus, or take an alternate route on the other side of the checkpoint. The delay caused by the administration in crossing the border at least twice a day, particularly during rush hours, was a source of frustration for the local residents of Shenzhen and a cause for local protest.[23]

Further contention arose when the government applied two sets of laws within the municipality. Different policies that dictated the social and political practices of the SEZ and non-SEZ portions of Shenzhen created inconsistencies in administration and permeated the Second Line's ideology of division throughout the city. For example, fines for running a red light were significantly higher in the SEZ due to the density of the city and the more stringent urban administration. Moreover, the taxi system was divided into red- and green-painted cars. The "in-zone" red cabs, equipped with drivers who had SEZ passes, mainly served the area inside the Second Line, though they were able to move across the border freely. The "out-zone" green cabs could only service the non-SEZ part of Shenzhen and were not allowed within the SEZ. The sparseness of development and longer traveling routes gave the green cab drivers a distinct disadvantage. These administrative and de facto apparatuses of segregation exacerbated the long-standing rural-urban divide, here in a fully urbanized context.[24]

By building the line, the central government not only quarantined a space of economic experimentation but also modeled the legitimacy of internal borders. The inequalities that existed across the line were not accidental by-products of political policies but specific details within the larger set of differentiating effects planned through its implementation. The Second Line was built as a tool for the production of two interrelated states of exception. The first was temporal—the broad national process of reform as a detour in time.

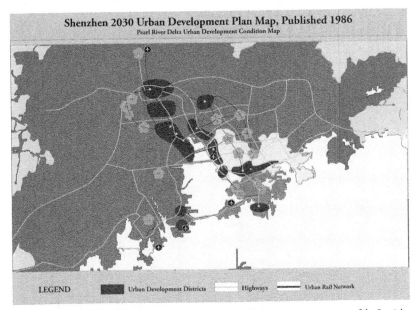

FIGURE 1. 2030 Shenzhen Urban Development Plan Map. Published in the 1986 iteration of the Special Zone Master Plan, this version imagines the future Pearl River Delta as primarily rural. Within Shenzhen, only the inner districts, Shekou (today Qianhai), and the riparian coast of Bao'an were designated for urban development. Note the planned use of bridges to connect Dongguan and Shenzhen to Guangzhou across the Pearl River and plans for five regional airports.

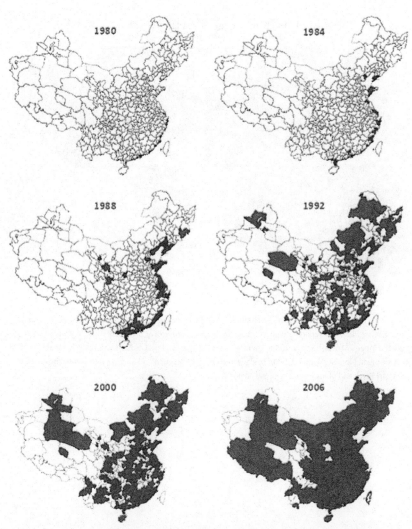

FIGURE 2. The geographic evolution of the Special Economic Zone experiment. *Source:* Jin Wang, "The Economic Impact of Special Economic Zones: Evidence from Chinese Municipalities," *Journal of Development Economics* 101 (March 2013): 137. Reprinted with permission. Note that Wang uses SEZ in the broadest sense to include all variations of zonal preferential treatment to attract foreign direct investment. The shaded area on the map shows municipalities with one or more forms of zone.

FIGURE 3. Map of Xin'an County, reproduced from the *Xin'an Xianzhi* [Xin'an County Gazetteer], c. 1819. The star indicates the location of the area's earliest county seat at the "center" of Xin'an County, which geographically speaking was located on the banks of the Pearl River, in the west of the city.

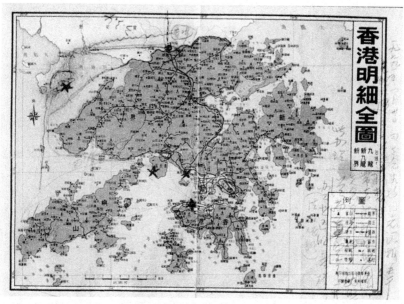

FIGURE 4. Map of Hong Kong, including Kowloon and the new territories, with hand-drawn notations in red crayon, black pen, and pencil, c. 1978. This is the famous "circle map" popularly associated with Deng Xiaoping, but in fact it was notated by Li Xiannian in a meeting with Yuan Geng and used to determine the location of the Shekou Industrial Zone. Reprinted with permission from the Shekou China Merchants Archive.

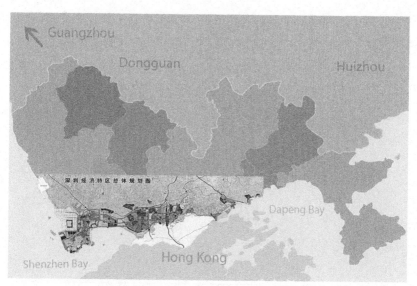

FIGURE 5. The 1986 Shenzhen Special Economic Zone Master Plan overlaid over a map of Shenzhen territory. Note that the majority of Shenzhen was unplanned.

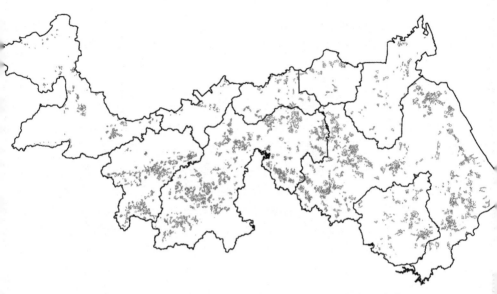

FIGURE 6. Location of Shenzhen urban villages, with approximate borders of individual villages, c. 2000.

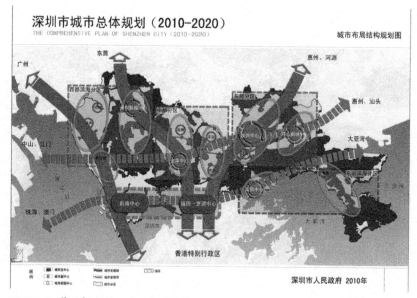

FIGURE 7. Shenzhen Comprehensive Plan, 2010–20.

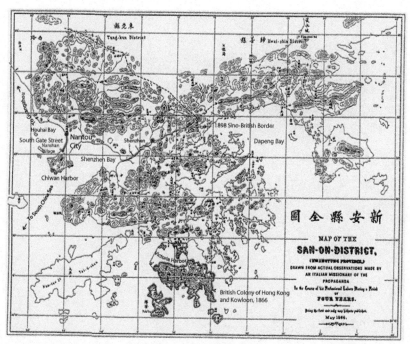

FIGURE 8. Map of Sun-On-District [Xin'an County], 1866. This map combines both Chinese and Western cartographic notations and shows the geographic location of the Xin'an county seat at Nantou and its distance from Shenzhen Market. In 1898, Shenzhen Market became the first town on the Chinese side of the Sino-British border. In 1913, it became a border station on the Kowloon-Canton Railway.

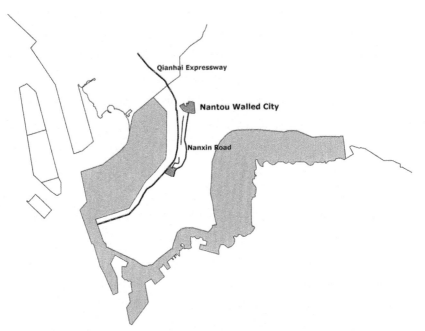

FIGURE 9. Map of concentric occupations of the Nantou Peninsula. The shaded area shows the extent of land reclamation. The broken line between the Qianhai Expressway and Nanxin Road marks the road that connected the Nantou (former Xin'an county seat) to villages on the peninsula. Image by Mary Ann O'Donnell, 2013.

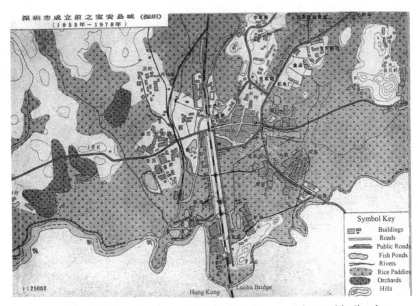

FIGURE 10. Map of Bao'an county seat, 1953–78. Before 1953, this area was the site of the Shenzhen Market, which was elevated to the Bao'an county seat in 1953 to take advantage of its railway connections.

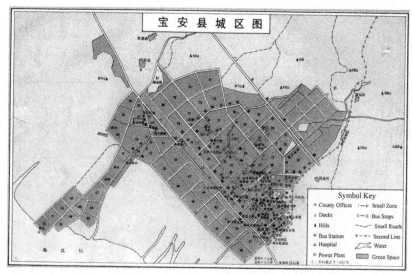

FIGURE 11. Map of the new Bao'an county seat, c. 1992. This map shows that the planning for territory outside the Special Zone, including the county seat of New Bao'an County, was to be treated as an industrial park with limited social development. The map notes that the county seat was "1 kilometer from Nantou, 9 kilometers from Shekou, and 23 kilometers from Shenzhen."

FIGURE 12. The 1996 Shenzhen Master Plan highlighting the location of Baishizhou.

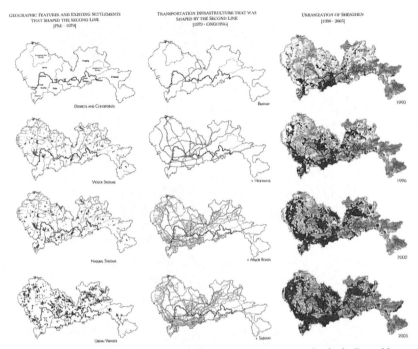

FIGURE 13. The Second Line and the urbanization of Shenzhen, 1979–2005. Graphic by Emma Ma.

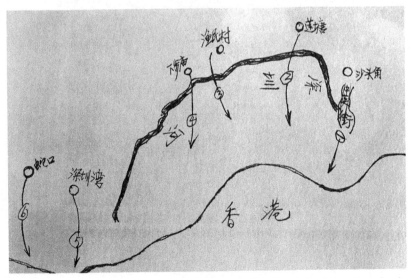

FIGURE 14. Hand-drawn sketch of major smuggling routes to Hong Kong before Reform. Reproduced from Chen Hong, *1979–2000 Shenzhen Zhongda juece he shijian minjian guangcha*, Wuhan: Changjiang Wenyi Press, 2006, 5.

FIGURE 15. Photograph of an agricultural gate on the Sino-British border, c. 1978. Image courtesy of the Sino-British Street Museum, Shenzhen.

FIGURE 16. Caiwuwei Village, Shenzhen, seen from the Diwang skyscraper, c. 2008. The MixC Mall is part of the complex under construction at the top left. Photograph by Jonathan Bach.

FIGURE 17. Museum model of Shenzhen, with no urban villages, c. 2008. Detail. Photograph by Jonathan Bach.

FIGURE 18. Aerial photograph of Tianmian Village, c. 1998, before redevelopment began. Courtesy of Yun Tai (Shenzhen) Developers.

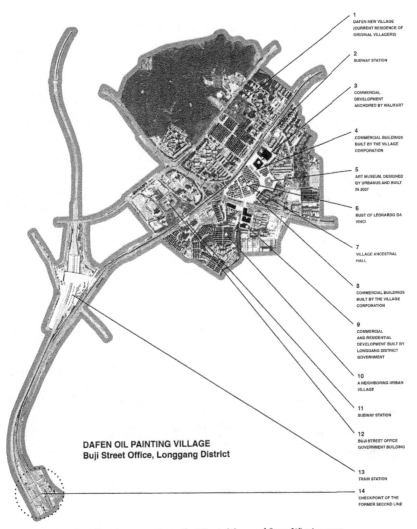

FIGURE 19. Dafen village in context. Image by Winnie Wong and Story Wiggins, 2015.

FIGURE 20. The built environment of the Dafen painting trade. Photo collage and composite image with section drawings. Image by Winnie Wong, Ettore Santi, and Story Wiggins, 2015.

FIGURE 21. A 2012 photograph of the construction site at Terminal 3, Shenzhen Baoan International Airport, Shenzhen. Photograph by Max Hirsh.

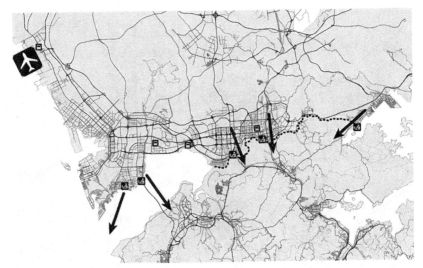

FIGURE 22. A map of Shenzhen displays the location of the city's airport, airport check-in terminals, and major border crossings. The dotted line demarcates the land boundary between Hong Kong and Shenzhen, while the arrows indicate the primary cross-border, or Shen Kong, transport axes. Map by Max Hirsh, 2014. Base map courtesy of Dorothy Tang.

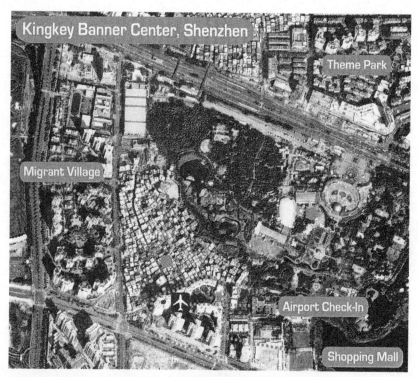

FIGURE 23. Kingkey Banner Center in Shenzhen's Nanshan District hosts one of the city's four cross-border airport check-in terminals. Map by Max Hirsh, 2011.

FIGURE 24. The border between Shenzhen and Hong Kong. In the background, a two-level pedestrian bridge connects the Futian border control point, located in Shenzhen, to its Hong Kong counterpart, Lok Ma Chau, located on the southern banks of the Shenzhen River. Photograph by Max Hirsh, 2011.

The second was spatial—the segregated territory of the SEZ itself. In the late 1970s, the Chinese state capitalist project of economic liberalization was presented as a temporally limited policy. The Second Line had been able to perform the complex topological movement of turning itself inside out, transforming China's interior into a new frontier for development. As a precedent for this, reformers pointed to Lenin's New Economic Policy implemented in the Soviet Union in the 1920s, with the implication that a certain stage of economic maturity was required before the economy could be returned to a more mature and equitable form of socialism.[25]

Unsanctioned Arrangements along the Second Line

As the Shenzhen SEZ was a site of economic experimentation, the lack of experience and clear legislative structure allowed for informal practices and unexpected uses of the specific geography of the Second Line. The territory of the Second Line itself operated as a space of undefined potential because the exact location of the Second Line was up for grabs, especially during the two-year period of the construction of the line between 1982 and 1984. The line was planned to cut through village lands, and village settlements would fall on one side or the other. Naturally the villagers would take advantage of their land use rights on the other side of the line and build strategically either into or out of the SEZ so that they could occupy both sides of the line through unsanctioned arrangements.

For example, one typical arrangement might be for landowners who had property just outside the line to develop their parcel, and offer border guards bribes to "open a door" in the line, thereby increasing the desirability of their property. In other cases, if a landowner had a plot outside the line, he or she might negotiate to have the address included within the SEZ. If the negotiations turned out to be successful, the line would effectively move. This would allow the landowner to reap the economic benefits and incentives of the SEZ and thereby increase their property value. Such practices constantly affected the specific geography of the Second Line, and its dimensions and porousness on the ground would continually evolve until 2003–4.[26]

The most iconic of these unsanctioned arrangements occurred where the Second Line entered and exited village lands. These areas were whimsically called "flower arrangement" plots or villages (literally "inserting flowers" plots or villages). These flower arrangement villages were situated immediately on, or adjacent to, the Second Line, and they were a special instance of the urbanized village. The villagers would expand their settlement across village lands that were located on the other side of the line so that the line would now pass through the middle of the village settlement. This would give the villagers a foothold in the SEZ. Large clusters of such villages developed around Meilin and Buji checkpoints.[27]

These unsanctioned arrangements were possible because there was no administrative structure in place to insist otherwise. Villagers' strategic activities thus deepened the discrepancy between the planned, political line drawn on the map and the physical, constructed border. In essence, by actively moving the planned position of the boundary, villagers created autonomous spaces that refused incorporation into the SEZ or the outer district government (then New Bao'an County). However, these villages created their own social services, including schools, clinics, utility lines, and sanitation services. One of the most profitable uses of these areas was the early establishment of informal border crossings. At the border crossings, local villagers smuggled migrants who were seeking factory work into the SEZ for a fee and allowed illegal residents or residents who had overstayed their passes to leave the SEZ undetected.[28]

In 2003–4, the municipal government of Shenzhen identified the flower arrangement villages as a "blind spot" or "black hole" in the administrative system and began a process of ex post facto criminalization. For example, it accused villages of harboring illegal migrants, diverting infrastructure, building substandard housing, and creating "black schools" for their children.[29] In so doing, this discourse specifically invoked the language of the "three ungoverneds" (*san bu guan*), which had previously been used to justify the razing and forced evacuation of the Kowloon Walled City in Hong Kong. The three ungoverned activities were ostensibly prostitution, gambling, and drugs, though the term came to represent the absence of state governance in general.

In September 2004, the municipal administration initiated a mobilization for a "thorough clean-up" of the flower arrangement villages. The procedure set out by the municipal government included the complete evacuation and demolition of illegal structures, investigation of past mistakes of jurisdictional neglect and implementation of appropriate disciplinary actions, increase of management, encouragement of adherence to new laws, and increase of cooperation between all parties involved in the administration of the villages.[30] The jurisdiction of illegal developments in central Shenzhen was assigned to Luohu District. The Luohu government would necessarily take responsibility for the quality of life and lawfulness in all aspects of social and physical support in the neglected developments. A part of the initiative included the assessment of schools in the area. Those that were below the standard would be shut down, their students dispersed to surrounding public schools.[31]

The Legacy of the Second Line

The official attention to criminalizing the flower arrangement villages was part of a larger governmental debate over the fate of Second Line. As Bach has detailed in chapter 7 of this volume, by 2004, Shenzhen no longer had any official "villages." Hence the distinction between collective (rural) and state-owned (urban) property within the city no longer needed to be demarcated by the Second Line. By 2003–4, the checkpoints of the Second Line were no longer regularly enforced, but the official debate over its fate continued until 2010 when the entire city was designated an SEZ.

This issue was first officially raised in 1998 at the "Two Sessions" meeting of Shenzhen.[32] At this meeting, the problematic issue of the "One City, Two Policies" nature of Shenzhen's governance was raised, and the Second Line was criticized for dividing the city and preventing its development as a coherent whole. Critics argued that Shenzhen as a bifurcated entity could not move forward from the SEZ period toward greater urbanization and economic goals. The central government responded with a State Council survey of Shenzhen in June 2000. The investigation concluded that the Second Line did not hinder the economic and urban development of Shenzhen. The conflict reached a climax in the national Two Sessions meeting of 2003. Major issues of note pointed to the traffic congestion at the busy passageways of Nantou, Tongle, Meilin, and Buji. It was argued that the primary cause of the issue was that the checkpoint system was the only method of access between SEZ and non-SEZ areas. The problem was expected to be resolved if the checkpoints and the time-consuming inspection process were eliminated. It was also noted that the management and upkeep of the Second Line also cost tens of millions of RMB every year. As Shenzhen was politically subjugated to Guangdong Province but economically subjected to Beijing, the cost of operation had a negative influence on the national budget. The matter was once again put to rest in January 2008, when the State Council concluded that the Second Line would not be annulled before the next general election of Hong Kong in 2017, thereby keeping the line in place for at least another decade.[33]

Nevertheless, at the same time, the process of entering the zone was relaxed significantly, and in 2003, "visitors" from the interior could simply purchase a pass at the border for RMB 2 for a week-long stay. Government-issued personal identification cards would be accepted during rush hours to streamline the border-crossing process, and workers within the zone could cross with their temporary residence permits. Eventually, even the temporary pass fee was waived in 2005, and checking border passes became less and less stringent.

Residents and visitors began to often enter the zone with only their personal identification cards until the border checkpoints were finally decommissioned in 2010.

The decommissioning of the Second Line finally occurred as a result of the 2009 review of the Shenzhen Comprehensive Reform Experimental Plan. The review stated that the existence of the line affected the "scope, land, and financial situation of Shenzhen," with each item pending further investigation. In Shenzhen's transition from an SEZ of China to an international megalopolis, the lag in speed of urbanization outside the zone made Shenzhen less competitive as a coherent municipality on both the national and international level. The decision to officially annul the Second Line was approved as a part of the 2009 Reform Plan. The formerly non-SEZ districts of Bao'an and Longgang, along with the two new districts that were cut out from them, Guangming and Pingshan, became incorporated into the SEZ in 2010. The effort added 1,553 square kilometers to the existing 395 square kilometers of the original SEZ, forming an area of 1,948 square kilometers—nearly twice the size of Hong Kong and the New Territories combined. According to the 2009 plan, the Second Line was to be moved in its entirety to the northern boundary of the city, between Bao'an and Longgang Districts (of Shenzhen) and the adjacent cities of Dongguan and Huizhou. While this was a hard administrative boundary, no walls or checkpoints were built there.[34]

As of 2014, practical questions remained with respect to the treatment of the former site of the Second Line. For the time being, the infrastructure of the checkpoints and fences remain, though some are being dismantled. Many checkpoints still function as bus terminals, as the original bus routes were planned so that passengers disembarked on one side of the line, passed through the checkpoint for inspection, and then reembarked on a different bus. New subway lines finished in 2010 are one step toward the unification of the Shenzhen public transportation system, as they pass through the Second Line without stopping, but the taxi system and all roadways that were initially divided by the Second Line still await integration. The city has begun this transformation by removing signs requiring slower speed limits near checkpoints. As of 2013, plans were being made to redevelop the 9.1 hectares of Buji checkpoint into a mixed-use development, including a 0.23 hectare public square and 0.5 hectares for a multimodel bus and train terminal.[35] The urban and social problems caused by the line, including traffic congestion, disparities between what are now known as the inner and outer districts, and the flower arrangement villages are still being addressed.

The key dimension of the Second Line was its unique status as an internal borderline. In this regard, the construction of the line was true to the experimental nature of the SEZ: by quarantining a space of economic experimentation, the central government also conducted an unprecedented experiment in the creation and function of internal borders for producing differences in economic value. Once the line itself was accepted, a whole suite of other, more subtle forms of bordering could be integrated into emerging forms of governmentality. While hardly foreseeable at the time of its creation, the Second Line became an important apparatus in the Chinese state's strategy of bordering, which produced a patchwork of overlapping and almost unbroken exceptional spaces, beginning with the SEZs and moving to open coastal cities, open economic regions, special municipalities, Special Administrative Regions, and autonomous regions.[36] The construction of the Second Line was a physical monument to the State's legitimacy as a zone of competitive and differentiated spatiality. As such, it was emblematic of the spatial innovation of Chinese reform: by enabling a controlled space for foreign market circulation, the government turned adjacent differences of capital, land, and labor into the motor of one of the longest sustained eras of economic growth ever seen. Yet, as unsanctioned practices and construction like the flower arrangement villages demonstrate, as inexorable as this process may seem in hindsight, the logic of exception also created unplannable contingencies that turned Shenzhen into the complex experiment it continues to be.

Notes

1. See Mary Ann O'Donnell, chapter 2, and Weiwen Huang, chapter 3, in this volume.

2. Kwan Yiu Wong, ed., *Shenzhen Special Economic Zone: China's Experiment in Modernization* (Hong Kong: Tai Dao Publishing, 1982), 16–19.

3. Steve Tsang, *Government and Politics: A Documentary History of Hong Kong* (Hong Kong: Hong Kong University Press, 1995).

4. Verna Yu, "Veterans Who Fled Mainland for Hong Kong in 1970s Tell Their Stories," *South China Morning Post*, January 6, 2013, http://www.scmp.com/news/hong-kong/article/1120852/veterans-who-fled-mainland-hong-kong-1970s-tell-their-stories.

5. He Huifeng, "Forgotten Stories of the Great Escape to Hong Kong," *South China Morning Post*, January 13, 2013, http://www.scmp.com/news/china/article/1126786/forgotten-stories-huge-escape-hong-kong.

6. Apparently, Wu noted that that the villagers on the Shenzhen side had an annual income of only 134 yuan, whereas their counterparts on the Hong Kong side had an annual income of 130,000 yuan. Chen Hong, *Shenzhen zhongda juece he shijian mingjian guancha 1979–2000* (Beijing: Changjiang Literature Art Publishing House, 1991), 8.

7. See O'Donnell, chapter 2 in this volume.

8. John Friedmann, *China's Urban Transition* (Minneapolis: University of Minnesota Press, 2005), 28.

9. Robert Kelinberg, *China's "Opening" to the Outside World: The Experiment with Foreign Capitalism* (Boulder, CO: Westview Press, 1990), 77.

10. Checkpoints sometimes have different names on the different sides of the border in addition to different Cantonese and Mandarin romanizations.

11. See Max Hirsch, "Stimulating Global Mobility," in this volume.

12. See Dong and Cheng, chapter 8 in this volume.

13. Chen, *Shenzhen zhongda juece*, 5.

14. See Huang, chapter 3 in this volume.

15. "Shenzhen erxian guan shi zenme lai de? [How Did the Second Line Come to Be?]," last modified June 8, 2010, http://jt.sz.bendibao.com/news/201068/210926.htm.

16. "How Did the Second Line?"

17. "How Did the Second Line?"

18. "How Did the Second Line?"

19. Ou Ning, "Bianjie jihua [Border Plan]," last modified December 13, 2007, http://www.china-up.com/special/wenhua/exhibit/pdf/边界计划.pdf; Shen Ying, "Shenzhen Anniversary: Second Line," last modified September 9, 2010, http://www.360doc.com/content/10/0909/15/22953_52389176.shtml.

20. The major checkpoints were Nantou , Tongle, Baimang, Meilin, Buji, Shawan, Yantian (named after the district), and Beizaijiao.

21. The minor checkpoints were located at Lishan, Liuxiandong, Niucheng, Baimang, Dawangkeng, Makan, Dakan, Dayin East, Fulin Village, Changlingpi, Changlingpi East, Buji Poultry Market, Zhangshekou, Xinwuxia, New Village Orchard, Dawang, Xintianzai, Hengpailing, Xikeng, Leibei, Xiaosanzhou, Dashuikeng, Anlekou, and Anleyong.

22. "How Did the Second Line?"

23. "Zhiyi Shenzhen tequ erxian guan [Questioning the Shenzhen SEZ Second Line]," *China Reform* 5 (2002): 21–22.

24. "Shenzhen tequ yanshen fang'an shangbao Guowuyuan: guannei guanwai heyi [Submission of Shenzhen SEZ Plan of Expansion to the State Council: Unification of Regions within and without the Second Line]," last modified January 29, 2013, http://www.szol.net/vip/article/2009-11-18/24040-1.htm.

25. Ezra F. Vogel, *One Step Ahead in China: Guangdong under Reform* (Cambridge, MA: Harvard University Press, 1989), 128.

26. Weiwen Huang, in discussion with the author, April 2012.

27. "Luohu quanquan jieguan erxian guan: chahuadi sanbuguan shidai zhongjie [Luohu Takes Jurisdictional Responsibility of the Second Line: The End of Flower Arrangement No-Man's Lands]," *Southern Metropolitan Daily*, October 8, 2003, http://www.szlh.gov.cn/icatalog/a/2003/j08/a1187_587401.shtml.

28. "Luohu Takes Jurisdictional Responsibility."

29. Jin Yong, "Shenzhen 'erxian' chahuadi: shishi 'kong lou xingdong' keburonghuan [Shenzhen Second Line Flower Arrangement Villages to Implement 'Evacuation Movement' Immediately]," *Shenzhen Special Economic Zone Daily*, September 6, 2004, http://sz.focus.cn/news/2004-09-06/76047.html.

30. Feng Yiju, "Shenzhen shi quanmian qingli 'erxian' chahuadi weifa jianzhu [Shenzhen Municipality to Thoroughly Clear Illegal Construction in 'Second Line' Flower Arrangement Lands]," *Shenzhen Special Economic Zone Daily*, September 2, 2004, http://www.southcn.com/news/dishi/shenzhen/shizheng/200409020325.htm.

31. "Luohu Takes Jurisdictional Responsibility."

32. "Shenzhen Second Line Flower Arrangement Villages."

33. "Shenzhen tequ ni che guan kuoda mianji jiang jiejin liang ge Xianggang [Shenzhen SEZ to Increase in Area through the Removal of the Second Line: New Area to Be Almost Twice That of Hong Kong]," *Qilu Evening News*, May 21, 2009, http://news.sina.com.cn/c/2009-05-21/072315660067s.shtml.

34. Ou Ning, "Shenzhen chahua di yu erxian guan [Shenzhen's Flower Arrangement Lands and the Second Line]," *Ouning's Blog*, November 18, 2005, http://www.alternativearchive.com/ouning/article.asp?id=34; "Shenzhen SEZ to Increase in Area through the Removal of the Second Line."

35. Jinwen Xie, "Meilin guan jiancha tongdao sheshi jinqi jiang chaichu: Buji guan ni gaizao [Meilin Checkpoint Passage Facilities Planned to Be Demolished in the Near Future; Buji Checkpoint Plans to Make Renovations]," July 11, 2013, http://www.jingme.net/content/2013-07/11/content_8288681.htm.

36. See Bach, chapter 1 in this volume.

7

"They Come in Peasants and Leave Citizens": Urban Villages and the Making of Shenzhen

JONATHAN BACH

When at dusk the neon lights come on in Shenzhen, China, they look from the top of its tallest building like small fires dotting a mist-shrouded landscape. The massive city of dreams lends itself to naturalistic metaphor—the eye scans skyscraper groves amid fields of striated worker housing and middle-class mid-rises like dragons playing—according to one interpretation—or like the undulating peaks of Guilin, interrupted by the clusters of tight, jumbled, low growth of the so-called urban villages (*chengzhongcun*). Sixty-nine stories below the cityscape sparks, oozes, and shudders along rivulets of light refracted in the humidity, the inky black of Hong Kong's New Territories across the border accentuating the lustrous Leviathan that is Shenzhen. It all fuses together in urban rhythm and meter, reminding me of a description I encountered in a government-issued coffee-table book about Shenzhen's central business district: "Construction is a poem, written by poets, who wrote them with steel and cement."[1] A hoary slogan, but with echoes of early twentieth-century fascinations with urbanization and form, from Russian constructivists to Bauhaus. Construction as poem captures Shenzhen as a paean to modernity, for it is, in its own way, the very equivalent of 1920s Berlin or New York in its urban intoxication, its inexorable nowness, its multiple overlapping fantasies of progress, promise, and peril. A luxury building under construction in downtown Shenzhen speaks to this lineage with a meter-high, block-long wraparound advertisement that starts with images of "1940 New York: Rockefeller Center" and, via images of Tokyo and Hong Kong, ends with Shenzhen "Now: City Crossing."[2]

Roland Barthes famously referred to the city as a poem that is not "tidily" structured around its subject but instead steadily unfolds its signifier.[3] This unfolding provides the grist for the ethnographer's mill and enables the city to be approached as a text—not in the service of uncovering secret codes or concealed meanings but in the struggle to give expression to the surfeit of meanings inherent in any text. Mary Ann O'Donnell reminds us that taking the city as text too literally risks flattening its multiplicity and emphasizing the agency of the individual author, as in de Certeau's famous "walker in the city" who also begins by looking down from the towering heights of the city's tallest building.[4] Yet the late Barthes's sense of text as an irreducible plural allows the multiple symbolic fields of the city to be woven together in a polyphonic ethnography.[5]

This polyphony renders legible the process by which this spectacular urban formation emerges from the workings of globalization and the innovative appropriation of socialist legacies in recently introduced market economies.[6] Shenzhen's component parts—its factories, settlements, migrants, managers, skyscrapers, villages, and parks—are often bundled together in a tidy narrative of progress and embrace of market forces, hurtling ahead at "Shenzhen speed."[7] As the city grew at an astonishing annual rate of more than 27 percent from 1980 to 2006, the city felt, as the son of white-collar migrants put it, "like a fever." The city's mythic quality lies in its meteoric ascent from villages and fields to a top-ranked world city in size and infrastructure, high technology and skilled manufacturing, airfreight and container ports.

In this process, Shenzhen, like most of China, has been shaped by the opposition of urban and rural and by the expression of this opposition through the terms city and village. At its most simplistic, this opposition functions as a rhetorical master code. In conversations with professionals and officials in Shenzhen, through advertisements and in official exhibits, websites, and newspapers, Shenzhen often appears as the proverbial opening to the imagined West—modern and market oriented, entrepreneurial and well ordered, atomized and universal. Conversely, whether in popular perception or musealized forms, the former villages of barely one generation ago represent the past, the (closed) East, the feudal, the mythic, the collective, the particular. If Shenzhen is a narrative about speed, progress, and civilization, its villages serve as the narrative's other, its measure of progress.[8]

Urban Villages and the Making of Shenzhen

When local white-collar residents talk about "villages," they are referring not to historic tourist sites (such as Nantou's "Ancient City" or Dapeng Fortress) but to the distinct warrens of narrow streets, illegal buildings, and concentration of rural migrants that interrupt the monotone cityscape of the high-rise city (see figure 16). There are 241 of these "villages in the city" that persist discursively and spatially long after their legal status was forcibly changed from rural to urban.⁹ The villages go back hundreds of years and were organized under Mao into collective farms prior to the 1980 establishment of the Special Economic Zone (SEZ) and city of Shenzhen, which literally grew up around them. This chapter argues that Shenzhen's urban emergence must be understood as a coevolution of the villages and the city that draws its force from the villages' legal and discursively ambiguous status within the city.¹⁰

In a technical, legal sense, the villages disappeared in 1992, the year they were fully "urbanized." For twelve years they had retained privileges from their days as collective farms that included the ability to pass on the land via inheritance and being exempt from the nation's One Child policy.¹¹ These privileges were lost in 1987 as the city government began to slowly transform villages into neighborhoods in a process that one city official described to me as "not popular but necessary." Once transformed to urban land, property became subject to "use rights" that expired in seventy years, whereas before it was inheritable.¹² Yet although the villages are, in a strict sense, long gone—their fields and ways of life absorbed into the physical and legal infrastructure of the urban—they continue to have an emotional and evocative power in linking Shenzhen's land and history to a different space and time.¹³ Former farmer families still refer to themselves as "villagers" even though they hold city residency, migrants still seek cheap housing in the crowded tenements on former village land, and middle-class professionals in business and government look down on the villages, ranking them "lower than migrant laborers" on their "moral or political map," as Helen Siu writes of analogous urban villages next door in Guangzhou.¹⁴ The village is a discursive category that lives on in many ways long after shedding its original function and form.¹⁵

The symbolic representation of the village is key to understanding Shenzhen's emergence because it makes legible discursive distinctions in which the villages appear simultaneously as the city's condition of existence and perceived obstacle to progress, its recognized heritage and its hidden past, its location for menace or entrepreneurial exuberance. The legal homogenization of the rural and urban is countered by its continued symbolic separation. But this separation is not merely a binary in which the village is disadvantaged by a contemptuous urban professional class, although this is often enough the case.¹⁶ The villages play neither a heroic role of resistance nor a submissive role of subservience to the city. The coevolution of city and village occurs precisely because both city and village use the ambiguous space that Shenzhen created between the urban and the rural to take advantage of each other in the context of transition to a market economy.

To capture this transformation of urban space in Shenzhen, I borrow semiotic concepts from Barthes to slow down our reading of this city of speed. The city's legal narrative of homogenization erases the villages from the picture, often literally. The particular of the village is subsumed into the universal of the city in what, following Barthes, I call a "readerly" narrative—the plot is tightly controlled by a "god" (the author, the state) and deciphered by the "priest" (the critic, the scholar, the official).[17] Here, interpretation takes place solely within a given "structure of signifieds" set by the state. Against this, I seek a "writerly" narrative that presents a "galaxy of signifiers" that provide different entry points into an ensemble animated by recombination of rural and urban forms. Through my ethnography of the villages and the city, the urban-rural divide appears as both essential and complicit in each other's continued production and effacement. Shenzhen is defined less by its teleology of urban speed than by the collective production of a contradictory, even cacophonous, identity that is both city and village.

"The First City with No Villages in China"

MODEL CITY

In the echoing, modernist Shenzhen city history museum, the visitor eventually comes upon the obligatory illuminated scale model of the city. Like all such models, it is graciously devoid of traffic, people, and pollution, offering a cartoon image of a glowing metropolis (see figure 17). An explanatory plaque draws one additional missing feature to your attention. It reads: "The urbanization of Bao'an and Longgang Districts made Shenzhen the first Chinese city with no villages, allowing for harmony among urban economy and society as a whole and the establishment of a firm basis for sustainable development." Indeed, the model shows no villages, no knots of jumbled buildings marking the cityscape. Instead, everything is regimented, orderly, planned. The "disappearance" of the villages in this ideal model reflects the omnipresent narrative of progress, civilization, and urbanization: Shenzhen is a civilized city, and civilization is urban, urbane, orderly. Villages are uncivilized, messy, and disorderly, and removing these "dirty, chaotic, and backward" spaces is akin to removing "the city's cancers," as Siu remarked on a similar situation in Guangzhou, and as Li Zhang detailed in her account of the destruction of Zhejiangcun in Beijing.[18] In the twenty-first century, where the Shenzhen government projects the city as the "global pioneer city of sustainable development,"[19] the villages are openly presented as a visual embarrassment, a challenge to government control, an impediment to planning, a vestige of a feudal past.

The classic official narrative of Shenzhen thus ends with the vanishing of the villages into the order of the urban. But a stone's throw from the museum, "villages" still pulse, making clear that the tension in the official narrative—to speak in textual terms—comes from the question of how this vanishing will be achieved. Barthes sees in these kinds of situations a narrative structure of suspense: once the conclusion is stated (here that Shenzhen became the first city with no villages in China), this sets up the object of the story (Shenzhen without villages) as an enigma, as a source of anticipation, because although we know the conclusion, we don't know how we will get there from where we are now. Therein lies the narrative tension. Such narratives "must set up *delays* (obstacles, stoppages, deviations) in the flow of the discourse."[20] The story of Shenzhen is also the story of the elimination of the villages and the way in which the villages "set up delays" to this outcome. But importantly, the "hermeneutic code" that sets up this scenario does not trouble the oppositions (village/city) but rather sets them in place. This makes the enigma here double. Together with the city, we ask, pragmatically, "What needs to happen to make the villages disappear?" but we suspect, thirty years since Shenzhen's inception, that the real hermeneutic question is why the villages persist.

EXCEPTION TO THE EXCEPTION

Beginning to address this enigma takes us to the very beginning of Shenzhen, when it was a rural setting with market towns and scattered collective farms cultivating lychee orchards, farming oysters, raising pigs, and growing bamboo. Deng Xiaoping chose this area abutting Hong Kong to develop the most ambitious of five SEZs designated in 1980 to experiment with capitalism and develop an export economy.[21] Thinking back to the early years of the zone, the wife of a former village head expressed her gratitude to Deng, for she had been working since she was ten years old, rising at three or four in the morning and getting to bed at midnight, eating bad food and working hard until the zone, as the cliché goes, changed everything. She was recounting these hardships to me over her regular sumptuous midmorning tea with seven other village wives in a private room of the "Tai Chi" restaurant in the village's four-star hotel, with steamed buns filled with barbecued pork that melts in your mouth and luscious shrimp dumplings wrapped in translucent rice flour. "Now," she said with satisfaction, "we have nothing to do but play mah-jongg and the stock market." They meet in the park every day for exercises at 7:30 a.m., have tea (yum cha) at 8:30 a.m., and go home at 10 a.m. to "do chores." The wives were preparing for a shopping trip to Japan, and they travel abroad once or twice a year. They have traveled to six countries in Europe, to South Africa, to southeast Asia and Taiwan, and all over China, including five trips to the Yellow River. On their first trip to Europe, the village head's wife spent RMB 40,000 (about $5,800) on gifts; the cost of the trips are covered by the village, which also pays for the village children's private education, including that of her son who studied in England, returned to the village, and now works in a foreign investment company.[22]

Although at first glance a rags-to-riches story in keeping with the master narrative of reform and progress, corresponding to the image of Shenzhen as a city of limitless opportunity, the fact of the villagers' wealth demonstrates the limits of centrally planned transformation rather than its fruit. In the short-lived original scheme for the SEZ, the villages would romantically continue to supply vegetables to workers. As fields yielded to factories, however, the villages became enclaves of a most curious sort. Rather than being absorbed into the formal city structure, the villages held on to forms of "rural" land status until late in the development of Shenzhen. Even after losing the legal designation as "rural," they retained the discursive and spatial imprint of the "village," giving rise to the term by which they are most known today—"village-amid-the-city" (*chengzhongcun*).[23] By 1992, when urban residency was forced on villagers, about 15 percent of Shenzhen's land remained under the control of village committees, later turned into shareholding corporations. In the more central districts, the villages occupy an area roughly equivalent to the core village size plus additional land on which villagers could collectively pursue small-scale farming and build homes. In the outlying districts, such as Bao'an, 80 percent of the land is "owned" by the village collectives.[24] The formation of Shenzhen thus initially institutionalized a rural-urban dichotomy within its borders. The city of Shenzhen became legally urban, while the villages, even when located in the heart of the rapidly expanding city, retained rural privileges.

The significance of the distinction between urban and rural cannot be overestimated: social control, resource allocation, biopolitical interventions from reproduction to health care, education, relative social status, and mobility all came to hinge on this distinction. In the first several decades of the People's Republic, rural-urban dichotomies were reshaped into the equivalent of a new class distinction between rural and urban identity through state policies restricting mobility and access to resources. As Dorothy Solinger puts it, by "barricading the cities against the peasants," the state not only turned the peasants into reserve labor but also discursively penned the peasant as "docile, disposable trespasser, and drudge."[25] The city dweller has access to rental housing, education, doctors, and hospitals; can move with increasing ease from city to city; can buy apartments and apply for passports; can become trendy, sophisticated, harried—in a word, modern. Their rural brethren are considerably constrained in all these pursuits, the city only available to them through the paths of formal employment or illegal migration that reduces them to the ghostly margins even as it invites their labor power. But rural residents could have more than one child, their villages collectively administered their own land, and they had the right of inheritance. This relative autonomy over the disposition of their own land stands in contrast to city folk, who have only limited "use rights" of property that they purchase.[26]

As long as rural and urban can be clearly delineated, these distinctions hold the promise of order and control. But economic prosperity has famously engendered massive rural-urban migration, and reform policies have diluted the original Communist structures of work unit (urban) and commune-collective (rural). The distinction still holds, but in recent years it has become absorbed into an opaque lattice of traces, legacies, laws, mores, and norms. O'Donnell writes that Shenzhen's formation contained a revolutionary premise that rural and urban China could interact on equal footing within the borders of the zone.[27] Yet the result was ambiguity because of the lack of any adequate administrative category "by which [the villages] might be re-incorporated into the Chinese state apparatus."[28] In the new SEZ, China suddenly confronted two forms of excess that could not be contained or disciplined within the state apparatus: global capital, whose flow was the raison d'être of the zone, and the villages, whose suddenly anomalous existence became the exception to the exception.

This exceptional state of inclusive exclusion supplemented village identity with the stark urban-rural identities forged under Mao, with the result that "village" continues to symbolically, spatially, and functionally exceed its legal category. Barthes has written famously of how signifiers (e.g., the cultural imprint of "village") remain while the signified (e.g., the buildings, legal status, etc.) change. This transience gives the signified the quality of an "empty space" around which a city can be organized.[29] Shenzhen's villages also function thus—they define Shenzhen yet resist signification, they are "empty" in that they exert an enigmatic power that cannot be adequately signified yet they remain central to the city's existence. They are there yet cannot be there. Thus we continue to speak of villages even after their legal status says they are urban.

AMBIGUITY AS OPPORTUNITY

What befell the people who lived there, the former farmers' families? With little or no formal education and with no city household registration (*hukou*), the villagers seemed consigned to become the "losers" of the transition to a market economy. Their holdings were reduced to ever-smaller plots of land, and they often had to seek unskilled and possibly illegal work in the surrounding city.[30] As the SEZ grew up around them, the local peasants were functionally excluded from better-paying employment beyond their village enterprises, often considered undesirable even for construction jobs. Although nominal efforts were made to provide some employment, the new local government was neither capable nor overly interested in villagers, hoping they would use their shrinking allocations of land to continue some small-scale agriculture and develop small-scale light industry to ameliorate unemployment.[31]

Left largely to their own devices, the villagers soon found that their ambiguity in the new environment contained the key to their eventual success: becoming landlords. With Shenzhen transforming from fields to a teeming metropolis of millions in barely one generation, the need for construction labor was unparalleled in China. Rural migrants sneaked or were smuggled into the zone to scramble up scaffolds and dig deep into the earth. The influx of factory workers and managers created a demand for subsidiary services, from food hawkers to taxi drivers to prostitutes. Forbidden from existing legally in the zone, excluded from housing and food and subject to arrest and removal, migrant workers found (and still find) a place to live in the villages, where the city had limited jurisdiction and even less enforcement ability regarding building codes, police registration for residents, business registration, fee collection, or taxes. Even "legal" migrants who had temporary permission to reside in the city sought both the lower prices and the relative protection of the village from extortion and harassment from local authorities.[32]

The villagers were quick to realize that more money can be made from rent than from farming. Shortly after the zone was created, villagers received allowance to build up to three stories on plots for their own homes. Instead, farmers began to build as high as they could without elevators, usually eight or nine stories, with open concrete shells for shops on the first floor, often initially reserving one floor for themselves and renting the rest out to migrants. With limited space to build, the substandard apartment buildings bend in on each other, reducing the sky to a thin line barely visible above layers of dripping air conditioners and caged windows at their narrowest. These "handshake" buildings (also known as "kissing" or "can't get a coffin in" buildings) form clusters of record-breaking population density.[33] They also generate revenue. In the village of Tianmian, where the eight-story buildings are of a relatively high quality, each floor contains three to four apartments that rented by the late 2000s for RMB 1,000 to 2,000 ($150 to $300) per month. Each village family owns one such building, whose rent in this case generates considerably more than $80,000 per year, not counting commercial income. Beginning in 1985, the collective farms formally started to be turned into joint-stock holding companies, with all villagers as shareholders. With this change, villagers profited from not only housing rents but also light industrial enterprises built on collective land in each village and, later, from deals with developers for office towers, middle-class and luxury residences, hotels, and commerce.

Thus the urban villages perform a double inversion of the Mao era: the city now encircles the countryside, and the peasants become landlords. In the far western edge of town in Nanshan District, hard against the flooded lowlands of the Pearl River Delta, a young migrant doctor from Shandong Province runs an informal clinic out of a 4 m2 space, garishly lit and barely high enough off the narrow street to avoid the flash floods that appear without warning during the rainy season. He pays RMB 550 ($80) a month for the space, which is more than 50 percent of his income. He rented from an absentee landlord who recently sold the building to another villager. The smart villagers invest their rent money, he said, and let a management company deal with the rental while they apply their skills to other jobs. Those villagers with no skills, like his landlord, manage their own buildings and make money only off the rent.

Out of the rain came a grandmother with her two-and-a-half-year-old grandson, his nose running and throat sore from the persistent dank of the rainy season. Wailing, he promptly received an injection in the rear—no disinfecting of the area—with a needle clearly used many times before and that the doctor had just used to fill multiple bottles. The shot cost RMB 47 ($7). A craggy-faced man from Anhui Province with nowhere to go nodded approvingly. "Hospitals are a waste of time and money," he declared, "because they make you take tests. But here," and he gestured at an old intravenous drip contraption affixed to the cement wall, "you're in and out—get your shot, pills, and two-hour drip—if you're seriously ill you can get the drip for two or three days." What's in the drip? "Vitamins."

As the doctor implied, and the wealthy wives of Tianmian personified, only the unlucky or unskilled villagers make their money from rent alone. The more unified and disciplined villages use the joint-stock company format to expand into other income-generating activities. Also known as a village enterprise corporation, this crucial development allowed villages to profit from land that they were not able to legally sell or convert to urban use because the land was subject to "rural" regulations governing land transfer. In Shenzhen, the advent of villagers forming a joint-stock company to share wealth was pioneered in the 1980s by a villager named Pan Qiang'en from the village of Wanfeng on the outskirts of Bao'an District. Pan Qiang'en has been described as a "phenomenon" who turned a farming village into a multinational enterprise following his "theory of joint ownership," which combines "the collective interests closely with the individual interests into an indivisible whole." In 1984, Pan Qiang'en convinced village party members and Wanfeng's sixty-two village households to contribute more than RMB 250,000 ($30,000) to establish the Wanfeng Stock Company.[34] By 2008, it had grown into Shenzhen Wanfeng Group Stock Company Ltd., with total assets of nearly RMB 1 billion ($130 million), one hundred twenty subsidiaries funded by foreign investment, and sixty thousand employees in sixty-nine factories making, among other things, electronics, plastics, clothing, hardware, paper, and toys.[35]

GLOBAL VILLAGE

Wanfeng Village's scale is unusually large but typical in how the village's success consists of taking advantage of not only rural land status in regard to property rights and land use but also collective identity and connections to the global economy. With Shenzhen next to Hong Kong, every village has someone who lives across the border, which had been open until 1951, and where farmers were allowed up through the early 1990s to tend fields on the other side. The farmers were and are well known for smuggling across the porous Hong Kong border, which maintained a flicker of trade during the most difficult years of the Great Leap Forward and the Cultural Revolution. Today, goods, drugs, and people still cross the border. "They are still smuggling," pointed out one Shenzhen professional about the farmers, "but now out of China—pirated CDs, clothing that exceeds quotas, pigs and food, diesel oil." But legal loopholes, not smuggling, is the source of wealth. The clan-like nature of the village is intimately connected to the transnational quality of village business, which relies on extensive networks in Hong Kong and relatives who are overseas Chinese across the globe.[36] All villagers are shareholders, even those who live abroad. Children who move to Hong Kong join extended family and use their dual residency status to move between the two systems to negotiate and manage business deals. What some Shenzhen officials referred to derogatorily as "feudal" kinship structures are, in turn, strengthened and spurred by the village joint-stock organization structure that sustains the transnational networks.[37]

The villages used their new livelihoods to embed themselves in translocal and transnational circuits in a double sense: Domestically, the villages enmeshed themselves in the vast network of inner-Chinese labor migration by using individual land plots to build rentals to which migrants streamed by the millions from all over China. Globally, the villages used their diasporas to attract overseas investment by using the remnants of collective land to provide foreign companies with land in exchange for sharing income and administrative costs.[38] Wang and colleagues argue that the villages de facto "started the industrialization process" even before the local government settled on clear plans for the zone's development.[39] By adeptly exploiting ambiguity in property rights, social relations, border privileges, and rural-urban distinctions, Shenzhen's villages became as much an experiment with the market as the zone itself.

"Responding to the Era's Call"

THE SECRET OF SUCCESS

The above section indicates how the villages are extremely capable of reconfiguring and adapting to changes in the local and global economy. It is through these reconfigurations that the villages coconstitute the city, providing the conditions for Shenzhen's growth and existence and giving villages a very different role than their symbolic status as vestiges of a backward past stuck anomalously in the modern city. The migrants who flock to Shenzhen with and without permission to take part in its economic miracle are more than just tangential to its growth. Without the villages, the city would never have grown with the famous "Shenzhen speed"—the city could never have sustained the cost of housing and supporting so much labor. The density that is the scourge of planners and the very image of squalor is also an answer to the city's housing problem. In central Futian District, more than 50 percent of the population (approximately 800,000 people) lives in its fifteen urban villages even though they make up only four square kilometers of land.[40] Seen in one way, this concentration is a catastrophic concatenation of health hazards, illegal buildings, slum-like conditions, and a classic breeding ground for crime and vice. Seen in another, the city of Shenzhen offloads its costs for 50 percent of its population in Futian onto the villages. The villages play a role in infrastructure for the zone beyond housing. As Webster and colleagues detail, one village alone in Bao'an District in the early 1990s invested more than RMB 30 million ($3.4 million) on roads, water, and power infrastructure.[41] The villages provide or invest in private schools, private security and, above all, housing. Village property management manages city-provided sewage, water, and electricity. Were Futian's villages to be destroyed overnight—something theoretically within the power of the government to do—the city would need to provide housing for nearly a million people. As Wang and colleagues write, village practices "enabled the municipal government to avoid taking responsibility for social, economic and infrastructural development in these villages."[42]

Two observations follow from this. First, the de facto reliance of the city on the informal provision of housing and services is less an unintended consequence than part of what one city official told me was "Shenzhen's secret of success"—the implementing of policies without paying for them, or as he phrased it, "building the city at no cost." In the earliest days of the city, each province from all over China was invited to be represented in the zone, and they sent and paid for work units from home to build their part. In this way, the city was able to avoid paying directly for much of the infrastructure. Through the unique terms of the SEZ's charter, Shenzhen was politically subordinate to the provincial capital of Guangzhou but reported directly to Beijing for economic policy. This resulted in unusually autonomous economic decision making and a range of incentives to local Chinese governments to set up shop in Shenzhen. Because Beijing was wary of southerners, they gave land to government ministries to develop. This favoritism from Beijing earned the resentment of Guangzhou, whose mayor in 2006, as some Shenzhen officials noted dryly, insisted on referring to Shenzhen as "that city next to Hong Kong."

Second, what appears as "backwardness" is closely linked to the villages' contemporary success. The village functions as a collective, but invariably the village leadership is run by a male village head who is usually also the village party secretary. This system, often decried by the government and urban observers as "feudal," lends a particular efficiency rooted in family and social ties. These ties not only make the villages prone to bitter power struggles but also bind the village internally and to other villages. This dense intervillage network frustrates government efforts to divide and conquer.[43] On top of this, rural migrants find the rural patterns of power structure and authority in the villages familiar and instantly recognizable. This enables the very migration that is simultaneously necessary and "illegal."

The "premodern" social structure also allows the villagers to enforce security and order better than the local police could do. Even the location has an analogy to rural days—the villages were laid out to enable an easy walk to the neighboring fields. Now their compact and central locations afford an easy commute to nearby office buildings and commercial areas where many migrants work. Keeping these old settlement patterns allows menial laborers to be more fluid and to follow jobs without high housing or transportation costs. As in the past, the villagers' livelihood and place of residence remain the same even though rents have replaced crops as the source of income.

The government struggles with precisely this tension between the benefits of displacing social costs onto the village and the costs of allowing villagers to exercise relative autonomy. In the last several years, the "usefulness" of the villages to the city government has reached, as one informant put it, "a tipping point." Concerns about security and safety trump the list, but "civilization" is the core issue. Having once played a useful role out of necessity, the villages are now seen as preventing the city from developing in a "controlled manner" (as one city official put it). The villages are, literally, largely "out of control" in the dual sense that their power structure is outside of the municipal administration and their nonvillager population is, as the term goes, "floating"—transitory, highly mobile, escaping the regulations about registration and official residency permission. Its semantic content, as Barthes might put it, exceeds and subverts the calculative impulses of planners and officials who find themselves continuously frustrated by the villages' slippage in and out of their grasp.[44] "Villagers exceeded everyone's expectations," explained a city official, "and we have no theory for explaining or dealing with them."

RELATIVE AUTONOMY

How does this interstitial excess manifest itself spatially? Even many years after the switch in legal status from rural to urban, villages remain spatially distinct and visually distinguishable for the most part by the sudden appearance of "handshake" buildings. Roads leading into the villages are usually gated and guarded by village security, providing both surveillance and revenue generation from parking: any entering vehicle has to get a ticket and return it on exit. Short stays, such as for taxis, are free. In this way, vehicular access is strictly controlled, and entrance and exit is monitored. When you walk through or past the gates into the village, you usually pass an inscription in verse that, as Mary Ann O'Donnell notes, connects village history to its new life after the creation of the SEZ, as in "By following the trend and developing, Shangsha Village responds to the era's call, and a prosperous Shenzhen is constructed."[45]

Once inside the gate, you enter narrow streets lined with concrete shells containing commerce, ranging from small, neat storefronts to dim jumbles of provisions stacked pell-mell selling cheap clothing, flimsy luggage of all sizes, brightly colored pails, used televisions and electronics, cell phones, alcohol, all manner of food, and used furniture. Some storefronts are workshops while others are clinics, cleaners, barber shops, and massage parlors. Density and closeness are omnipresent: of people, stores, aromas, buildings—all hovering between the comforting vibrancy and the vertiginous confinement of proximity. The better-organized villages, such as Tianmian, Shangsha, or Huanggang, dismantle these stereotypes through development of commercial housing, offices, or new layouts, but the poorer ones, like Shawei and Gangxia, approach the stereotype of urban slum, with ever-narrowing streets and mixtures of shabby, haphazard buildings and slowly decaying original stone structures from the pre-Shenzhen era.

In the village of Xinzhou, the image of the village as blight is taken to an extreme, where for one half-kilometer stretch, old village structures gape and list as sparse ruins in a desolate moonscape, and pedestrians walk surreally through curving paths cut through piles of garbage and waste dumped illegally on the site. More typical is the juxtaposition of Gangxia, where the village has been split in two by a major nonvillage thoroughfare. Behind the Bang and Olufsen audio and the high-end Japanese TOTO bath and kitchen fixture stores, you can just barely discern the curved roofs of an old village compound lilting gracefully above a cement wall, the carved stone eaves shimmering with phosphorescent green moss as garbage and laundry offer themselves side by side to the summer sun.

Spatial differentiation is supplemented by an attendant discourse of danger and exoticism. The poorer villages are infamous for illegal activities, such as Xiawei, described by a city official as "Hong Kong's Tijuana" because of its mix of drugs and prostitution that caters to clients from across the border. For the same reason it is also commonly referred to as "Second Wife Village" for all the Hong Kong (and presumably Shenzhen) men who "keep" a woman in Xiawei. One particular corner is famous as the "Hong Kong drop-off" for taxis coming from the border station. Prostitution fronts masquerade weakly as barbershops, massage parlors, karaoke TV joints, or saunas. "Want a beautiful woman?" ask female touts desultorily, milling outside nondescript storefronts.

With security provided by the village, there are tales of profitable relations between some villages and organized crime. This lends the sketchier villages an aura of deliberate but restrained violence. Crime is one of Shenzhen's biggest preoccupations—it has a deserved reputation as a transit point for drugs and human trafficking, stolen cars and goods from and to Hong Kong and Macao, abductions for ransom, car jackings, random violence, and all manner of petty and grand theft. These stories are a steady backbeat. A friend's secretary was robbed in the Central Park at lunch time; an acquaintance was stabbed seemingly at random coming out of a bar at night in Shekou; a friend of a friend was reportedly kidnapped by a taxi driver, held in an apartment, and escaped by jumping out the bathroom window, breaking his leg. The newspaper reports how a whole subway escalator was stolen, in broad daylight, as it was being installed outside the municipal government. A couple was brutally murdered in the model middle-class housing complex of Meilin.

Villages are hardly the only locus of crime, but they fit all too simplistically into narratives where crime exists as the dystopically privileged purview of migrants, premodern ways, and poverty. Villages can present a security and public safety problem in myriad ways due to overcrowding and illegal buildings, but the inability of the government to intervene effectively in the village security apparatus serves to reinforce the autonomous ability of the villages to enforce their own codes of conduct and justice. Police stations are often set up directly outside a village, and while they do undertake raids, there is little to no regular police presence in the villages. Although jurisdiction plays a role, the small police presence is also question of money. Police are allocated in the city based on population, and the official population of the villages is significantly smaller than the many tens of thousands who can actually be living there. The result is a misallocation of police resources that rests on the fiction of formal household registration.

More than anything, this fiction of registration gives the villages a sense of spatial exception outside of state control, especially for the rural migrants for whom this is advantageous. Countless residents of the villages remain unregistered. Many landlords simply fail to report their tenants to the police. The city has tried to implement stricter regulations and send "caretakers" door to door to enforce registration but finds, unsurprisingly, that those without "proper income" fail to cooperate and provide false information. Landlords, according to a cautiously worded government report, also create a "certain degree of difficulty" and "are incapable" of supporting caretakers. The report's understated conclusion is that "timely monitoring of the move-in and move-out of the tenants is impossible and the updating of information is slow."[46]

All this creates a potentially volatile situation for the city government: eliminate the "villages" too abruptly and risk both resistance from the villagers and responsibility for the displaced social problems, or tolerate them and risk fomenting autonomous spaces outside the state apparatus. The villagers, one government official said, want the benefits of both their exceptional status and their city services. The government wants to avoid social unrest and to minimize acknowledging the limits of their abilities. This leads to a complicated dance between village and city—a series of moves and countermoves that circle, cautiously at times and viciously at others, around the villages' increasingly valuable land.

THE COMPENSATION DANCE

Local government had sought to exert control over the housing in the villages since the SEZ was created. Each successive attempt by the city to curb illegal structures achieved the opposite effect.[47] In 1982, the government first gave each household a small plot of land for building a house, and in response, extended families innovatively divided themselves into more discrete households to claim more plots. Four years later, the government sought to curtail construction by limiting height and reducing house size. This spurred even faster development and "unauthorized buildings ran out of control in many villages in the following years."[48] When the government responded by further reducing plot size, it only pushed up building height and density, leading the government to take the drastic measure of stopping all approvals for house building. Yet this only resulted in flagrant illegality, and most all-new or rebuilt buildings became even larger and higher than what had previously been allowed. From the late 1990s onward, 80 percent of new buildings were double to seven times over the height limit.

The government changed tactics in 2001 and offered an amnesty: they would recognize the buildings as legal if households paid a penalty and agreed contractually to transfer the land from rural (i.e., collective administration with inheritance rights) to urban (i.e., "use rights" limited to seventy years). The penalties proved a form of moral hazard: they were so low compared to the profit to be made from renting that they only encouraged more building. The shift to "use rights" from collective land ownership was unpopular but did not have an immediate effect on slowing the practice of illegal building.[49]

The most straightforward way for the city to eliminate the villages would be to take their land, but this requires compensation. If the land is technically rural, compensation is formally calculated by figuring the average farmland output over a prior three-year period and multiplying that value by up to ten—yet clearly this makes little sense for land that has not been farmed in decades. If, instead, the villagers hold use rights, then compensation must still be paid, but this is determined more loosely by local governments (one main reason the government pushed for this).[50] Most villages work out deals with the municipal government and developers.[51] Usually compensation is paid to the collective, but it can also be paid to individuals. This works both ways—the government can seek to divide and conquer by appealing to individuals if the collective is driving a hard bargain, but individuals can also take advantage of this option.

The most famous case in the Western media was the so-called nail house in Chongqing, where from 2004 through 2007, owners Yang Wu and Wu Ping held out against the development of a shopping mall, refusing more than $100,000 above the house's estimated value and defying court orders, holding up a $38.79 million project.[52] The case drew attention to the competing values of money versus lifestyle.[53] It also drew attention to the power of citizens to rally the media, especially online, and to base their claims on rights in the Chinese Constitution.[54] Inspired by the Chongqing case, and concurrent with the heated debate about a new sweeping property law that went into effect in October 2007, a couple living in the Shenzhen urban village of Caiwuwei decided to resist the powerful and politically connected developer Kingkey.[55] Even though the collective had agreed to compensation for the site, located in the heart of downtown and the future location of one of Shenzhen's tallest buildings, Zhang Lian Hao and Cai Zhu Xiang refused to give up their six-story building, which they rented out, built in 1996 for about RMB 1 million ($134,000). The husband told reporters, "The new law says that private property has the same rights as public property. What they were trying to do was to tear down our apartments forcefully and it was against the law and wasn't for any public benefit. It was totally for commercial activity."[56] After a campaign fought through the Internet and news media, the couple eventually settled on compensation of RMB 12.58 million ($1.8 million). Throughout and after the process, the couple was intimidated by the developers; hounded by villagers, extortionists, and gold seekers; and eventually went into quasi-hiding to avoid the publicity. When Cai once sought protection from the authorities, an unsympathetic police official reportedly warned, "98 percent of all nail house owners die in car accidents."[57]

SENDING A SIGNAL

Compensation is one form of exploiting ambiguous property rights. Another is what can be termed a land grab, as attempted in 1994 by Yunong Village, near the border of Hong Kong. Yunong's villagers, apparently, had sought to significantly expand their holdings by erecting thirty-seven new high-rise apartments—some seventeen to twenty stories tall—on adjacent land beginning in August 2004, hoping that if they established facts on the ground, they would be able to claim the entire area for either village income or future compensation. The villagers, according to a city official, "hoped the government would ignore" their "grab" for land that nonvillage developers also laid claim to. They built the buildings at a furious pace, adding one story every three days in an echo of the original "Shenzhen speed" of the 1980s. The government, however, saw this provocation as the perfect opportunity for a display of strength. As part of a new campaign called "Action Combing," launched in 2004 to "renovate" forty-two urban villages by replacing all the illegal buildings, the Shenzhen government set a powerful example by spectacularly blowing up sixteen of the new high-rise buildings. It symbolized their commitment to the destruction of illegal buildings (even if in reality they would fall far short of their goals), it sent a signal to the villagers that they meant business, and it was an unsubtle reminder of state power. In all, fifteen buildings with a combined area of 62,000 m² tumbled like toys in a Godzilla film under the impact of nine hundred kilograms (nearly two thousand pounds) of explosives. Villagers, media, police, curious onlookers, and a four-hundred-strong demolition crew from two subcontracted companies were on hand as the state media trumpeted "China's Number 1 Blast."[58] One building stubbornly remained standing and needed to be demolished by bulldozer. A district official was quoted in Chinese media as declaring that the "blast would completely end Yunong Village's status of 'one family, one illegal building.'"[59]

The tactics of demolition or compensation in the form of an offer that can't be refused are often successful, but they invite confrontation and do not serve the purpose of successfully "integrating" the villages into the city, which some officials see as the ideal outcome. Creating "neighborhoods" based on social relations rather than villages based on "blood relations" is the goal, as one official put it to me. Emphasizing that this was not government policy, he sketched the outline of an internal debate about how to allow the villagers to maintain their quality of life in an urban environment as an alternative to removing the villages by buying out and tearing down. The trick is to enlist, co-opt, and otherwise engage the villagers in the process of "renovation" of the villages such that, as this official put it, they "come in peasants but leave citizens."

COMPROMISE THROUGH DEVELOPMENT

For government advocates who seek integration and not eradication, the "win-win" outcome would consist of villages working with developers to visually, spatially, and administratively "integrate" their villages into the "civilized" modern urban landscape. This necessitates a certain empowerment of the very village structures that the government hopes to weaken. It requires generational patience, for it seeks to reap the benefits of turning hereditary rural land rights into expiration-based urban use rights. Because use rights for residential housing expire in seventy years, village property transformation will be spread out over at least two to three generations.

Yet the hope is that long before the expiration of use rights, market forces and the benefits of city integration will irrevocably change the villagescape.

A poster child for this type of transformation is Tianmian, a village of about two hundred villagers and approximately eight thousand nonvillager residents. During the first two decades of the zone, it was a ramshackle village located in Futian District just beyond the western edge of Shenzhen's downtown area, surrounded by lychee orchards and fields. The 1996–2005 master plan for Shenzhen proposed building a new central business district

(CBD) directly to the west of Tianmian, which would house the municipal government, museums, office towers, high-end housing, hotels, shopping malls, public library, and a convention center. Looking from above, almost like the classic layout of the Forbidden City, the CBD was separated from the old downtown by a new Central Park. By 1995–96, Tianmian Village hung off the eastern side of the CBD like a small fruit, extending into the park along a narrow river. In 1998, Tianmian became the site for one of the most successful "renovations" of an urban village, in part because of its visible location along the central east-west axis connecting "old" Shenzhen to the CBD, its green yet central location, and its relatively compact layout (see figure 18).

As part of a government-approved plan, developers negotiated with villagers to comprehensively redesign what had been a hodgepodge of old and illegal buildings. Although in other villages the villagers were either compensated through relocation to substandard housing or given cash that they used to build new structures on their own, in Tianmian, the municipal government brokered a deal between the village and the developer, Han Guo. If the village shareholders let the developers build a luxury housing complex along with a combination office tower and luxury hotel, giving them exclusive rights to market and profit from these buildings, they would build new improved "handshake" buildings for the families to continue to rent out, along with some middle-class housing (Tianmian Gardens) and commercial space. It was essentially a barter arrangement with government blessing—let the developers renovate your village at no cost if you allow them to reclassify and use part of village land for profit.

Han Guo then brought in the second largest real estate company in China, World Union, to market the new luxury development, called Millennium Oasis. It was a "great location," according to one of the World Union real estate executives: its nine eighteen-story buildings overlooked Central Park with views that could not be blocked, it had good fēng shui being next to water (the river is still polluted but less "stinky" than a decade ago), and it was directly next to the planned CBD. When it opened in 2001, with its private underground garage, health club, swimming pool, private garden, expansive views, and exclusive aura, the 130–40 m^2 (1,400–1,500 ft^2) apartments were, at the time, among the most desirable in the rapidly expanding city. At its 2001 opening, an apartment cost about 30 percent more than the average Shenzhen apartment at the time and has almost tripled in price since then. Three years after Millennium Oasis opened, Han Guo completed construction on the thirty-story office building and four-star Grand Skylight Garden Hotel.[60]

Some villagers moved to the luxury buildings, while others took over full floors in the new rental buildings or became absentee landlords. Tianmian

Gardens, the eight-building, high-rise, middle-class development, was aimed at middle-class management, while the forty eight-story former handshake buildings continued to rent to migrants. The migrant area was spatially segmented in the "back" of the village, remained formally "unapproved" by the government (and hence subject to future demolition), and the buildings remained closed together. But they now bore the hallmarks of modern planning—the buildings were built in a tight rectangular grid, all the buildings were identical and clad in pink tile, and despite their closeness they allowed for more air circulation and emergency vehicle access. They boasted elevators and security guards, and the village ran their own fire department. The apartments rented in 2008 for up to RMB 2,000 ($300) per month.

These improvements earned Tianmian accolades as a "civilized village," and it continues to be a model. The desire to make citizens out of peasants seems to indeed bear fruit here—when I asked shareholders what they were most proud of in the village, they replied that they were most proud of the social identity of being "city persons" and "not being looked down upon." "Now we do nothing and earn money," one village wife pronounced with satisfaction. With such a comment, however, the difficulties of turning "peasants into citizens" is also manifest, for their pride in making money by "doing nothing" is the source of widespread resentment against villagers-turned-landlords among the broader population. The resentment seems greatest not from the rural migrants who live in the villages but from middle- and upper-class Shenzhen residents who are both contemptuous and jealous of their unexpected success.

Civilization Is a Taste

Civilized Shenzhen, Warm Homestead
SLOGAN ON A CITYWIDE BUS STOP POSTER

In our trajectory of Shenzhen's villages, we discern Barthes semic and symbolic codes, where villages are symbolically differentiated by the rural-urban dichotomy, yet their spatial, discursive, and material forms constantly suggest additional meanings that reinforce and undermine the core binary separating the rural and urban. These modes of signification form a tight weave with the central cultural code of the post-Mao era, that of civilization (*wenming*). Together with the phrases "harmonious society" (*hexie shehui*) and "quality" (*suzhi*), *wenming* is the key ingredient in "building a civilized city," and the villages-in-the-city gain legibility in the context of this cultural code.[61] The honor of being designated "civilized" is both one of the greatest tools of leverage the state can bring to bear on its population and a coveted designation—Ann

Anagnost relates the story of a family who stole their neighbor's "civilized household" plaque to save face when a potential marriage partner came to visit.[62] Being civilized is not only a goal for Shenzhen but also its raison d'être. Accordingly, what villagers, managers, migrants, and executives share is the promise of civilization as opportunity that is metonymic with Shenzhen.

Recall the village head's wife at the beginning of the chapter who gave her thanks to Deng for delivering them from a time when they were "so bitter and getting thinner and thinner." This loyalty to Deng takes form as a "civilizational" meme that structures the city, which, like no other in China, is predicated on Deng Xiaoping's exhortation: "To get rich is glorious." The notion of being civilized is wholly inextricable from the promise of the market, and Shenzhen is all about the market. "Time is money, efficiency is life," proclaimed an infamous billboard from the 1980s, an icon to Deng's optimism and can-do spirit that thoroughly infuses the city of migrants seeking fortune at every level of the spectrum.[63] Shenzhen venerates Deng like nowhere else; his only statue in China is poised in midstride on a hill overlooking the CBD and an oversized billboard with his face greets all who head west on the wide Shennan Boulevard from the old city center. When Deng died, Shenzhen trembled, remembering that other party secretaries in 1992 had criticized Deng's trip to Shenzhen with comments like "The flag is no longer red" and "We should stop this, it will change China." A friend's mother, on hearing the news, cried, "What will happen to Shenzhen?"

The economic reforms for which Shenzhen was the "test bed" elevated the discourse of civilization to a "national strategy for radical social transformation."[64] Peasants in particular gained a prominent role as agents capable of progression from backwardness to civilization, with those unready for civilization disturbing the national narrative of progress.[65] Peasants were to be made ready to receive the "evaluative gaze of foreign capital" or suffer the ignominy of failing to "embody international standards of modernity, civility, and discipline."[66] Shenzhen's urban villages fit this sense of the "inappropriate other" (a term Anagnost borrows from Gyan Prakash), and are thus key spaces for the act of civilizing through all necessary means, from absorption into municipal administrative structures to the various forms of renovation explored above.[67]

FROM VILLAGE TO MALL

In 2005, Shenzhen was named one of China's top ten "civilized cities," a designation that indicates the correct environment for proper living. During the city's campaign for this national distinction, Carolyn Cartier observed how subdistricts that wished to be considered officially civilized had to be free of informal structures and how demolishing illegal migrant housing was a central part of the campaign to improve the conduct, outlook, and "correct values" of zone residents.[68] In the southern part of Caiwuwei Village, once the location for Communist Party headquarters during the era of collectivization, one can see the application of the civilizing discourse along a continuum from the removal of informal buildings to their replacement with one of Shenzhen's premier malls, the vast MixC Mall, Shenzhen's urban answer to the old "rural" market town of Dongmen, which lives on as a cacophonous warren of shops a brief walk away.[69] MixC boasts an ice-skating rink, Gucci and Häagen-Dazs stores, and a gourmet supermarket with organic products. As the village literally recedes, the mall gains iconic power. In summer 2008, the mall advertised itself with a large poster, backlit in purple with three black circles containing, respectively, a T-shirt, a handbag, and a plate with a knife and fork. In the center of each object glows the English word *civilized* in white letters. Underneath, the shopper is informed, in Chinese, that "civilization is a taste" (*Wenming ye shi yizhong pinwei*).

This advertisement expresses succinctly how "civilization" shapes bodies and behavior and becomes connected to "new" spaces replacing the "old" villages. The term *civilization* is what Barthes refers to as a "seme," or a proper name that groups together other signifiers. These other signifiers (in this case, the handbag, Western fork and knife, T-shirt, or the mall itself) must be repeatedly connected to the proper name such that its invocation is implicit whenever these subordinate signifiers are mentioned in context. In this way, the "seme" connotes meaning and, writes Barthes, is thus "literally an index: it points but does not tell . . . it is both the temptation to name and the impotence to name."[70] This characterization of temptation and impotence reveals a tension within the advertisement, for the Chinese word *pinwei* has connotations that exceed the English word *taste*. The kind of taste it implies reflects a discerning palate, one attained through proper upbringing rather than through money alone. It implies the ability to sort the genuine from the fake (a practical challenge in China, especially in Shenzhen) and alludes to a graceful quality that exudes nobility without snobbishness. But of course, the point of the advertisement is precisely to imply that you can acquire civilization through proper consumption, which requires significant amounts of money. Thus the temptation and the impotence—the desire to earn and spend to become a "civilized" person of taste, and the mimetic inferiority of always trying to attain a sophistication that can never entirely be yours through this method.

CIVILIZATION BY DESIGN

While the mall signifies civilization through taste, the collectively owned industrial areas of villages are also "building civilization" by refashioning themselves as centers for creative industries.[71] As the new millennium began, the city had more than sixty thousand designers generating 4 percent of GDP, with hopes for a tenfold increase in the years to come.[72] The government branded Shenzhen "the birthplace of modern Chinese design" due to the comparatively high number of designers who helped businesses compete in a market environment, and in 2008, Shenzhen was designated a UNESCO "City of Design" (COD). In the lead up to this designation, which incorporated Shenzhen into a global network of "creative cities," Tianmian repurposed a corner of the village where light industry had been set up to produce paper boxes and sweaters. In 2007, the dingy factories were gutted, enclosed in glass, and repackaged as a COD using the English name and acronym, drawing on the campaign for the official UNESCO designation. With close ties to the government, the company that manages the COD, named Sphinx, provided connections and competitive rents. It began by attracting seventy companies to the almost 50,000 m², 3.5 acre site, from China, Germany, Austria, Japan, and the United States. By 2014, there were two hundred thirty companies listed at the COD.[73]

Most striking about the COD is not its Bauhaus-style renovations and its public art but the way in which it begins to erode the difference between village and city through aesthetics. With some financial backing from the municipal government and incentives from the village, it presents itself as part of the process of "building a civilized city." Civilization beckons at the interface of production and consumption, display and use, design and identity. The COD wine bar doubles as a showroom for wine export. It is deeply upscale, sporting a plush, dark wood interior, walls lined with wine, a cigar humidor, and a waitress from Anhui who studied English and business before migrating. Thomas Friedman's *The World Is Flat* and a book on Hillary Clinton nestle in a book nook near the entrance.[74] Nearby in building eleven, a café called Book Bar is open until two in the morning serving espresso, tea, and cigars, while designers bent over IKEA "Grundton" tables work wirelessly on their laptops. COD hired graffiti artists, reportedly including some from Hong Kong, to decorate two outlying buildings that were left unrenovated to give them a gritty urban look á la New York. The theme is Shenzhen, with a skateboard reference, anime-type figures, and distorted urban landscapes. Nearby on a lawn large, colored geometric shapes sit incongruously, and public art displays rotate in the open plazas. On the side of one building loom two four-story-tall pixilated photographs of Einstein and (best anyone there can tell) 1998 Nobel Laureate in physics, Daniel Tsui. Local workers outside the building had no idea who he was—Deng Xiaoping, they offered, authoritatively, automatically. A Porsche dealership occupies a far corner. Unconfirmed rumors made the rounds in 2008 that Tianmian was going to get its own subway stop and its own Starbucks, certain markers of "civilization" (neither have materialized).

Conclusion: Village-City

The overall impression of Tianmian, with all its contradictory cultural and economic gulfs between former farmers, migrant workers, white-collar residents, and designers, is that the village transforms the city as much as the city transforms the village. This is not because of any specific government policy or clever scheme but because shareholders, businessmen, and migrants alike are adapting the legacies of socialist organization and management to a market environment. Webster and colleagues remind us that socialist management was always more ad hoc than Western observers imagine it to have been, and that the urban villages are recombinant adaptations that express an older Chinese concept of "cities within a city" in modern form.[75] Accordingly, the villages are part and parcel of the new urban spatial order. This spatial order is also always a social order, one mediated by the village leadership. Acting collectively, the village leadership moves deftly between officialdom and informality, staking out a role as the social group that makes local control possible for the government in nebulous times of great transformation.[76]

A growing number of architects and city officials I spoke with in Shenzhen view the villages less as an urban antithesis and more as entanglements of urban essence. In comparison to the disciplined rows of high-rises and housing developments that occupy most of Shenzhen, they wonder if the urban villages' "spatial vitality" could be deployed to control their own subjectivity.

This question of approaching the symbolically "rural" part of cities as something other than a space to be wholly assimilated or physically excised is a key challenge for the rapid urbanization happening around the globe. Shenzhen is one part of this phenomenal trend: the urban population will add two billion people globally by 2040, and much of this growth will take place where rural and urban alter and encounter each other. Already by the early 2000s in Asia nearly a majority (43 percent) of urbanites live in variations of informal settlements; in Africa this figure is more than 70 percent.[77] In China the projected growth to eight hundred million urbanites in the coming decades means, as a World Bank report once put it, the equivalent of adding a new million-person city every month for nearly the next half century. In this context, the gray zones of quasi-informal and quasi-legal communities are inextricable from urban texture, raising questions of displacement and replacement of populations and of tactics and trajectories of agency, accommodation, recombination, and reappropriation.[78] These spaces engender not only municipal hostility but also co-optation through gentrification, especially where land value is high, such as in Shenzhen or Mumbai, or where nostalgia for "rural" traces leads to the simultaneous erasure and production of heritage.[79]

Writing of Hong Kong, Shenzhen's big (br)other across the border, Ackbar Abbas writes that there is no such thing as a "systematic reading of the city, only a compendium of indices of disappearance."[80] My reading of Shenzhen has sought to index the city's narrative of progress with the symbolic valuation of villages, which both disappear and remain in plain sight. In this way, one can see how the "villages" are both a key locus for China's urban civilizing mission and the lump in its urban throat. Urban villages play a complicated role in China's transvaluation of rural and urban. Under Mao, the rural was rhetorically privileged in party doctrine although subordinated to the city in practice; since Deng, it has been the urban that marks progress.[81] The urban villages interrupt this transvaluation, forming spatial inconsistencies to supplement what Michael Rowlands has labeled "temporal inconsistencies" under globalization, where tradition and modernity are coextensive.[82]

For a city famous for speed, by slowing down the reading of Shenzhen through a gentle application of Barthes, I hope to trouble the legally accurate but symbolically misleading official narrative in which Shenzhen's villages have vanished and are visible only as vestiges. The chapter began by noting how the official narrative of the city functioned like a good murder mystery: the conclusion to the story is known—the narrative tension comes only from figuring out how to get there from here. (How will villages disappear? How will they resist disappearance?) Barthes contrasts this hermeneutic form of emplotment with what he calls, after Aristotle, the proairetic—narrative tension is not located in the unfolding of a foregone conclusion but transpires from an emergent sequence of motions that leaves us eager to know what happens next.[83] The sequences are neither arbitrary nor deterministic. They follow a recombinatory logic in which social practices emerge around use rather than follow a tight script.[84] To evoke Barthes's metaphor that started the chapter, the village and the city do not form a poem structured "tidily" around its subject but rather navigate between hierarchy and assimilation, the binary images of a backward remnant or an anonymous modern development, the slum or the mall. As spaces of navigation, the urban villages simultaneously contest state claims to what urban life should be while functioning as laboratories for the coevolution of contemporary Chinese spatial organization and technologies of governance.

In the Shenzhen experiment, villagers and migrants form a kind of circuit with the state that enabled the city to take shape. Along this circuit travel the contradictions inherent in China's transition to a market economy: ambiguous property laws interrupt the assignment of market value, overlapping jurisdictions result in competing forms of social control and regulation, and ambiguous signals from government and village (compensation, demolition, integration, cooperation, resistance) allow for wide negotiating space with contradictory outcomes. In all this, the rural-urban dichotomy is simultaneously reinforced and undermined, creating interstices that both village and city seek to exploit. Shenzhen's villages are thus ultimately neither the city's other nor solely its history but its accomplice in the creation of the urban.

Notes

This chapter appeared originally in *Cultural Anthropology* 25, no. 3 (2011): 421–58. I would like to thank Mary Ann O'Donnell and Yukiko Koga for their inestimable guidance. My gratitude also to Hugh Raffles, Winnie Wong, Chang Tianle, Yang Chen, Michael Gallagher, Lara Luo, Li Kefu, Wang Jia, and many more in Shenzhen who were extraordinarily generous with their time, friendship, food, and insights. I wish to gratefully acknowledge research support from The New School Faculty Development Fund, The New School's India China Institute, and Brown University's Watson Institute for International Studies.

1. *Futian* (Shenzhen: Shenzhen Municipal Government, 2005), 14.

2. In observing similar advertisements in Beijing, Anthony King and Abidin Kusno note how relocating "marginal" places in relation to floating global symbols of modernity raise the expectation that "the accumulation of signified meaning, attached to familiar signs, will in some way drain off into the one that is unknown." Anthony King and Abidin Kusno, "On Be(j)ing in the World: 'Postmodernism,' 'Globalization,' and the Making of Transnational Space in China," in *Postmodernism and China*, ed. Arif Dirlik and Xudong Zhang (Durham, NC: Duke University Press, 2000), 159.

3. Roland Barthes, "Semiology and the Urban," in *The City and the Sign*, ed. Mark Gottdiener and Alexander Lagopoulos (New York: Columbia University Press, 1967).

4. Mary Ann O'Donnell, "Vexed Foundations: An Ethnographic Interpretation of the Shenzhen Built Environment," in *Shenzhen: On and Beyond China's Fastest Growing City* (New York: The New School, 2008); Michel de Certeau, *The Practice of Everyday Life* (Berkeley: University of California Press, 1984).

5. Roland Barthes, *Image, Music, Text*, trans. S Heath (New York: Hill and Wang, 1977), 159; Stephen A. Tyler, "Post-Modern Ethnography: From Document of the Occult to Occult Document," in *Writing Culture*, ed. James Clifford and George E. Marcus (Berkeley: University of California Press, 1986). For contemporary approaches to city as text, see also Tay Kheng Soon and Robbie B. H. Goh, "Reading the Southeast Asian City in the Context of Rapid Growth," in *Theorizing the Southeast Asian City as Text*, ed. Robbie B. H. Goh and Brenda S. A. Yeoh (Singapore: World Scientific, 2003); Tracey Skillington, "The City as Text: Constructing Dublin's Identity

through Discourse on Transportation and Urban Re-Development in the Press," *British Journal of Sociology* 49, no. 3 (1998); and James Donald, "This, Here, Now: Imagining the Modern City," in *Imagined Cities: Scripts, Signs, Memories*, ed. Sallie Westerwood and John Williams (New York: Routledge, 1997).

6. On the impact of globalization on Chinese cities, see, in particular, essays in Li Zhang and Aihwa Ong, eds., *Privatizing China: Socialism from Afar* (Ithaca, NY: Cornell University Press, 2008); Terrence McGee et al., eds., *China's Urban Space: Development under Market Socialism* (New York: Routledge, 2007); Fulong Wu, Jiang Xu, and Anthony Gar-On Yeh, eds., *Urban Development in Post-Reform China: State, Market and Space* (New York: Routledge, 2007); Fulong Wu, *Globalization and the Chinese City* (New York: Routledge, 2006); Laurence Ma and Fulong Wu, eds., *Restructuring the Chinese City: Changing Society, Economy and Space* (New York: Routledge, 2005). On postsocialist encounters with the market, see David Stark and Laszlo Bruzst, *Postsocialist Pathways: Transforming Politics and Property in East Central Europe* (Cambridge: Cambridge University Press, 1998). On the introduction of neoliberalism in Buenos Aires, see also Emanuela Guano, "Spectacles of Modernity: Transnational Imagination and Local Hegemonies in Neoliberal Buenos Aires," *Cultural Anthropology* 17, no. 2 (2002): 181–209.

7. The phrase "Shenzhen speed" was coined in 1985 for the then amazing feat of building one floor a day on the fifty-three-story World Trade Tower. It is now found in many business names and even as the title of a popular song.

8. This exploration of urban formations echoes Ann Anagnost's analyses of the concepts of *wenming* (civilization) and *suzhi* (quality). Ann Anagnost, *National Past-Times: Narrative, Representation, and Power in Modern China* (Durham, NC: Duke University Press, 1997); and her "The Corporeal Politics of Quality (*Suzhi*)," *Public Culture* 16, no. 2 (2004): 189–208.

9. This figure is from Yan Song, "Housing Rural Migrants in China's Urbanizing Villages," *Lincoln Institute of Land Policy: Land Lines* (2007): 1–7. The exact number of the total villages varies mildly depending on the source. Often one "administrative village" will contain two or more historical villages. The villages are further divided into ninety-one inside the original SEZ (inside the First Line) and one hundred fifty in the outlying districts. See Ya Ping Wang, Yanglin Wang, and Jiansheng Wu, "Urbanization and Informal Development in China: Urban Villages in Shenzhen," *International Journal of Urban and Regional Research* 33, no. 4 (2009): 959.

10. I draw on fieldwork conducted over three visits to Shenzhen in 2006, 2007, and 2008, including interviews and discussions with villagers, village residents, developers, intellectuals, city officials, planners, architects, artists, and workers.

11. The outlying district villages were urbanized only in 2003–4.

12. Hang Ma, " 'Villages' in Shenzhen: Persistence and Transformation of an Old Social System in an Emerging Mega City" (dissertation, Bauhaus University Weimar, 2006); O'Donnell, "Vexed Foundations"; Shenzhen City Government, *Futian District Report on Cities among Villages* (Shenzhen: Shenzhen City Government, 2005).

13. For a discussion of this process in the very different context of rice in Japan, see Emiko Ohnuki-Tierney, *Rice as Self: Japanese Identities through Time* (Princeton: Princeton University Press, 1993), 132.

14. Helen F. Siu, "Grounding Displacement: Uncivil Urban Spaces in Postreform South China," *American Anthropologist* 34, no. 2 (2007): 334.

15. Urban villages are a phenomenon across China, as rural areas are being incorporated into cities at ferocious speed. Shenzhen's villages are particularly interesting as they were not

"THEY COME IN PEASANTS AND LEAVE CITIZENS"

incorporated into an existing city but became the space from which the city emerged. Li Zhang's important ethnography of an urban village known as Zhejiangcun in Beijing describes the rise and fall of an urban village in the capital city. As in Shenzhen, local villagers rented to migrants, but in Zhejiangcun, migrants sought to spatially separate themselves from villagers in rudimentary compounds (*dayuan*) to form close, quasi-autonomous communities. "Migrant entrepreneurs" practiced new market-based norms while contesting the system that sought to control the definition of "proper" urban space and navigating across different forms of bureaucratic and private power. In Shenzhen it is the local villagers who seem the more aggressively entrepreneurial, accruing to themselves a mediating role between unruly and official urban spaces. Li Zhang, *Strangers in the City: Reconfigurations of Space, Power, and Social Networks within China's Floating Population* (Stanford, CA: Stanford University Press, 2001). On women migrants in Beijing, see also Tamara Jackra, "Finding a Place: Negotiations of Modernization and Globalization among Rural Women in Beijing," *Critical Asian Studies* 37, no. 1 (2005): 51–74. On female migrants in Shenzhen, see especially Pun Ngai, "Women Workers and Precarious Employment in Shenzhen Special Economic Zone, China," *Gender and Development* 12, no. 2 (2004): 29–36. On the urban village phenomenon in China more generally, see Daniel B. Abramson and Samantha Anderson, "Planning for the Urban Edge in Quanzhou, Fujian: Foreshadowing an Enablement Approach to Village Urbanization," *Projections: The MIT Journal of Planning* 5 (2006): 9–26; Deborah S. Davis, Richard Kraus, Barry Naughton, and Elizabeth J. Perry, *Urban Spaces in Contemporary China: The Potential for Autonomy and Community in Post-Mao China* (Cambridge: Cambridge University Press, 1995); John Friedman, *China's Urban Transition* (Minneapolis: University of Minnesota Press, 2005); and Chris Webster, Fulong Wu, and Yanjing Zhao, "China's Modern Gated Cities," in *Private Cities: Local and Global Perspectives*, ed. G. Glasze, C. J. Webster, and K. Frantz (London: Routledge, 2005). On urban villages in Shenzhen, see Mei Fangquan, "Report on Power Structures in Mingxing Village, Shenzhen City, Guandong Province," *Chinese Sociology and Anthropology* 36, no. 4 (2004): 44–65; Ma, " 'Villages' in Shenzhen"; O'Donnell, "Vexed Foundations." For an extensive treatment of villages in neighboring Guangzhou, see Peilin Li, *Cunluo de Zhongjie—Yangchengcun de gushi* [The End of Village—Stories of Yangcheng Village] (Beijing: Shangwu Yinshuguan, 2003); Siu, "Grounding Displacement: Uncivil Urban Spaces in Postreform South China." See also Gregory Eliyu Guldin, ed. *Farewell to Peasant China: Rural Urbanization and Social Change in the Late Twentieth Century*, Studies on Contemporary China (Armonk, NY: M. E. Sharpe, 1997).

16. Siu, "Grounding Displacement."
17. Roland Barthes, *S/Z*, trans. Richard Miller (New York: Hill and Wang, 1974), 174.
18. Siu, "Grounding Displacement," 335; Zhang, *Strangers in the City*.
19. See Bach, chapter 1 in this volume.
20. Barthes, *S/Z*, 75. Emphasis in the original.
21. In 2002–3, the SEZ was no longer a formal administrative category for Shenzhen, making it a "normal" city, but still largely defining its identity. The first five zones were Shenzhen, Zhuhai, Shantou, Xiamen, and the island of Hainan. For overviews of Shenzhen's development, see Mee Kam Ng, "Shenzhen," *Cities* 20, no. 6 (2003): 429–41; and Mary Ann O'Donnell, "Becoming Hong Kong, Razing Baoan, Preserving Xin'An: An Ethnographic Account of Urbanization in the Shenzhen Special Economic Zone," *Cultural Studies* 15, no. 3/4 (2001): 419–33. For a discussion of Shenzhen's development in the context of China's SEZs, see George T. Crane, " 'Special Things in Special Ways': National Economic Identity and China's Special Economic Zones," *Australian*

Journal of Chinese Affairs, no. 32 (1994): 71–92. For an overview of China's urban development in the reform era, see Friedman, *China's Urban Transition*; Li Zhang, "Conceptualizing China's Urbanization under Reforms," *Habitat International* 32, no. 4 (2008): 452–70.

22. An apparent issue of some contention within villages was whether these types of trips and education costs should come from income from collectively managed land (e.g., industrial or commercial income), as seemed to be the case here, or from pooled resources from family-controlled residential income.

23. When the zone was created in 1980, the village cores were left largely intact, although they lost their farmland. In compensation for the lost fields, the remaining village land was given special status for residential and commercial purposes. Generally speaking, there are three types of villages located, respectively, within the inner city (*Chengzhongcun*), at the interface of the city and suburbs (*Chengbiancun*), and in the suburban districts (*Chengwaicun*). Wang, Wang, and Wu, "Urbanization and Informal Development in China," 959–60. This chapter focuses on villages within the inner city.

24. Webster, Wu, and Zhao, "China's Modern Gated Cities."

25. Dorothy Solinger, *Contesting Citizenship in Urban China: Peasant Migrants, the State, and the Logic of the Market* (Berkeley: University of California Press, 1999), 36–45. The Chinese word for peasant (*nongmin*) can also be translated as *farmer*. Although farmer is perhaps more accurate in the case at hand, in English it is often romantically identified with heroic individualism. I use peasant in this essay and in the title to convey the English connotation of a backward class subject to transformation.

26. On the question of household registration (*hukou*) and the status of rural migrants to urban areas, see, among others, Solinger, *Contesting Citizenship in Urban China*; Kam Wing Chan and Li Zhang, "The Hukou System and Rural-Urban Migration in China: Processes and Changes," *China Quarterly*, no. 160 (1999): 818–55; C. Cindy Fan, "The Elite, the Natives, and the Outsiders: Migration and Labor Market Segmentation in Urban China," *Annals of the Association of American Geographers* 92, no. 1 (2002): 103–24; Song, "Housing Rural Migrants"; Linda Wong and Huen Wai-Po, "Reforming the Household Registration System: A Preliminary Glimpse of the Blue Chop Household Registration System in Shanghai and Shenzhen," *International Migration Review* 32, no. 4 (1998): 974–94.

27. O'Donnell, "Vexed Foundations."

28. O'Donnell, "Vexed Foundations," 6.

29. Barthes, "Semiology and the Urban"; Roland Barthes, *Empire of Signs*, trans. Richard Howard (New York: Hill and Wang, 1982).

30. This is very different from how the introduction of private property into former Soviet collective farms led to impoverishment in Russia. See Jessica Allina-Pisano, *The Post-Soviet Potemkin Village: Politics and Property Rights in the Black Earth* (Cambridge: Cambridge University Press, 2009).

31. Wang, Wang, and Wu, "Urbanization and Informal Development in China."

32. Of the officially registered 8.76 million official residents in Shenzhen in 2008, 74 percent were temporary migrants (6.496 million) and only 26 percent (2.28 million) had Shenzhen as a primary residence—that is, held the Shenzhen *hukou* (city registration; see "Overview," Shenzhen Government Online, [2010]). Rural migrants were allowed to register as temporary residents in 1985 to deal with the need for labor.

33. Shenzhen's population density is reported as 17,150 people per square kilometer, making it the world's fifth most crowded city after Mumbai, Calcutta, Karachi, and Lagos. Xiao Qi,

"Shenzhen Fifth Most Crowded City in the World," *China Daily*, February 25, 2010, accessed February 26, 2010, http://www.chinadaily.com.cn/china/2010-02/25/content_9504088.htm. The Shenzhen municipal government, however, takes exception to the method of calculation, claiming that if the maximum area was taken into account, the density would drop by more than half. See *Shenzhen Daily*, "Shenzhen Ranking Doubted," December 19, 2007.

34. This figure, although based on the official history of Wangfen Village, is a little hard to accept. To raise RMB 250,000 ($30,000), each household would have given an average of around RMB 4,000 ($500). In 1984, this would have been an extraordinary amount for a village family. If the amount is correct, it implies that families had access to funds from Hong Kong or from other sources.

35. Qiang'en Pan and Jiabao Yu, *Zhongguo nongcunxue* [Study of Chinese Farming Villages] (Beijing: Zhonggong Zhongyang dangxiao, 1999); Wanfeng Group, "Wanfeng Group Home Page," http://www.szwanfeng.com. Wanfeng's eagerness for industry did not emerge without context, as villages had a long history of Township and Village Enterprises starting in the 1950s from commune- and brigade-run industries during the Great Leap Forward. See Xia Jun, Shaomin Li, and Cheryl Long, "The Transformation of Collectively Owned Enterprises and Its Outcomes in China, 2001–5," *World Development* 37, no. 10 (2009): 1651–52; James Kai-Sing Kung and Yi-Min Lin, "The Decline of Township-and-Village Enterprises in China's Economic Transition," *World Development* 35, no. 4 (2007): 569–84.

36. Aihwa Ong, *Flexible Citizenship: The Cultural Logics of Transnationality* (Durham, NC: Duke University Press, 1999).

37. On urban transnationalism, see Nina Glick Schiller, "Transnational Urbanism as a Way of Life: A Research Topic Not a Metaphor," *City and Society* 17, no. 1 (2005): 49–64; Andreas Huyssen, ed. *Other Cities, Other Worlds: Urban Imaginaries in a Globalizing Age* (Durham, NC: Duke University Press, 2008); Ong, *Flexible Citizenship*; Michael Peter Smith, "From Context to Text and Back Again: The Uses of Transnational Urbanism," *City and Society* 17, no. 1 (2005): 81–92.

38. These approaches echo the role of other interfaces that seek ways to reinvent themselves in a context of constant flux. In his study of the suburb of Pikine in Dakar, Senegal, AbdouMaliq Simone writes of the neighborhood's negotiation with "new forms of livelihood that transverse local places." See AbouMaliq Simone, *For the City Yet to Come: Changing African Life in Four Cities* (Durham, NC: Duke University Press, 2004), 28.

39. Wang, Wang, and Wu, "Urbanization and Informal Development in China," 964–65.

40. Shenzhen City Government, *Futian District Report*.

41. Wang, Wang, and Wu, "Urbanization and Informal Development in China," 159.

42. Wang, Wang, and Wu, "Urbanization and Informal Development in China," 967.

43. But relations between villages are not always friendly alliances despite (or because of) extensive intermarriage. When I asked about cooperation with Shangsha (Upper Sand Village), a high village official in the successful Xiasha (Lower Sand Village) shot back, "No, no, absolutely no. About eight hundred years ago we were brothers." But the villages were only separated in 1953, joined again in 1968, and then separated again in 1978. The competition today is fierce: Xiasha is closing the last of its factories and building a five-star hotel, offices, and a shopping center. Its future lies in attracting textile and industrial designers. See also Fangquan, "Report on Power Structures," 44–65, which offers an insightful perspective on village dynamics in a study of Mingxing Village in Shenzhen.

44. Barthes, "Semiology and the Urban."

45. O'Donnell, "Vexed Foundations," 24.

46. Shenzhen City Government, *Futian District Report*, 24.

47. Wang and colleagues expertly detail the back and forth described in this paragraph, from which this is partly drawn. See Wang, Wang, and Wu, "Urbanization and Informal Development in China," 960–64.

48. Wang, Wang, and Wu, "Urbanization and Informal Development in China," 961.

49. In 1982, all households received a plot of one hundred fifty square meters, of which eighty square meters was designated for a house up to a total building size of two hundred forty square meters. In 1986, a height limit of three stories and size limit of forty square meters per person was instituted, and in 1993 the plot area was reduced to one hundred square meters, and floor space was further capped. All approvals for housing were suspended in the mid-1990s, and in 2001, the amnesty was instituted. Details are in Michael Gallagher, "Consequences of Urban Village Redevelopment without Consideration for Where Migrant Workers Will Go" (Shenzhen: Shenzhen Institute of Urban Planning and Design, n. d.); Wang, Wang, and Wu, "Urbanization and Informal Development in China," 960–64.

50. See the 1998 Land Administration Law, Article 47. Siu notes how, in Guangzhou, "it seemed ironic that while rural populations at large were trying their best to shed rural status . . . the villagers here were obsessed with protecting their *hukuo* in the collective." Siu, "Grounding Displacement," 340.

51. See You-tien Hsing, "Brokering Power and Property in China's Townships," *Journal of Pacific Review* 19, no. 1 (2006): 103–24; and You-tien Hsing, *The Great Urban Transformation: Politics and Property in China* (New York: Oxford University Press, 2009).

52. Estimated house worth was RMB 2.47 million ($319,414), and media reported that developers offered RMB 3.5 million ($452,612). Figures and quote from Rui Zhang, "First Test Case for Newly Approved Property Law?," *China.org.cn*, March 23, 2007, accessed March 8, 2009, http://www.china.org.cn/english/China/204173.htm.

53. Yang Wu insisted on "a same-sized apartment at the original location on the same floor and with same exposure to the sun, as well as a temporary residence and shop space." Zhiyong Wang, "Detailed Compensation Pace for 'Nail House' Revealed," *China.org.cn*, April 4, 2007, accessed March 8, 2009, http://www.china.org.cn/english/China/205952.htm.

54. A 2004 amendment to the Constitution states, "Citizens' lawful private property is inviolable." In the end, the couple received a new apartment and additional compensation for the loss of their business. See "'Nail House' in Chongqing Demolished," *China Daily*, April 3, 2007, accessed April 18, 2009, http://www.chinadaily.com.cn/china/2007-04/03/content_842221.htm; Zhiyong Wang, "Detailed Compensation."

55. The CEO of Kingkey is the niece of former Vice Premier Zeng Qinghong. See Catherine Jiang, "Chinese Homeowners Nail down Their Rights," *Asia Times*, November 2, 2007, accessed April 18, 2009, http://www.atimes.com/atimes/China_Business/IK02Cb01.html.

56. Jiang, "Chinese Homeowners."

57. See coverage of the events in "The Shenzhen Nail House," *EastSouthWestNorth*, October 4, 2007; "Nailhouse Blues: Unanswered Questions," accessed March 2, 2009, http://zonaeuropa.com/20071004_1.htm; *Mutant Palm*, April 11, 2007, accessed March 11, 2009, http://www.mutantpalm.org/2007/04/11/nailhouse-blues-unanswered-questions.html.

58. The companies were Haofengda and Shenzhen Heli Blast Technology Engineering Co., Ltd.

59. Yan, "Fifteen Old Buildings Demolished in Shenzhen," *News Guangdong*, May 23, 2005; and "Border Bridge to Close for Blast," *News Guangdong*, May 13, 2005.

60. In 2001, these apartments cost about RMB 8,000 per square meter (about $150,000 for a 135 m2 apartment), in 2008 they cost RMB 20,000 per square meter (about $400,000 for a 135 m2 apartment). Data in this section comes from interviews with World Union employees. For an overview of the mechanics of China's commodified housing market and especially the role of intermediary agents, see Bo-Sin Tang, Siu-Wai Wong, and Sing-Cheong Liu, "Property Agents, Housing Markets and Housing Services in Transitional Urban China," *Housing Studies* 21, no. 6 (2006): 799–823. For a good explanation of the role of the government in the housing market, see You-tien Hsing, "Socialist Land Masters: The Territorial Politics of Accumulation," in *Privatizing China: Socialism from Afar*, ed. Li Zhang and Aihwa Ong (Ithaca, NY: Cornell University Press, 2008).

61. On *suzhi*, see Anagnost, "Corporeal Politics of Quality"; Andrew Kipnis, "Suzhi: A Keyword Approach," *China Quarterly*, no. 186 (2006): 295–313.

62. Anagnost, *National Past-Times*, 86.

63. Yuan Geng, the creator of the Shekou Industrial Zone in the western part of Shenzhen, erected the billboard with its famous slogan at the zone entry. Its appropriateness was debated all the way up to the People's Congress, although today a local Shenzhen newspaper regards the controversy as quaint, calling the slogan "bland and self-evident these days."

64. "Records of Comrade Deng Xiaoping's Shenzhen Tour," *People's Daily*, January 18, 2002; Anagnost, *National Past-Times*, 81.

65. Anagnost, *National Past-Times*, 77.

66. Anagnost, *National Past-Times*, 76.

67. Erik Harms shows the workings of the "civilization" discourse in postsocialist Vietnam, where, as in China, it is linked to "market-based notions of right to ownership" and security and hygiene. In this context, Harms shows how civilization is historically linked to spatial order in the service of privatization. In China, Yukiko Koga shows ethnographically how civilization is closely linked to space in the case of marketing colonial inheritance in Harbin. King and Kusno point to an important shift in how civilized modernity is being increasingly signified by environmental indicators of "quality of life" rather than architectural monumentality. This is in keeping with civilization as a primary mode for cities "to accept, and contribute to, the same urban symbolic language, to participate in the same symbolic economy, to speak in the same architectural and spatial terms, as exists elsewhere in the global economy." See Erik Harms, "Vietnam's Civilizing Process and the Retreat from the Street: A Turtle's Eye View from Ho Chi Minh City," *City and Society* 21, no. 2 (2009): 186–206; King and Kusno, "On Be(j)ing in the World," 59–60; Yukiko Koga, "'Atmosphere of a Foreign Country': Harbin's Architectural Inheritance," in *Consuming the Entrepreneurial City: Image, Memory, Spectacle*, ed. Anne M. Cronin and Kevin Hetherington (New York: Routledge, 2008), 221–53.

68. Carolyn Cartier, "Transnational Urbanism in the Reform-Era Chinese City: Landscapes from Shenzhen," *Urban Studies* 39, no. 9 (2002): 1527.

69. MixC is short for "Mix City." The Chinese name for the mall is "All Earthly Phenomena" (*Wanxiang*). It is interesting to contrast this kind of shopping center with the older, traditional commercial center of Dongmen, and also the newer, more chaotic shopping district of Huaqiangbei. On the latter, see James Jixian Wang and Jiang Xu, "An Unplanned Commercial District in a Fast-Growing City: A Case Study of Shenzhen, China," *Journal of Retailing and Consumer Services* 9, no. 6 (2002): 317–26.

70. Barthes, *S/Z*, 62.

71. See Wong, chapter 9 in this volume.

72. "Shenzhen, a Booming City of Design," *Shenzhen Daily*, June 10, 2008.

73. Information comes from interviews at the Sphinx company's COD office. See also "City of Design Home Page," *Sphinx*, http://www.cityofdesign.com.cn/2008/english.

74. Thomas Friedman, *The World Is Flat* (New York: Farrar, Straus and Giroux, 2005).

75. Webster, Wu, and Zhao, "China's Modern Gated Cities," 165.

76. In this way, the villagers play a role similar to the migrant "bosses" that Li Zhang analyzes in Zhejiangcun, where migrant bosses who control housing and markets maintain local order by mediating between self-described bureaucratic (*quanli*) and local (*shili*) forms of power. See Li Zhang, "Strangers in the City," 10 and chapter 4. Given the strength of villagers' associations in Shenzhen, local control is mediated in a nested series of brokers from the villagers down to subunits of migrant organization. Both village and migrant forms of organization are networked within China (to other villages or home villages) and abroad (through the diaspora). Thus social order is maintained and replicated both vertically and horizontally, even as specific villages or settlements disappear. By looking at the mediating roles of the villagers and the migrants, as in Zhang's study, we gain a fuller picture of the emerging forms of social order in rapidly urbanizing China. Zhang, *Strangers in the City*.

77. In Latin America, the Middle East, and Northern Africa, the corresponding figures are 32 percent and 30 percent, respectively. See Barney Cohen, "Urbanization in Developing Countries," *Technology in Society* 28 (2006): 64.

78. Partha Chatterjee, *The Politics of the Governed* (New York: Columbia University Press, 2004); Mike Davis, *Planet of Slums* (New York: Verso, 2006); Robert Neuwirth, *Shadow Cities: A Billion Squatters, A New Urban World* (New York: Routledge, 2005); Simone, *For the City Yet to Come*; AbdouMaliq Simone, *City from Jakarta to Dakar* (New York: Routledge, 2009).

79. O'Donnell has detailed this process for the case of Shenzhen and Hancock for the case of Indian cities. See O'Donnell, "Becoming Hong Kong, Razing Baoan, Preserving Xin'An"; O'Donnell, "Attracting the World's Attention: The Cultural Supplement in Shenzhen Municipality," *Positions* 14, no. 1 (2006): 67–97; and Mary Hancock, *The Politics of Heritage from Madras to Chennai* (Bloomington: Indiana University Press, 2008). See also Jonathan Shapiro Anjaria, "On Street Life and Urban Disasters," in *What Is a City? Rethinking the Urban after Hurricane Katrina*, ed. Phil Steinberg and Rob Shields (Atlanta: University of Georgia Press, 2008); Arjun Appadurai, "Deep Democracy: Urban Governmentality and the Horizon of Politics," *Environment and Urbanization* 13, no. 2 (2002): 23–43; Vyjayanthi Rao, "Slum as Theory: The South/Asian City and Globalization," *International Journal of Urban and Regional Research* 30, no. 1 (2006), 225–32.

80. Ackbar Abbas, *Hong Kong: Culture and the Politics of Disappearance* (Minneapolis: University of Minnesota Press, 1997), 73–74.

81. See Solinger, *Contesting Citizenship in Urban China*.

82. Michael Rowlands, "Temporal Inconsistencies in Nation-Space," in *Worlds Apart: Modernity through the Prism of the Social*, ed. Daniel Miller (New York: Routledge, 1995); Simone, *For the City Yet to Come*, 233.

83. Barthes, *S/Z*, 18.

84. David Grahame Shane, *Recombinant Urbanism* (Hoboken, NJ: Wiley, 2005); David Stark, *The Sense of Dissonance: Accounts of Worth in Economic Life* (Princeton: Princeton University Press, 2009).

8

Sex Work, Migration, and Mental Health in Shenzhen

WILLA DONG AND YU CHENG

Sex Work in Shenzhen

One Friday night near the Shuiwei checkpoint, all the booths in the karaoke bar were filled with middle-aged men who took turns crooning ballads to the entire bar. The hours passed with the rattle of dice in tumblers, the clinks of small glasses of watery beer, and conversations shouted in Cantonese under the lazy spin of a disco light. After midnight, one of the men signaled to the "mommy,"[1] who conferred with him at his booth. She made calls on her cell phone, and within a few minutes, a series of young women came in one by one through the front door and were introduced to him. The last girl sat down at the booth and he lit her cigarette.

One Wednesday afternoon, several women sat on couches and chairs at a beauty salon watching a talent competition rerun on TV. The conversation took place entirely in Sichuanese. A middle-aged man entered the salon and motioned to one of the women. The entire time, he kept his eyes forward and strode directly to the back. The woman he pointed to drew out a condom from a little jar and headed out the back.

It was close to midnight one Saturday night in a clothing stall at a night market, and the salesgirl was trying to sell a tight green dress to the mommy, telling her that in her line of work, it was perfectly acceptable to dress flamboyantly.

One Sunday evening, sandwiched between a pizza restaurant and convenience store, a woman sat in a "hair salon," where the front room was empty except for a couch and a strip of red neon light.[2]

Similar to the development of various Special Economic Zones (SEZs) globally, the opening of the Shenzhen SEZ has been accompanied by the reappearance of commercial sex after its near eradication between the late 1950s and 1970s. Well into 2012, sex work continued to be woven into the everyday landscape of Shenzhen and other Chinese cities despite cycles of crackdowns after its reemergence in the 1980s.[3] In 2000, the Joint United Nations Program on HIV/AIDS, citing Chinese Public Security Bureau sources, estimated that there were between four and six million sex workers in China.[4] With the rise in sexually transmitted infection rates and the advent of the HIV epidemic, public health efforts targeting sex workers has focused on these areas.[5] However, like other migrants, sex workers have a range of occupational and everyday health needs. Given sex workers' unique position of encountering high levels of stigma and vulnerability due to criminalization, mental health remains underaddressed as a matter of social policy, but it also provides us with an important perspective on the challenges faced in their lives, as well as those of other Shenzhen migrant workers.

Though all urban centers in China have witnessed a dramatic increase in their populations due to internal migration, Shenzhen remains unusual for its unprecedented pace of urbanization and for being composed of an overwhelming majority of migrants.[6] In addition, Shenzhen is also distinguished as one of the first sites in which sex work reappeared; by 2001, Shenzhen was recognized as a hotspot for sex and sexual services. One study estimated that from the Luohu checkpoint alone, ten thousand men entering China engage in "sex networking" daily.[7] For these reasons, understanding the mental health needs of female sex workers (FSWs) in Shenzhen provides an informative case study for interrogating the mainstream narrative of public health development surrounding sex work and its purported achievements.

In this volume, the marginalization of migrant workers through administrative and spatial apparatuses such as *hukou* and the Second Line have been detailed, as has the recognition of their persistent—if not legally permanent—position in the city of Shenzhen. Moreover, Florence and Wong have shown in this volume how the valorization of migrant workers became necessary in official and public discourse in Shenzhen from the mid-1990s through the 2000s. At the same time, some migrant workers remain a problematic or even main site of negativity for public and official discourse,[8] and FSWs in Shenzhen are therefore a crucial element of this larger development that, in many ways, runs counter to the accepted ways in which *some* forms of labor have been incorporated into a new Shenzhen identity.

This chapter represents one study of FSWs in Shenzhen based on the results of twenty-seven semistructured, in-depth interviews conducted in Shenzhen, China, between March 2012 and February 2013, supplemented with fieldwork.[9] The interviews were conducted in sex workers' workplaces, including hair salons, massage venues, and karaoke bars in an effort to protect the informants' privacy by avoiding meeting in public or in their homes.[10] The venues where sex work takes place in Shenzhen are also part of our analysis of the spatial, or place-based, concepts of sex workers' mental health.

Sex Work and Shenzhen Urbanization

If each Chinese city were to be affixed with a label, the best label for Shenzhen would be "Reform." This sort of "reform" is reflected not only in the rapid economic growth but also in significant shifts in lifestyle and values. Under Mao, campaigns to eradicate sex work were carried out through closing brothels, punishing venue owners and managers, and attempts at "reeducation" for sex workers.[11] While the historical purpose of sex was to produce a male heir to continue the family line, this orientation was especially emphasized during the Cultural Revolution.[12] Sex was policed through the enforcement of informal codes of conduct against nonmarital sexual and romantic relationships and official silence toward sexuality.[13] Since economic reform and opening up, sex work in China has developed into an enormous industry.[14] Sex work has thrived in confluence with the expanding scale of the mobile population. First, most FSWs in China are rural-to-urban migrants.[15] Like many migrants, these women's *hukou* (household registration) remain in their rural hometowns, which means that they cannot access many social and health care services. Likewise, their children's *hukou* also remain in these rural areas, and they also have limited access to services, including education. This migration is due to the concentration of public resources in cities, as well as widening urban-rural income disparities. The majority of older sex workers are those who have migrated to the cities to participate in sex work in order to bolster their families' income. Most of these women will return to their home villages after earning enough money in the cities and will use the accumulated funds to subsidize family expenses or to start a business. Younger sex workers, in addition to making money, are also drawn by the urban lifestyle, such as using the Internet, going to the cinema, and experiencing nightlife.[16]

Large numbers of male migrants may increase demand for sexual services. During migration, many married male migrants are separated from their wives, who often remain home to raise the family, and this separation may result in sexual frustration.[17] As a result, many migrant men may engage in commercial sex while away from home. It has been hypothesized that large populations of surplus men—defined as men who are young, single, and poor—are encouraged to migrate by the lack of economic opportunities available in their hometowns and high bride prices, which can be earned through working in the city.[18] The demand for sexual services may be facilitated, as these men are far from their homes and less constrained by the norms of their hometowns.[19] Additionally, residence in single-gender dormitories attached to a work unit in the cities may also impede male migrants from meeting noncommercial female sexual partners.[20]

Economic reform, and the ensuing social inequality, has given rise to another class of clients, deemed "mobile men with money." In contrast to "surplus men" clients, this group consists of government officials and wealthy businessmen.[21] Though the current market economy offers opportunities for entrepreneurs, access to resources still depends on government officials. A set of practices for facilitating these relationships and building *guanxi* have emerged out of this context, including banqueting, singing karaoke, and patronizing sex workers.[22] Given Shenzhen's earliest purpose to attract foreign investment and expertise, transnational businessmen from Taiwan and Hong Kong also take part in these practices, which often extend to taking on second wives, who provide intimacy and companionship often in exchange for living expenses and housing.[23]

In Shenzhen, the resurgence of sex work can be divided into three stages based on ethnographic fieldwork and interviews conducted in 2005, 2006, and 2008 throughout the city by Yu Cheng.[24] These stages are based on both the scale of the sex industry and trends in its development:

1. *1979–89:* Though China had officially embraced economic liberalization and opening up, strict control over population mobility did not completely disappear during this stage. Thus urban professionals who arrived in Shenzhen to start businesses and Hong Kong residents who crossed the border constituted the main part of the mobile population in Shenzhen. Sex workers began to trickle into the city at this time, primarily those who worked as second wives providing sexual services to Hong Kong men who crossed into Shenzhen regularly, such as businessmen and truck drivers. During this time, sexual services were concentrated in points near the Luohu crossing, including Liantang, Chunfeng Road, and Badun Street.
2. *1990–2006:* This period was the "Golden Age" for Shenzhen's flourishing sex industry. The Southern Tour speech delivered by Deng Xiaoping in 1992 entrenched the process of reform and opening up in Shenzhen during a time when it was still under debate. As a result, these economic forces further encouraged the influx of rural women (who make up the vast majority of sex workers in China) and male migrant workers, who formed a large pool of potential clients and sellers for the burgeoning sex industry in Shenzhen. During this time, FSWs mostly fell into three groups: those

working in upscale entertainment venues and restaurants, providing services for businessmen or government workers; FSWs who provided services for Hong Kong residents crossing into Shenzhen; and FSWs who provided services for male migrant workers. While sex workers could be found throughout the city, the most well-known red light districts were concentrated near the Shuiwei checkpoint in Futian District, which included Shangsha, Xiasha, Shazui, and Shuiwei. By November 2006, the sex industry in Shenzhen reached its summit. At this point, it was estimated that there were 110,684 FSWs in the city.[25] In the neighborhood of Shazui alone, there was estimated to be no less than ten thousand FSWs. In a survey of sex workers conducted prior to November 2006, it was found that 95 percent of the women were under thirty years of age, 72.7 percent had a junior high school education or lower, and all were migrants to Shenzhen.[26]

3. *November 2006–present:* The sex industry in Shenzhen became bottlenecked during a series of crackdowns that culminated in November 2006. After arresting more than one hundred FSWs, police officers paraded these women through the streets to publically humiliate them. Overnight, the widely recognized red-light districts in Shenzhen (Shangsha, Xiasha, Shazui, and Shuiwei) collapsed, and sex work shifted to points further from the city center, including Guanlan in Bao'an District, Nantou in Nanshan District near the Shenzhen Bay checkpoint, and Shawan in Longgang District. Additionally, the structure of the industry began to change. There was an exodus of FSWs, especially of street-based women who provided services for migrant workers and low-income men from Hong Kong, due to the threat of the crackdowns. Many of these women moved to other cities in the Pearl River Delta or to other inland cities. However, Shenzhen also witnessed an increase in the number of FSWs working in upscale entertainment venues since 2006. Another crackdown occurred in 2011 in advance of Universiade, an international sporting event for college athletes.

The red light district of Shazui in particular exemplified these changes in a comparison of its past and present landscape. Prior to 2006, every street was lined with sex workers who could be observed publicly soliciting clients. Outside of the sex industry, the neighborhood only consisted of businesses supporting the sex industry, such as restaurants catering to sex workers and beauty salons. Today, it is rare to catch sight of the figure of the provocatively dressed sex worker, and the solicitation of clients is not performed in public anymore. Businesses unrelated to sex work, such as clothing stores and Internet cafés, have become established in this area.

Official Interventions and Absences

Public health agencies in Shenzhen often viewed migrants as vectors of disease, and interventions targeting FSWs have focused on preventing the transmission of HIV and other STIs.[27] In 1988, Shenzhen established sentinel surveillance for HIV among key affected populations, including FSWs, and the first cases of HIV among this population detected by this system were diagnosed in 1998.[28] After 2000, the terms "female sex worker" (*nüxing xinggongzuozhe*) and "commercial sex worker" (*shangyexing xing gongzuozhe*) emerged in the literature. In the public health literature with regards to Shenzhen, the term *nüxing xinggongzuozhe* was first used in a journal article in 2004 by Dr. Yang Fan.[29] This shift corresponded to governmental action prioritizing disease control work among FSW groups. In a behavioral intervention program among high-risk populations in Shenzhen in 2005, FSW groups were included in HIV prevention. This program established a supervision team to implement behavioral interventions at the city, district, and neighborhood levels in Shenzhen. During this period, several international programs aimed at FSW groups conducted in Shenzhen achieved satisfactory results, including collaborations with the European Union and with the Global Fund. After 2006, much of the FSW population dispersed to other cities and commercial sex activities became more covert; consequently, health interventions and disease surveillance for FSWs became difficult due to the hidden nature of the population. One example of a program after these crackdowns was implemented by the Shenzhen Family Planning Research Institute and AIDS Care (a Hong Kong-based NGO) in Xiasha, which featured a developed sex industry. Here, a health station targeting sex workers was created to provide health education from 2008 to 2011 in addition to activities such as condom distribution and safer sex workshops.[30]

Sex workers face many health issues arising from their unique social and occupational niche, including repetitive stress injuries, violence, alcohol and other drug use, stress, and infectious diseases.[31] A series of studies in Guangxi found that 14 percent of FSWs report suicide ideation, 8 percent report suicide attempts, and 30 percent have elevated depressive symptoms.[32] When compared to female migrants who did not sell sex, those who did had higher rates of depression.[33] The high burden of negative mental health outcomes among FSWs indicates the need for the inclusion of mental health in public health efforts serving this group.

The specific circumstances of sex workers' work, life, and social context are linked to mental health outcomes. Several factors related to life and work conditions are significantly associated with suicidal thoughts, including being deceived or forced into sex work, alcohol intoxication, dissatisfaction with life conditions, and having one or multiple stable partners.[34] Stigma against sex work is also linked to mental health. Sex workers who reported high levels of stigma were significantly more likely to have elevated depression, suicidal thoughts, and suicide attempts compared to those who reported perceiving low levels of stigma.[35]

Rather than arising from the inherent nature of sex work, these findings suggest that sex workers' mental health status is influenced by the conditions of their life and work.[36] Thus the World Health Organization's conceptualization of mental health beyond solely the absence of disease in an individual is utilized here: "A state of well-being in which every individual realizes his or her own potential, can cope with the normal stresses of life, can work productively and fruitfully, and is able to make a contribution to her or his community."[37]

Migration and Place

The following case studies provide concrete examples of how everyday stressors and anxieties are directly linked to migration for FSWs in Shenzhen. Here, the concept of "place" is resonant for exploring these connections. "Place" is defined as "the sum of resources and human relationships in a given location."[38] Place therefore refers to not only the notion of a geographic location at the center of one's daily life but also the social milieu of that location and "represent[ing] the nodes of the life biography" through one's interaction with a place.[39]

Accordingly, these women's perceptions of the place of Shenzhen were prominent features in setting expectations for migration and in navigating the stressors of urban life, specifically as a migrant woman selling sex. Migrating was a tactic for addressing the problems originating from their hometowns; the time in Shenzhen was an opportunity for economic and social advancement. However, these cases also illustrated the difficulties these women faced in this particular place due to their lack of an urban *hukou*, the illegal and highly stigmatized sex trade, the high cost of urban living, and the difficulty of maintaining relationships with their families across long distances.

In Fullilove's framework, displacement interrupts the attachment of an individual in his or her original place (i.e., hometown) and causes disorientation, homesickness, and alienation.[40] Similarly, the framework proposed by Li and his colleagues attributes poor mental health experienced by rural migrants to the consequences of stigmatization mediated by the psychological processes of social isolation, alienation, reduced self-efficacy, and lowered self-esteem.[41] As Florence shows in this volume, the feeling of "belonging" in the city was mobilized as a key figure of migrant workers' acceptance of their precarious status by official and public discourse in the 1990s.

Venues and Workers

The venues identified as sites of sex work in this study included three hair salons, two massage venues, and one small karaoke bar, which were located within the SEZ, in Luohu, Futian, and Nanshan Districts. All three salons were located on side streets alongside other small businesses and varied to the extent that they maintained an appearance as a hair salon. For example, one venue, described in the vignettes, had peeling lettering that advertised the prices for haircutting, styling, and washing services that were not provided. Inside, mirrors atop counters crammed with beauty products and styling chairs ran alongside one wall, parallel to a couch and coffee table set. Next to the mirrors were dorm rooms for the sex workers, and in the back were rooms where sexual services were provided. In contrast, the glass pane front of one of the three hair salons had neat lettering in a bold black font that described various beauty services that were also not actually provided, a sparse anteroom, and dorm rooms and rooms for sexual services behind a partition.

The small karaoke bar in this study differed from other karaoke venues, where guests rent out private rooms for their group. Instead, it was a small bar where each party gathered at high leather booths surrounding low tables and took turns singing songs that were played in the main bar area. Additionally, there was a wet bar in the back as well as a private room for larger parties. The non-sex-worker bar staff, who were women in their thirties, would accompany the guests in drinking, singing, and playing dice.

One of the two massage parlors in this study was located in the same area as the hair salons. A large sign made up of strips of rainbow-colored lights decorated the entrance and the interior was covered with a worn maroon carpet. The dorm rooms for the masseuses fit three metal bunk beds covered with thin mattresses and a small toilet. In contrast, the other massage parlor, which charged double the rates, featured hotel-like private rooms, marble floors, and expensive light fixtures.

The sex workers interviewed in this study all worked in these venues and varied in their income and status.[42] Most of the informants had been at the venue for less than a year, but the length of time in Shenzhen as well as job history and reasons for migrating varied. While some had worked previously in factories or in the service industry, others had migrated to Shenzhen expressly to work in sex work after learning about it from hometown friends. Additionally, economic needs were cited as a reason for migrating, including repaying debts, financing a family member's health care expenses, or supporting a child or themselves after a divorce. A total of sixteen of the twenty-seven informants had left children behind in their hometowns. Most did not plan on remaining a sex worker after these obligations were fulfilled.

All informants stated that their involvement in sex work was a choice, albeit one that was sometimes made under difficult circumstances. Only one informant described controlling practices, where her passport was withheld while she was working in Indonesia. Their entry into sex work was often facilitated by a friend or someone from their hometown who had been involved in sex work. The views of the informants on their work varied widely. Some viewed sex work as *bu guangcai* (not honorable) and felt shame and inferiority with regards to their involvement. Others felt that sex work was a job just like any other and that they were making a contribution to society. Most felt that the only benefit of this job was the ability to earn money quickly, while other informants cited the flexibility and independence afforded by sex work, as well as the relative ease compared to other low-wage work as positive aspects. Drunk and disrespectful clients were brought up as a significant source of workplace-related stress.

Several women discussed the stigma they experienced from the surrounding community and society at large. These expressions of stigma included being called names by neighborhood children and having strangers give them disgusted looks, particularly when they were in the company of an older man. Many women also actively hid their work from family members and *laoxiang* through tactics such as avoiding socializing or not giving them their Shenzhen mobile phone number. A few participants, especially in the higher-end venues, stated that sex work was not seen as negatively in the cities as it was in the conservative countryside.

While most of the women described their mental health as *zhengchang* (normal), they listed several sources of stress, including their physical health, the high cost of housing, maintaining relationships with their husbands and children, hiding their work from friends and family, earning enough money to send home, dealing with societal stigma and feelings of inferiority, and their children's welfare. Many women experienced symptoms such as sleeplessness, sadness, and feelings of immense pressure. The women in turn would cope with these symptoms and stressors in varying ways. Some would discuss these problems with friends or a family member while others would hold these feelings inside and cry or scream. However, none of the informants viewed seeking mental health care services as a viable option, citing the high cost and the perception that these services were reserved for those with severe mental illnesses.

Because many of these women viewed sex work as temporary, similar to the short-term migration undertaken by many internal migrants, they described a variety of plans for the future. Some women planned on going back to their hometowns and starting a small business with their savings, while others sought out opportunities in Shenzhen, including one woman who wanted to start selling Amway or Mary Kay products. Finally, besides the economic opportunities, a few of the informants also discussed their hopes of eventually getting married, while two informants already were engaged and in the process of exiting sex work.[43]

Ah-Chen[44]

Ah-Chen is a thirty-four-year-old woman who is from a rural village in Shandong. At the time of the interview, she worked as a masseuse providing sexual services in a few large massage venues. Prior to migrating, she worked as a secretary in her hometown, but her wages were not enough to support her father's medical expenses. For this reason, in 2001, she left home for the first time to come to Shenzhen, where she sent home her earnings from sex work as a hostess in a nightclub. Around 2003, after her father's death from health complications, she returned home to care for her mother. During this time, she began a romantic relationship with a man from her hometown. Over the course of this two-year relationship, they attempted to start a hair salon (that did not provide sexual services) in their hometown, pouring RMB 150,000 of her earnings into the venture while her boyfriend contributed little. However, since most of her savings went into the business, she was left with little when the hair salon went bankrupt. Additionally, Ah-Chen's fiancé possessed a violent temper and frequently beat her. As a result, she left the relationship. Because of the losses incurred from her hair salon business, she migrated again to Shenzhen to work. Consequently, Ah-Chen described herself as preoccupied with money to the point of obsession.

Ah-Chen also attributed much of her preoccupation with money to her experience in securing funds for her father's health care expenses. After her father was diagnosed with uremia following renal failure, he required RMB 4,000 a month to pay for dialysis. Prior to her migration, Ah-Chen's family asked to borrow RMB 1,000 from each of her father's four siblings, but they were refused because of the siblings' view that Ah-Chen and her sister had few economic prospects due to their gender and would not be able to repay the loan.

During her first stint in Shenzhen, Ah-Chen worked as a masseuse and did not provide sexual services, but switched to selling sex in nightclubs and hair salons when she heard that she could earn money more quickly; eventually, she sent money home in monthly installments of RMB 10,000. Early into her first stint in sex work, Ah-Chen described the nightly need to secure housing as a major stressor. She initially lived day-to-day off her earnings, where not being able to earn money for the day meant that she could not rent a room for the night. As a result, she described waking up every day preoccupied with the need to make money.

At the time of the interview, Ah-Chen's worries had shifted to her future. Along with getting married to her new fiancé, she was also searching for an apartment amid high housing prices. Ah-Chen also planned on buying the apartment in her name as an important form of insurance in light of recent changes to China's Marriage Law. With the changes to this law, after a divorce, marital property is transferred to the person in whose name the apartment was registered and who took out the mortgage. For her, Ah-Chen stated that financial independence as a woman—and for women in general—is a necessity given her experiences with dealing with her father's health care expenses. Additionally, she saw owning a home as a responsibility to the future child she planned to have sometime after she got married. Ah-Chen wanted to ensure that this child would have a better future. Highlighting the inequities in education and life opportunities between rural and urban areas, she stated, "Can I change my identity as a rural person? This is my destiny, which is the reason I came out. I want to change my destiny as well as my child's because I do not want my child to suffer in life as I did."

Sister Zhang

Sister Zhang is a thirty-six-year-old woman from a rural village in Hunan. At the time of the interview, she worked full time at a "hair salon." She left Hunan in December 2010 around the time of her divorce, which left her alone to support her daughter. As Sister Zhang knew wages in coastal areas were higher than those in Hunan, she migrated to Shenzhen and began working as a server in a restaurant for the first six months, earning a little less than RMB 2,000 per month (she would only be able to earn a few hundred RMB a month in inland areas). However, this income was not enough to support herself as well her child's expenses. Her child lived in the home of a caretaker in her hometown, as Sister Zhang no longer owned a house there. Additionally, her parents both died and her sisters had their own families and could not support her child. Because of her financial burden, she began working as a sex worker in a hair salon. She also supplemented her income by working as a nanny. She planned on staying in Shenzhen for at most another year in order to pay for her daughter's educational, living, and care expenses and then on returning home to start a business.

Like some of the other informants, Sister Zhang had a negative view of sex work that touched on the tradeoff between easing a financial burden and confronting societal stigma. Besides earning more money than in her restaurant job, she did not feel that there were any positive aspects of this job. Similarly, she believed that other sex workers were here primarily for a clear financial purpose—either supporting a child or lessening their family's economic burden—and that she would not be involved in sex work if she were from a well-off family. She also felt significant pressure to hide this work from her family and friends she made while working at the restaurant because they would discriminate against her. Additionally, like a few of the other informants, she felt significant shame and *zibeiguan* (inferiority) because of her involvement in sex work and stigmatization by society. However, she stated that she had "no other options except sex work" because of her need to support her child.

Sister Zhang immediately rated her mental health as poor because of the inferiority she felt about herself, the stress of the stigma she experienced from her community and society, the pressure of hiding her work from her friends and family, and the general pressure she felt in her everyday life. Additionally, she described experiencing symptoms such as lack of appetite, frequent worrying, and insomnia that she attributed to the pressure of needing to earn enough money. She stated that "it [was] impossible to work in sex work and not feel pressure," as she did not feel that sex work was honorable work.

Besides her finances, Sister Zhang also frequently worried about how her daughter's welfare, specifically regarding her academic performance and health. However, she was unable to travel back to see her daughter because of the distance and the fact that she no longer had a place to live in her hometown.

Ah-Lin

Ah-Lin is a thirty-year-old woman from a rural part of Chongqing. She worked part time at the same hair salon as Sister Zhang. In 2007, she migrated to work in Shenzhen outside of the SEZ, including stints in electronics, toy and clothing factories, as well as working in restaurants. In addition to the low wages, her paychecks were sometimes delayed, as in the case of the clothing factory where she had to wait three months before getting paid. Additionally, the work was not steady, as some of the factories went bankrupt. Later that year, she was introduced to a man from her hometown who she eventually married. Soon after, they migrated to Xinjiang to find work, but could only find work in construction. However, she could not handle the physical demands of construction work. Eventually she got pregnant and gave birth to their daughter. After her daughter's birth, she and her husband would frequently argue, so they divorced. Her daughter was ultimately placed into the custody of her ex-husband due to poor representation by her attorney.

When she returned to Shenzhen, she first started working as a housekeeper and in other low-paying jobs. One day, when she was washing blankets at another "hair salon," a client noticed her and felt attracted to her. From then on, she began working at that venue whenever she needed extra money. At the time of the interview, she rotated between different low-wage jobs and sex work at various hair salons.

Like some of the other informants, her stressors included securing housing and worrying about her child's future. She did not have a regular residence, as she found renting to be expensive. She also declined to live with her parents or her brother and his wife in order to better conceal her work. She sometimes stayed with a man that she was somewhat involved with at his apartment. In addition to sending money to her parents, who had also migrated to Shenzhen to work, she also sent money to her ex-husband in her hometown for her daughter to enroll in extracurricular English classes. As she no longer had custody, it was even more difficult for her than the other informants to remain in touch with her child. She worried that her child was not treated well by her ex-husband and his wife, but was unable to spend much time with her child during her visits home due to the custody arrangement.

Overall, Ah-Lin was positive about sex work. While she described stressful aspects of the job, such as dealing with intoxicated clients and the frustration of not earning money regularly, she viewed this job as allowing her to be independent and well-rested. This aspect of sex work was directly contrasted with her experience of factory work.

Unlike the other informants, Ah-Lin was uncertain about how long she planned on staying in sex work and lived on a day-to-day basis financially, with large fluctuations in her monthly income. She described herself as not having an objective. In the meantime, she had been studying to be a bookkeeper and nearly passed the certification exam a year ago.

Lan Hua

After arriving in Shenzhen from Hubei Province in 2001, Lan Hua, who is forty-seven years old and holds an urban *hukou*, began sex work in 2012 to repay a gambling debt. Prior to working at a high-end massage venue near the Nantou checkpoint in Nanshan District, she ran an Internet café. Unlike many of the other informants, Lan Hua came from a family who was relatively well-off, and before arriving in Shenzhen, she ran a small business in their hometown after being laid off from a government work unit. In 2008, she began playing the stock market and gambling and eventually lost more than RMB 1 million and had to sell off her stake in the Internet café. Currently she shares an apartment with her adult son, who she supports financially.

Overall, Lan Hua sees her work for both herself and other sex workers as temporary, stating that no one would do this work if they had other options, and that she herself had never done this work in her twenties. While she does not see this work as prestigious and does not particularly enjoy it, she has grown to accept her work and attributes her ability to cope with the negative aspects of sex work to a strength that comes from undergoing many different life experiences. Though she sees sex work as *mai kuli* (selling hard labor), she also simultaneously sees herself as being skilled, as she does not believe that clients would pay for nothing.

Migrant Sex Workers

For rural-to-urban migrants, navigating the place of Shenzhen was often defined by the endeavor to shape the environment into what Fullilove conceives of as a "good enough" place, with consequent impacts on their mental health.[45] The experiences of these informants problematizes the "ubiquitous narratives of success that form the foundational myth of [Shenzhen]" through highlighting the tensions of urbanization, migrant sex work, and health.[46] Many of the informants belonged to a specific profile—rural, older, divorced, working in a lower-middle-tier venue, and supporting at least one family member in their hometown. Though efforts were made to diversify the sample of participants through recruitment in various districts and venues, the post-2006 environment for sex work meant that very few women worked on the street. As a result, participants were recruited through the networks of community gatekeepers who were either based at a venue (such as the mommy and the manager) or performed outreach at specific venues (such as the physicians and the health educators). Women working in venues such as hair salons and massage parlors are often older and married and need to support their families back in their hometowns. For this reason, they are less willing to risk working on the street where they are more vulnerable to crackdowns and where the income is lower. As a result, the following discussion may not be applicable to other groups of sex workers in China.

On the surface, the migration undertaken by all the informants in pursuit of economic opportunities aligns with the image of the city. Shenzhen is imagined as a place of boundless opportunity fueled by the entrepreneurship of male newcomers.[47] In particular, since the mid-1990s, those migrants whose labor has been transmuted into the tall buildings that fill Shenzhen are celebrated by the mainstream press.[48] Rural women are systematically excluded from the entrepreneurial classes that may lead to a Shenzhen *hukou*.[49] Similarly, the stressors detailed by the informants are connected to their lack of this *hukou*. For example, the stress of maintaining relationships with their children back home originated in large part from this policy, which denies migrant children access to free public education and other services in urban areas, causing many children of poorer migrants to remain in their hometowns.[50] However, the contributions of migrant women who are sex workers are excluded from this narrative despite their physical and emotional labor that are essential to the business practices that not only take place in these buildings but finance the construction of these buildings in the first place.

In addition, migrant sex workers experience additional, specific forms of exclusion. Despite the integration of sex work into city life, many of the informants experienced varying levels of stigma from their communities.[51] Many informants attributed their migration and entry into sex work to a *laoxiang* (a social connection from same hometown), suggesting that the availability of sex work in Shenzhen is known in hometown networks as well.[52] However, sex work in China exists against the backdrop of societal stigma that persists to varying degrees both in the city and among migrants' hometown networks.[53] For example, Sister Zhang described hiding her sex worker status from the friends she made in her previous job in a restaurant in Shenzhen. Other informants hid their jobs from their *laoxiang*, which Ah-Lin cited as one reason for not living with family in Shenzhen despite her lack of a stable residence. As a result, stigma against sex work may decrease these women's access to the resources of their non-sex-worker hometown social networks.[54] Additionally, sex work is criminalized in China and sex workers are subject to arrest, imprisonment, public humiliation, and physical violence from various law enforcement agencies, while their clients are rarely punished.[55] Stigma against sex workers remains a significant occupational hazard.[56]

Sex workers' needs in Shenzhen cannot be addressed without examining the context of migrant women's work outside of the sex industry. Many of these jobs require long hours, tiring work, provide low wages, and offer little flexibility in scheduling. For many sex workers, these jobs do not allow them to sufficiently provide for themselves or their families. Some informants equated not being able to remit money with effectively consigning a family member in need of health care to an early death. The experiences of these informants—who were often older, divorced, and had low levels of education—illuminate the inadequacies of work available in the formal economy that would offer resilience rather than vulnerability to sudden changes. It is also necessary to look to the conditions of rural poverty in China. These cases have highlighted the uneven economic opportunities for migrant women not only in the cities but also in their rural hometowns. For many, their involvement in sex work stemmed from the need to pay for medical expenses that were not covered by China's weak rural health insurance. It is telling that many of the informants had stated that if their economic conditions at home were better, they would not be working as sex workers.

Beyond mainland China, the story of sex work in Shenzhen must also be examined in the context of globalization. First, the sex industry in Shenzhen that developed originally served men from Hong Kong crossing into the mainland. Additionally, the developed sex industry made Shenzhen an attractive location for businessmen and international sex tourists.[57] Consequently, Shenzhen is a particularly important example for other metropolises in China as the first city to adopt market-based economic reforms that are being implemented throughout the country. For example, after the Shenzhen crackdowns in 2006, many sex workers moved to Dongguan, which also experienced a similar pattern of crackdowns in 2010 after a period of tolerance of sex work.[58] In February 2014, another large-scale crackdown of sex work was carried out in Dongguan and a map[59] tracking the real-time movement of migrants during Spring Festival reflected an immediate exodus from the region.[60] As patterns of migration shift, other cities may attract sex workers and undergo similar dynamics in the intertwining of sex work and local economic development. The growth of both the economy and the sex industry are then often followed by crackdowns motivated by political reasons or to further urban renewal.[61] However, as the history of sex work in Shenzhen and recent events in Dongguan demonstrate, sex workers then move elsewhere; sex work is not simply eradicated. While the borders delineating the SEZ have produced the "exceptional" space of Shenzhen,[62] the continued exclusion from a legal status for migrants, longstanding societal stigma against sex workers, and crackdowns on sex work over the past decade demonstrate that Shenzhen is ultimately not an exception to the rest of China in these regards. Though excluded from the dominant narratives of migrant labor, sex work remains "the unexceptional, totally ordinary narrative of oppressed workers under neoliberal capitalism."[63] While sex workers' experiences of migrating to and living in Shenzhen exemplify many of the possibilities for creating a better life for themselves and their families, the alienation, stress, and other mental health concerns arising from these processes make evident the high human cost of this model of economic development and urbanization.

Notes

1. A "mommy" refers to a woman who acts as an intermediary between the client and sex worker in China. She receives a portion of the sex worker's fee for introducing clients to sex workers. In this case, the mommy owned the karaoke bar, which did not employ sex workers but allowed them to operate there. For more information on relationships between mommies and sex workers, see Huso Yi, Tiantian Zheng, Yanhai Wan, and J. E. Mantell, "Occupational Safety and HIV Risk among Female Sex Workers in China: A Mixed-Methods Analysis of Sex-Work Harms and Mommies," *Global Public Health: An International Journal for Research, Policy and Practice* 7, no. 8 (2012): 840–55, http://www.tandfonline.com/doi/abs/10.1080/17441692.2012.662991.

2. Based on fieldwork conducted by the first author between March 2012 and September 2012.

3. X. Ren, "Prostitution and Economic Modernization in China," *Violence against Women* 5, no. 12 (December 1999): 1411–36, doi:10.1177/10778019922183453, http://vaw.sagepub.com/cgi/doi/10.1177/10778019922183453; V. E. Gil, M. S. Wang, and A. F. Anderson, "Prostitutes, Prostitution and STD/HIV Transmission in Mainland China," *Social Science and Medicine* 42, no. 1 (1996): 141–52, http://www.sciencedirect.com/science/article/pii/027795369500064X.

4. Estimates of sex worker numbers vary. In 2010, Huang Yingying, a scholar on HIV and sex work in China, quoted numbers ranging from three to four million and up to ten million in a *China Daily* opinion piece. See Yingying Huang and Wenyan Liu, "Debate: Prostitution," *China Daily*, May 31, 2010, http://www.chinadaily.com.cn/opinion/2010-05/31/content_9909738.htm. For a further discussion of these figures, see the literature reviews by Yan Hong and Xiaoming Li, "Behavioral Studies of Female Sex Workers in China: A Literature Review and Recommendation for Future Research," *AIDS and Behavior* 12, no. 4 (July 2008): 623–36, doi:10.1007/s10461-007-9287-7; and Catherine Pirkle, Riswana Soundardjee, and Artuso Stella, "Female Sex Workers in China: Vectors of Disease?" *Sexually Transmitted Diseases* 34, no. 9 (September 2007): 695–703, doi:10.1097/01.olq.0000260989.70866.94, http://www.ncbi.nlm.nih.gov/pubmed/17507838; UNAIDS, "HIV/AIDS: China's Titanic Peril," 2001, http://www.hivpolicy.org/Library/HPP000056.pdf.

5. Hong and Li, "Behavioral Studies of Female Sex Workers in China."

6. Jonathan Bach, "Shenzhen: City of Suspended Possibility," *International Journal of Urban and Regional Research* 35, no. 2: 414–20, doi:10.1111/j.1468-2427.2010.01031.x.

7. J. T. F. Lau and J. Thomas, "HIV Risk Behaviours of Hong Kong Male Residents Travelling to Mainland China: A Potential Bridge Population for HIV Infection," *AIDS Care* 13, no. 1 (2001): 37–41.

8. See also Mason, chapter 10 in this volume.

9. These interviews were supplemented by fieldwork, including neighborhood strolls and participant observation at a karaoke bar, which began after an introduction to a mommy who owned the venue. The interview guide was designed based on a literature review, and with the input of public health researchers. Informants were conveniently sampled and recruited through the assistance of this mommy, physicians, venue managers, and health educators who worked closely with sex workers.

10. The interviews were audio recorded and were conducted in Mandarin by a native Mandarin Chinese speaker. Informants were compensated with a phone card worth RMB 100 for their time. This study was approved by the Institutional Review Board of the Department of Anthropology at Sun Yat-sen University.

11. C. Henriot, "'La Fermeture': The Abolition of Prostitution in Shanghai, 1949-58," *China Quarterly* 142 (1995): 467-86, http://journals.cambridge.org/abstract_S0305741000035013.

12. Suiming Pan, "Transformations in the Primary Life Cycle: The Origins and Nature of China's Sexual Revolution," in *Sex and Sexuality in China*, ed. Elaine Jeffreys (New York: Routledge, 2006), 21-42, http://www.sex-study.org/news.php?isweb=2&sort=49&id=864&classid=.

13. Emily Honig, "Socialist Sex: The Cultural Revolution Revisited," *Modern China* 29, no. 2 (April 2003): 143-75, doi:10.1177/0097700402250735, http://mcx.sagepub.com/cgi/doi/10.1177/0097700402250735.

14. M. Liu and James O. Finckenauer, "The Resurgence of Prostitution in China: Explanations and Implications," *Journal of Contemporary Criminal Justice* 26, no. 1 (January 2010): 89-102, doi:10.1177/1043986209350172, http://ccj.sagepub.com/cgi/doi/10.1177/1043986209350172; Ren, "Prostitution and Economic Modernization"; Gil, Wang, and Anderson, "Prostitutes, Prostitution and STD/HIV."

15. Pirkle, Soundardjee, and Stella, "Female Sex Workers in China."

16. Yu Cheng, unpublished fieldwork, 2006.

17. Yu Cheng, unpublished fieldwork, 2006.

18. J. D. Tucker, G. E. Henderson, and T. F. Wang, "Surplus Men, Sex Work, and the Spread of HIV in China," *AIDS* 19, no. 6 (March 2005): 539-47, http://www.ncbi.nlm.nih.gov/pubmed/15802971.

19. Xiushi Yang, "Temporary Migration and HIV Risk Behaviors in China," *Environment and Planning A* 38, no. 8 (2006): 1527-43, doi:10.1068/a3814, http://www.envplan.com/abstract.cgi?id=a3814.

20. Christopher Smith, "Social Geography of Sexually Transmitted Diseases in China: Exploring the Role of Migration and Urbanisation," *Asia Pacific Viewpoint* 46, no. 1 (2010): 65-80.

21. Elanah Uretsky, "'Mobile Men with Money': The Socio-Cultural and Politico-Economic Context of 'High-Risk' Behaviour among Wealthy Businessmen and Government Officials in Urban China," *Culture, Health and Sexuality* 10, no. 8 (November 2008): 801-14, doi:10.1080/13691050802380966, http://www.ncbi.nlm.nih.gov/pubmed/18975228.

22. Uretsky, "'Mobile Men with Money'"; Tiantian Zheng, "Cool Masculinity: Male Clients' Sex Consumption and Business Alliance in Urban China's Sex Industry," *Journal of Contemporary China* 15, no. 46 (February 2006): 161-82, doi:10.1080/10670560500331815, http://www.tandfonline.com/doi/abs/10.1080/10670560500331815.

23. Alvin Y. So, "Cross-Border Families in Hong Kong: The Role of Social Class and Politics," *Critical Asian Studies* 35, no. 4 (December 2003): 515-34, doi:10.1080/1467271032000147014, http://www.tandfonline.com/doi/abs/10.1080/1467271032000147014; Hsiu-Hua Shen, "The Purchase of Transnational Intimacy: Women's Bodies, Transnational Masculine Privileges in Chinese Economic Zones," *Asian Studies Review* 32, no. 1 (March 2008): 57-75, doi:10.1080/10357820701870759, http://www.tandfonline.com/doi/abs/10.1080/10357820701870759.

24. Yu Cheng, unpublished fieldwork, 2005, 2006, 2008.

25. Chen et al. 2008.

26. Yu Cheng, Li Han, and Yunshi Huang, "Health Needs and Health Risks among Chinese Female Migrant Workers: A Qualitative Assessment," in *Migrants and Urban Health in China*, ed. Bettina Grasnow and Daming Zhou (Berlin: LIT Verlag Münster, 2010), 40-51.

27. See Mason, chapter 10 in this volume.

28. Tiejian Feng et al., "1988-1998 Nian Shenzhen Shi Aizibing Liuxing Bing Xue Fenxi," *Zhongguo Xingbing Aizibing Fangzhi* 2 (2000): 100-103.

29. Fan Yang, Yongshuang Li, Hua Zhou, and Zhihua Wu, "Yingguang Dingliang PCR Jiance Nvxing Shangye Xing Xing Gongzuo Zhe Xing Chuanbo Ganran," *Zhongguo Mafeng Pifu Bing Zazhi* 20, no. 4 (2004): 334–35.

30. Yu Cheng, unpublished fieldwork, 2008.

31. Priscilla Alexander, "Sex Work and Health," *Journal of the American Medical Women's Association* 53, no. 2 (1998): 77–82; W. C. W. Wong and Y. T. Wun, "The Health of Female Sex Workers in Hong Kong: Do We Care?" *Hong Kong Medical Journal* 9, no. 6 (2003): 471–73.

32. Yan Hong, Xiaoming Li, Xiaoyi Fang, and Ran Zhao, "Depressive Symptoms and Condom Use with Clients among Female Sex Workers in China," *Sexual Health* 4, no. 2 (July 2007): 99–104, doi:10.1016/j.bbi.2008.05.010; Yan Hong, Xiaoming Li, Xiaoyi Fang, and Ran Zhao, "Correlates of Suicidal Ideation and Attempt among Female Sex Workers in China," *Health Care for Women International* 28, no. 5 (2007): 490–505.

33. H. Yang et al., "Workplace and HIV-Related Sexual Behaviours and Perceptions among Female Migrant Workers," *AIDS Care* 17, no. 7 (October 2005): 819–33, doi:10.1080/09540120500099902, http://www.pubmedcentral.nih.gov/articlerender.fcgi?artid=1949042&tool=pmcentrez&rendertype=abstract.

34. Hong et al., "Correlates of Suicidal Ideation and Attempt."

35. Hong et al., "Self-Perceived Stigma, Depressive Symptoms, and Suicidal Behaviors among Female Sex Workers in China," *Journal of Transcultural Nursing: Official Journal of the Transcultural Nursing Society/Transcultural Nursing Society* 21, no. 1 (January 2010): 29–34, doi:10.1177/1043659609349063, http://www.ncbi.nlm.nih.gov/pubmed/19820172.

36. I. Vanwesenbeeck, "Another Decade of Social Scientific Work on Sex Work: A Review of Research 1990–2000," *Annual Review of Sex Research* 12 (January 2001): 242–89, http://www.ncbi.nlm.nih.gov/pubmed/12666742.

37. World Health Organization, "Mental Health Atlas 2005."

38. Mindy Thompson Fullilove, "Psychiatric Implications of Displacement: Contributions from the Psychology of Place," *American Journal of Psychiatry* 153 (December 1996): 1516–23.

39. Anssi Paasi, "Deconstructing Regions: Notes on the Scales of Spatial Life," *Environment and Planning A* 23, no. 2 (1991): 239–56.

40. Fullilove, "Psychiatric Implications of Displacement."

41. Xiaoming Li, Bonita Stanton, Xiaoyi Fang, and Danhua Lin, "Social Stigma and Mental Health among Rural-to-Urban Migrants in China: A Conceptual Framework and Future Research Needs," *World Health and Population* 8, no. 3 (January 2006): 14–31, http://www.pubmedcentral.nih.gov/articlerender.fcgi?artid=2249560&tool=pmcentrez&rendertype=abstract.

42. Pan Suiming created a widely used seven-tier classification scheme for female sex workers in China based on factors including the income, venue, and sociodemographic characteristics of the workers. For more information, see Pan, *Cunzai Yu Huangmiu* (Guangzhou: Qunyan Chubanshe, 1997).

43. See Wong, chapter 9 in this volume, for further discussion of women migrants' desires for family life.

44. All names of informants have been changed.

45. Fullilove, "Psychiatric Implications of Displacement."

46. Bach, "Shenzhen."

47. Bach, "Shenzhen"; Mary Ann O'Donnell, "Path Breaking: Constructing Gendered Nationalism in the Shenzhen Special Economic Zone," *Positions: East Asia Cultures Critique* 7, no. 2 (1999): 343–75, http://muse.jhu.edu/journals/positions/v007/7.2odonnell.html.

48. See Florence, chapter 4 in this volume.
49. O'Donnell, "Path Breaking."
50. Zai Liang, Lin Guo, and C. Duan, "Migration and the Well-Being of Children in China," *Yale-China Health Journal* (April 2008), http://csda.albany.edu/imc/migration_and_the_well-being_of_children_in_china.pdf.
51. Cheng, Han, and Huang, "Health Needs and Health Risks."
52. Kongshao Zhuang and Shiling McQuaide, "AIDS/Other STIs Prevention in China: The Effect of Sex Worker Migration and the Organization of the Sex Industry," *Anthropology and Medicine* 20, no. 1 (April 2013): 36–47, doi:10.1080/13648470.2013.774935, http://dx.doi.org/10.1080/13648470.2013.774935.
53. Hong et al., "Self-Perceived Stigma, Depressive Symptoms, and Suicidal Behaviors."
54. See Tucker, Henderson, and Wang, "Surplus Men." The social networks of female sex workers are complex and heavily based on these hometown connections. This study provides an in-depth discussion of the characteristics and functions of female sex workers' social networks in the Pearl River Delta.
55. Gil, Wang, and Anderson, "Prostitutes, Prostitution and STD/HIV"; J. Tucker, X. Ren, and F. Sapio, "Incarcerated Sex Workers and HIV Prevention in China: Social Suffering and Social Justice Countermeasures," *Social Science and Medicine* 70, no. 1 (2010): 121–29.
56. Hong et al., "Self-Perceived Stigma, Depressive Symptoms, and Suicidal Behaviors."
57. Yu Cheng, unpublished fieldwork, 2005.
58. UNAIDS, "Protecting Sex Workers Key to Preventing Spread of HIV," September 9, 2011, http://www.unaids.org.cn/en/index/topic.asp?id=783&classname=PhotoStories&class=2.
59. http://qianxi.baidu.com.
60. "Hong Kongers Flee the Heat from a China Sex Raid," *Asia Sentinel*, February 11, 2014, http://www.asiasentinel.com/society/hong-kong-flee-from-china-sex-raid.
61. Yu Cheng, unpublished fieldwork, 2006.
62. See Ma and Blackwell, chapter 6 in this volume.
63. Sarah M., "To the Would-Be Sex Work Abolitionist, or, 'Ain't I a Woman'?," *Rabble*, February 3, 2012, http://rabble.ca/news/2012/02/would-be-sex-work-abolitionist-or-aint-i-woman.

PART III

Extensions (2004–Present)

9

Shenzhen's Model Bohemia and the Creative China Dream

WINNIE WONG

True Art and the New Creative Industry

In 1999, culture and propaganda officials in Shenzhen's Buji Township (later redesignated Street Office) learned from a Guangzhou newspaper of the existence of an "artists' village" right under their noses (see figure 19). According to the article, Dafen Village, then a rural village located outside the Special Economic Zone (SEZ) border checkpoint in an otherwise undistinguished and impoverished district (Longgang), was home to a large and thriving community where oil painters were working, living, and selling their artwork to buyers in Europe and America.[1] In the wake of the sudden international success of Beijing's East Village and 798, artists' villages on the outskirts of Beijing where Chinese artists had gained the attention of influential foreign collectors and curators, Shenzhen officials saw in Dafen Village a rare opportunity to put the city on the cultural map.

The painters of Dafen, however, turned out not at all like the Chinese contemporary artists who were setting auction records and exhibiting in art museums all around the world. First, they were mostly rural-born and uneducated migrant workers; most had never finished secondary school and few had had any formal art education at all. Second, it turned out that their paintings were compositions specified by their buyers, and, given their predominantly Western clientele, many of their paintings were imitations of famous masterpieces of European art, like Leonardo's *Mona Lisa* or Vincent van Gogh's *Sunflowers*. Third, since they painted dozens, hundreds, and even thousands of paintings per "order" (*dingdan*) for shockingly low prices (to the Western consumer), Dafen painters were seen by Western journalists as "forgers" or "assembly line copyists," producing "fake art" that was fooling the world's unsophisticated middle-class consumers. Even though imitating classical and canonical works has always been a central part of the Western art tradition, Western professional painters and journalists called on the Chinese government to halt the "copyright violations" of Dafen Village.

From an art historical perspective, and in contrast to these inflammatory portrayals, Dafen's painters' working methods were neither new nor exclusive to China. In fact, they worked much like professional artists in many periods of Western art, and as many professional painters have done throughout Europe, America, and elsewhere—painting according to their clients' (or patrons') demands, in their own homes and studios, while controlling their own materials, methods, and time of work. They painted in standardized sizes suitable to the domestic spaces where their paintings would be hung, and they were paid for each painting when their buyers were satisfied with them. With few exceptions, this is by and large how artists have worked throughout the history of art, and as such, Dafen Village largely did not inspire inflammatory reactions among all practicing artists. Rather, some even embraced it as part of a new "outsourcing to China" trend in the elite world of cutting-edge contemporary art.[2]

What was new in reform-era Shenzhen, though, was the scale of these practices. The size of the orders, the productivity of the painters, and the number of them that had gathered in one area of the city to work was indeed unprecedented. Some painters temporarily worked or lived in their bosses' "workshops" (*huafang*) and "factories" (*huachang*), but they usually came and went at the hours they wanted, painted as they would do in their own homes or studios, and were still paid by the piece for each finished painting. When working in their homes, their painting practices did not change substantially, except that they often appropriated the labor of their wives and girlfriends, or welcomed extended family members, co-villagers, or newcomers as apprentices and assistants, so great was the need for painting labor. When working independently, Dafen painters got their orders from bosses and other painters, and when their orders were too large, they passed them onto their small network of former apprentices, extended family, and accumulated associates. Similarly, if a client ordered a painting they could not paint, they would farm out the work to a friend or an associate, taking a small cut. With new clients, they took a 50 percent deposit on each order and would deliver the finished paintings in roughly two weeks in hopes of receiving the remainder if the paintings were acceptable.

To produce so much so quickly, Dafen painters did indeed work very hard (at least, harder than the popular and romantic image of the modern artist). In Dafen's heyday in the late 1980s and early 1990s, they often painted at astonishing rates of twelve to twenty oil-on-canvas paintings per day. In the words of G. S. Shieh, who studied Taiwan's simple subcontracting production practices (many of which were transferred to Shenzhen through Hong Kong and

Taiwanese clients in trades that include this one), Dafen's piecework painting trade encouraged workers to "exploit themselves" but also offered them social mobility through self-employment and entrepreneurialism.[3] In Dafen, painters who could save up a little capital or land a client or two would become small-time bosses (*getihu* or just *laoban*) and open their own galleries, studios, and workshops, or even ancillary businesses like frame shops, photo studios, art supply stores, art schools, cafes, and so on. The trade thus became characterized by innumerable small-sized firms, which rose and fell with the precarity of their orders. Painters passed in and out of "worker" or "boss" status frequently, constantly self-trained to develop new specialties, and eagerly sought market knowledge about what kind of paintings were in demand or what kind of new commercial opportunities they might pursue. It is said among Dafen painters that when the Buji party secretary visited Dafen's largest "factory" in 1999, he was so surprised by its small scale (probably housing no more than a few dozen painters at the time) that he remarked, "But this is just an underground workshop!" Dafen's "painting factories" were, in other words, not very different from the hundreds of unregulated, simple subcontracting workshops in and around Shenzhen that produced many kinds of consumer goods for export.

For their part, Dafen's foreign buyers also turned out not to be elite art collectors at all, but rather were largely bosses of small export wholesale and retail firms based in Hong Kong, South Korea, Taiwan, and Japan. The Hong Kong bosses dominated at first, and many came to put out orders and receive inventory every two weeks. They in turn sold and reexported the paintings to clients in Europe and the United States, who were based at major distribution points such as Los Angeles, Miami, Munich, Amsterdam, and so on. There the paintings often passed through several more hands before they were ultimately retailed in frame shops, souvenir shops, furniture stores, and sidewalk stalls and finally hung in restaurants, cafes, hotels, and homes. As Dafen's fame spread due to the Chinese government's propaganda promotion of it after 2004, bosses from Russia, France, Italy, Israel, Morocco, and many other global distribution points also made Dafen their main source of "hand-painted art products," and their stock was also reexported to other markets. Throughout its history, this was a trade that was transnational, but its firms were always small in scale. They operated through loose and informal networks and not as massive corporations or distinguished institutions.

It was estimated by a researcher at the US-China Chamber of Commerce in 2006 that 60 percent of Dafen's paintings ultimately ended up reexported (via Hong Kong, Taiwan, Europe, and elsewhere) to the United States.[4] But the networks of the transnational painting trade are so diffuse—and orders pass through so many hands—that such estimates are difficult to evaluate. Many Dafen bosses were told in the 1980s and 1990s that their largest orders were ultimately destined for big American retailers like Wal-Mart and K-Mart, yet as small subcontractors, they never had the means to verify where their paintings ended up. In the flexible production system that characterized Dafen's trade (but also many other Shenzhen industries of the period), no buyer can know who actually painted their painting, just as no painter can really know for whom his works are intended.

Importantly, unlike products for which the site of production would elevate the sense of authenticity associated with them, in the case of Dafen's oil paintings, the greater the number of middlemen the paintings passed through, the greater the chance that the identities of their painters (and especially their Chineseness) could be forgotten. At the sites of retail, whether on Cat Street in Hong Kong, in flea markets in Long Island City, in souvenir stalls at Mont Blanc in France, on the sidewalk of the Metropolitan Museum of Art in New York, or in the Piazza Navona of Rome, Dafen paintings were sold as the works of "local struggling artists," a "Russian art academy professor," "art students," or a "possibly world-famous Italian artist." The farther the paintings traveled through the extended global supply chain, the more their origins were forgotten, obscured, aggrandized, and mystified. Because this forgetting was not always intentional, we cannot accuse the whole trade of forgery or fraud. However, the more the origins of the paintings were blurred, the higher the price these paintings seemed to fetch, until Dafen painters joked about Chinese officials going all the way to Europe to buy expensive oil paintings only to discover upon returning home that they had bought Dafen products.[5] Like so many urban rumors in Shenzhen, such tales are both apocryphal and possible.

From a global and historical perspective, it is important to note that the basic procedures of the Dafen painting trade would seem to run exactly counter to the romantic and modernist myths of art and the artist's work. The threads of this rhetoric are so well known that they hardly need summarizing—indeed, they are often used by retailers to sell Dafen paintings after they have been taken far away: "Art" comes from gifted and rebellious individuals who struggle and strive alone in their studios without care for their patrons' demands. He—and it is almost always a "he"—rebels against his teachers, who have nothing to teach him. Women, especially his models, fall passionately in love with him, while his paintings of them are expressions of his innermost feelings. He shuns the market and all hints of interest in commercial concern in pursuit of his singular personal vision, which will ultimately bring him worldwide recognition, if only after his death. Even so, his "masterpieces" have an ineffable authenticity to them—they speak of his unique and individual persona and will someday be recognized by the most sophisticated connoisseurs and, eventually, the whole world.

But of course, such a discourse of "true art" is historically constructed as well, and it has been equally criticized as bourgeois, a product of the Cold War, a trope of Hollywood, or a form of what German artists once criticized as "Capitalist Realism." Dafen village, too, presents an uncanny contrast to the rhetoric of "true art." Since its operation in the early 1980s, Dafen Village has operated as a modest community of migrant workers who came to the fringes of a booming city seeking to work as painters and to secure a modest livelihood as independent artists. When painters began entering the trade in Shenzhen in the late 1980s, they were more than satisfied—in many cases, quite overjoyed—to have found a way to make a living as independent painters. Many were very proud that their paintings were appreciated by foreign buyers all around the world. It was only much later that they learned that the hierarchies of the world market excluded them, erased their names, and considered them factory painters, mechanical copyists, or worse—forgers. When Dafen became famous enough that local officials started mentioning that it was perhaps destiny that the name "Dafen Village" sounded just like "Da Vinci," some Beijing artists in the "true" artists' village in Song Zhuang wrote online that Dafen sounded just like "full of shit."[6] Thus, although Dafen's painters began with high aspirations that they would eventually be recognized as artists, they later came to realize that they were definitively excluded from the art world's institutions and markets by gatekeepers for whom Dafen painters were not true "artists." After all, Dafen painters did not claim to paint from their "hearts" or from their "minds"—they did not claim to be new or avant-garde, they did not claim to invent their compositions, and they had no dealers to "sell them." That is, these skilled migrant workers could not mobilize the discourse of modernity to lay authorial claim to their labor, even if their works might give every appearance of being "art."

To their supporters, Dafen spoke to the grit and optimism of China's migrant workers, given the insurmountable obstacles they faced in being recognized as artists. In sociological terms, Dafen could be regarded as not so different from the historical bohemias of Montmartre, Soho, Greenwich Village, East Village, or its contemporaries Beijing's East Village and Song Zhuang, where artists who were first considered marginal found themselves turned into the art business of the day. Like these now famous urban neighborhoods, Dafen Village was also as an arts district located on the periphery of a metropolitan city, emerging at a time of middle class prosperity and public and private interest in real estate development. Migrants young and old came to Dafen to reinvent themselves as artists. New arrivals saw the possibility of realizing their dreams to become artists, while jaded old-timers regularly complained that the place had already become too commercialized, too conservative, and too calcified. In the classic definition of bohemia so aptly compiled by César and Marigay Graña, Dafen may just as well be described as "a place of youth and disenchantment," a "tavern by the wayside on the road of life," and a "social mechanism for absorbing excess population until adequate status opportunities become available."[7] When such social patterns are acknowledged, Dafen is a manifestation of the ideal of art as a universally accessible endeavor—made possible by the faith that "genius" and "talent" would be recognized anywhere it was worked at.

Is making art a social privilege or a universal right? The socialist ideal to democratize art spurred the Chinese government, at various levels of administration, to take up its own representation of Dafen Village in and around 2004. In the contemporary Chinese cultural policy context, Dafen's painters embodied the two most heroized class types of socialist culture: the artist and the worker. As Mary Ann O'Donnell and Eric Florence discuss in this volume, the Chinese narrative for the development of Shenzhen puts into action an explicit rural peasant critique of elite urban culture. In this spirit, some Dafen painters banded together to call themselves "art workers" in order to emphasize inclusiveness and their service to society. This term expresses the egalitarian ideals shared by both Chinese and Western post-1960s anticapitalistic artistic movements, and it echoed quite perfectly the celebratory Shenzhen story of post-Mao transformation, the utopian ideals of the Reform and Opening policy, and the transnational leftist defense of migrant workers' rights and social justice. Dafen painters were regarded as migrant workers who had learned to paint the world's masterpieces (at least, well enough to sell to consumers in the world market), becoming an attractive symbol of social justice to both China and the West. After all, they were disadvantaged, rural-born individuals who—through self-perseverance, hard work, and self-skilling—had attained an independent, productive, and creative life in a global city. In Shenzhen and the Pearl River Delta, where so much attention had been focused on the plight of the exploited female migrant worker (*dagongmei*), here was a group of (mostly male) migrant workers whose lives echoed some of the most attractive success stories of world culture. Indeed, this positive rhetoric was aptly executed by the chief photojournalist of the *Shenzhen Economic Daily*, Yu Haibo, whose photographic series of Dafen Village won him a prize in the World Press Photo Awards from the organization in Amsterdam in 2006. The subsequent and widespread interest in Dafen in China and around the world similarly echoed this leftist and populist ideal.

Shenzhen officials, especially those of the Buji Street Office and Longgang District Propaganda and Culture Departments, thus vigorously took up the promotion of Dafen as a new model for creative industries, the global urban policy trend of the 2000s that, through Dafen, they would definitively embrace as both new and a continuation of socialist ideals. The speed and breadth of the adoption of a creative industries policy can be readily seen in the explosive administrative attention Chinese officials paid to Dafen. In 2004, the Chinese Ministry of Culture named Dafen Village a national model cultural industry, and Dafen was named the only "satellite site" of the China International Cultural Industry Fair, held annually in Shenzhen (see figure 20). (By 2010, there were thirty-six satellite sites in Shenzhen.) Local officials implemented cultural and labor policies, redeveloped infrastructure and public spaces, and created promotional activities to "elevate" Dafen into a "true artists' village." They addressed issues of *hukou* for Dafen painters by introducing policies that would allow them to gain Shenzhen *hukou* or permit their children to attend local schools, aided the development of new businesses like a Dafen auction house, and concertedly promoted Dafen throughout the official and public sphere. Through Dafen, they sought to show that even the lowliest Chinese migrant worker can achieve his or her dream of becoming an artist. When a national daytime television show on China Central TV Channel 1 (CCTV-1) aired a documentary about Dafen in 2008, it opened with this promise: "As long you come here, you can pick up a paintbrush. As long you can pick up a paintbrush, you can make your fortune."[8] As the show's six episodes pointed out, making money did not always mean being a successful or good artist, and vice versa. Dafen, it suggested, pointed to the critical dilemma of all aspiring artists—the relationship between commercial success and true creativity.

Though such longstanding issues of modern society remain unlikely to be resolved by propaganda, they did not prevent Shenzhen officials from attempting to remake Dafen Village itself—with all the promise of embourgeoisement and rise in real estate values such a transformation would bring. Throughout the 2000s, with village-, street-, and district-level investment and direction, Dafen's basic infrastructure was redeveloped along the same urbanization patterns taking place throughout Shenzhen with the exception that its dense urban village form was left largely intact.[9] Indeed, Dafen became the only urban village in Shenzhen to be the official object of "urban preservation." What Dafen's development throughout the 2000s demonstrated was the level of local-level government coordination that brought the village corporation, street-level office, and district officials in line with Shenzhen's and China's creative industry policies. Such efforts got the attention of national propaganda to the extent that the "Dafen model" was discussed in national ministry publications and implemented in cities throughout China, even in centers of power like Beijing and Shanghai.

At the village level, this governmental attention had a very clear impact: Dafen's sewage and water services were upgraded, its streets and alleys were repaved, large commercial buildings were constructed with government investment and ownership, and a Dafen subway stop was built. Eight thousand "painter-workers" (*huagong*) and six hundred firms were officially registered, an industry association was formed, and a public-private enterprise called Dafen Oil Painting Village was established to manage the village, which was immediately named by China's Ministry of Culture as China's first "Model Cultural (Art) Industry." By 2010, Dafen had its own museum of art, an officially sanctioned artists' association, an auction house, government-subsidized artist housing, and a cafe street with water fountains and public sculptures. It had been featured on the front cover of the *New York Times*, in the pages of *Artforum* (America's flagship contemporary art magazine), and countless foreign newspaper and TV news outlets. When China mounted its first World Expo in 2010, the pavilion that the city of Shenzhen proudly mounted featured exclusively the success story of Dafen Village. The pavilion touchingly focused on the dreams of Dafen's migrant-worker painters, portraying them as aspiring artists who came to the city to realize their creative aspirations. Organizers of the pavilion (the Shenzhen Urban Planning Commission and Urbanus Design) boldly named the pavilion, "Shenzhen, Frontier for China Dreams." This was a couple of years before the "China Dream" became the whole nation's rhetoric.

How the Dafen success story was circulated nationally is hence of consequence to the emergence of "dreams" as an ideological force that links and legitimates individual participation in rural-to-urban mobility and intensive urban development. From 2004 onward, individual government departments and state officials beyond Shenzhen had taken special interest in Dafen Village and promoted its attractive story as a model for creative industry throughout the nation. Popular Dafen's policies (for example, enabling painters' children to attend local public schools regardless of *hukou*) were adapted and adopted in the city of Xiamen in Fujian, where other neighborhood concentrations of trade paintings rebranded themselves as painting villages too. One Dafen boss—owner of the "Dafen Louvre"—opened a massive painting mall named the "Beijing Louvre" on the outskirts of Beijing, complete with replicas of Dafen's public sculptures (themselves replicas of classical European sculpture like the *Venus de Milo*). Lishui "Barbizon" Village (the subject of a *New Yorker* essay by Peter Hessler), the massive wholesale markets of Yiwoo city, and a creative district in Hangzhou were all created as painters' villages from which Dafen painters and bosses were directly recruited.

But beyond the new discourse of creative industry policy, the Chinese party-state's interest in promoting Dafen Village as a cultural phenomenon in the mid- to late 2000s also fits squarely within the far longer historical trajectory of the socialist avant-garde. We could broadly understand this as a literary and artistic legacy and movement more concerned with society and politics than with art's autonomy, with vernacular and popular forms rather than elite and private consumption, and with egalitarian and collaborative forms of art-making rather than the valorization of individual genius. Dafen Village was seen by some officials, propagandists, and its own painters as just such a socialist phenomenon—assembly-line production is a kind of collaborative creativity, the production of low-cost imitations a kind of popularization of art for the masses, and Dafen's transnational painting trade a means of providing hard-working migrant workers with the kind of knowledge and cultural work that would make them into global cultural citizens.[10]

However, like other efforts to construct a post-Mao socialist discourse in Shenzhen,[11] the representation of Dafen Village by local officials added up to a notable departure from Mao-era socialist precepts. In this case, officials adapted the socialist rhetoric of egalitarian culture to promote an export-driven, profit-led, creative industry—suggesting that the policies of creative industries themselves could be nationalistic, class-leveling, and profitable all at the same time. This neoliberal assemblage of a socialist-capitalist vision, an inheritor of Hu Jintao's "harmonious society" and a precursor to Xi Jinping's "China Dream," was most evident in the cultural narratives and events sponsored and produced by local officials, such as the twenty-episode television melodrama set in Dafen and produced by Guangdong television in 2008. In these television narratives, the socialist artistic legacy was transformed into arguments for the liberating potential of the art market, the positive role of patriotic but foreign-located Chinese entrepreneurs, the importance of contract obligations, the legitimacy of intellectual property law, and the legitimacy and positive effects of the party-state's social policies.

In other words, the success of Dafen Village was made into an emblem of Shenzhen's version of the China Dream. First, it represented a rural and populist challenge to the "capitalist" (class-based) monopoly on art and culture, one that excluded those who were provincial, uneducated, and lowborn. Second, it challenged the isolationist Mao-era doctrine that cultural work could not be profitable, commercial, or international in its outlook and forms. As a form of cultural criticism, Dafen presented an opportunity to think through some of the most protracted problems of modern artistic production in a society with egalitarian ideals. Those who championed Dafen saw it as emblematic of a new Chinese globalism.

Real and Fake

Dafen Village held different significations for contemporary artists in China and in the West despite the fact that both artistic cultures ostensibly trace their ideals to Marxist, leftist, and anticapitalist politics. These differences speak first to the cultural dynamics of authenticity raised by China's rise as the manufacturing center of the world in the post-Mao era and second to Western domination of the legal and cultural definitions of authorship and intellectual property. Furthermore, these differences show how concerns about global inequality in the area of labor and migrant labor, while shared by Chinese and Western publics, are expressed and imagined in different ways. The results were diametrically opposed representations of Dafen. Whereas Western journalists represented Dafen painters as "sweatshop workers" whose individual creativity was being suppressed by Chinese state censorship and the factory regime, Chinese journalists presented Dafen painters as hardworking peasants whose individual creativity was being unleashed by Shenzhen's embrace of world culture and the global market. Underlying both of these representations were misunderstandings about the way painting production was actually practiced in Dafen and the cultural, commercial, and legal values professionally attached to originality and copying.

The central international policy discourse that underlined the representation of Dafen Village in both China and the West was China's ascension to the World Trade Organization (WTO) in 2001 and the corresponding introduction of the intellectual property legal regime codified in the Agreement on Trade-Related Aspects of Intellectual Property Rights (TRIPS) and the Berne Convention. The introduction of a legal intellectual property regime is important for assigning the value of authorship to cultural products such as paintings. This basic power dynamic, however, enabled American consumers and politicians to accuse Chinese producers of imitation and intellectual property infringement of Western products and ideas, and Dafen was thus predictably seen as yet another case of rampant Chinese copying, maliciously and illegally mass producing the masterpieces of high art. This presentation is problematic in numerous ways and is perhaps most egregious from the point of view of Western art history.

Within its own confines, the Western artistic tradition, especially with respect to painting, has always recognized many modes of imitation as a central component of artistic development, production, and critique. *Imitatio* was a central artistic discourse of Renaissance Italy, and classicism—the imitation of ancient styles, works, and subjects—was the highest value of the French academy. In modern art, *influence*—often taking the form of pastiche, collage, satire, or referentiality—was a central component of the construction of the avant-garde. In the contemporary post–World War II period, *appropriation*— the intentional and direct reuse of existing artworks and images—is among the most celebrated of artistic strategies. Indeed, innumerable Western artists whose works are "copies" in the traditional sense have been securely accepted, praised, collected, and historicized as supremely "original" works. Though contemporaries of Dafen painters, these artists are not accused of being "copyists" at all—and when they are, they are vigorously defended by major art institutions. From an art historical perspective, there is little reason, formally, the paintings of Dafen could not be subjected to the same standard of analysis.

Strictly speaking, the accusation that Dafen painters are "forgers" is on the whole an inaccurate portrayal of the trade because the vast majority of paintings made in Dafen come from unrecognizable sources—that is, the source of most Dafen paintings is not known to the painter, the buyer, or the expert. In other words, the painting trade has so transformed the images that form the sources and inspiration of subsequent works that only a small percentage of them are even recognizable as copies of any single work. This means that, in a general sense, there is neither intentional nor knowable "fraud" in the mind of the maker or buyer of Dafen paintings. Finally, from a legal point of view, even in cases where Dafen's painting are painted from recognizable "masterpiece" works, most of these masterpieces are in the public domain, and many were made centuries before the notion of copyright was even codified or extended to visual art.

Yet, because of the high cultural status of "the artist" and oil painting held in both Chinese and Western traditions, the idea of Chinese-made "copies" of Western masterpieces was tantalizing. Leonardo da Vinci's *Mona Lisa*, however, is an excellent example of why the notion that a Dafen "copy" of the *Mona Lisa* infringes on Leonardo's intellectual property rights is logically absurd. First, Leonardo da Vinci never "owned" a copyright to his portrait because copyright was not even extended to pictorial, graphical, or visual works under the American Copyright Act until 1870 (never mind the fact that Leonardo was obviously not working in America and that France, where the painting is housed, has a different legal tradition of authorial rights). Second, legal history aside, the history of art is founded on innumerable instances in which artists copied and appropriated an image into a work of their own, and the *Mona Lisa* is a well-established example. Salvador Dali drew himself as the

Mona Lisa with a moustache. Marcel Duchamp took a postcard of the *Mona Lisa* and wrote a racy pun on it. These works are securely interpreted within the practices of "art" and accepted today as canonical works with no trouble by Western artists and historians. The difficulty, of course, comes when the same apparent practices are done outside of the West, where they are almost always considered "derivative" and "belated" of the Western tradition, or unconnected to the local (e.g., "native") tradition. The plight of Dafen painters is, in other words, an extreme version of what all non-Western artists have faced in trying to gain legitimacy as artists in a Eurocentric art world.

To the extent that all works of art must appear like previous works of art in order to be recognizable as "art" at all, a "true original" and an "exact copy" exist only in theory, for no paintings (even two painted at the same time by the same artist) can ever be physically or philosophically exactly alike. When Dafen painters paint *their* Mona Lisas, they are indeed making visible alterations of greater variety and difference than Dali or Duchamp did with theirs. Furthermore, when consumers purchase a *Mona Lisa* (in Dafen for about RMB 300, elsewhere for about US$250 to €90) in one of various sizes, on canvas, put into the consumer's choice of frames, with any alterations they might desire, they likely know that they have not purchased the *Mona Lisa* that hangs in the Musée de Louvre in Paris. Most probably, they have been told that they have purchased a painting by an art student or unknown artist, which might look familiar to them. All of these situations amount to neither forgery nor fraud unless one believes that there is a single, unalterable *Mona Lisa*; that Leonardo da Vinci owns all rights to its imitation, variation, duplication, or interpretation; and that all consumers believe that all paintings sold to them are probably stolen goods. If such assumptions were to govern our visual world, then neither the museum shop of the Louvre nor the Western painting tradition could exist "legally."

The *Mona Lisa* is an illustrative and extreme case, but the vast majority of Dafen paintings are in fact made after minor works whose origins are unrecorded by experts, dealers, or merchants and are so visually transformed that most art historians—let alone the casual consumer—would have great difficulty in identifying its source. Historical information about the image (e.g., artist name, title, or date) does not habitually accompany the source image as it is passed from client to boss to painters. Hence, with a few exceptions, Dafen paintings are usually painted after works that are unknown to the painter (who indeed has never seen the original), and they are bought by buyers who likewise do not know precisely what the original might look like. For these reasons, we must question whether Dafen paintings are "copies" at all—or rather, we must consider how much knowledge and its institutions must travel with images in order for the modern apparatus of "originality" to maintain its logic and coherence.

Beyond the modern Western discourse that always imagines itself the origin and the non-West an imitation, what the history of art has consistently shown is that "originality" is a historical concept born of a social process of evaluation and legitimation in which changing determinants of artistic authorship have been constantly negotiated and contested. *If* class, education, ethnicity, gender, and other social determinants of the identity of the artist were truly considered immaterial to the theoretical status of originality, then Dafen Village should be understood within this same spectrum of debate about the function of imitation and appropriation as a reaction to modern creations of "originality." Unfortunately, it has not, and it has instead been consistently rendered by both Chinese elites and the Western public as a case of copying *tout court*. In my opinion, this is very much the result of Shenzhen's place in both the Chinese and Western imaginary.

My observations about the art products of Dafen contradict prevalent perceptions of counterfeits, fakes, and *shanzhai* in Shenzhen in the Chinese and Western media. The manufacture of a wide range of counterfeit and imitation designer consumer goods, such as handbags, watches, golf clubs, sunglasses, mobile phones, and other branded products, has been strongly associated with Shenzhen. However, as in the case of Dafen paintings, media portrayals of "true" and "fake" reinforce a set of assumptions that obscures a complex set of issues surrounding authenticity. For example, at Shenzhen's Luohu Commercial City at the Luohu border checkpoint, where such goods have been sold since the earliest days of the SEZ (and before that on Sino-British Street),[12] one can buy a vast variety of handbags in many shapes and colors and with any number of designer logos attached (e.g., a Prada bag with a Gucci logo, and vice versa). With a great deal of effort, a discerning buyer can find an acceptable match to the authentic thing at a fraction of the price, while many others that bear little resemblance to the original are bought by consumers who aren't even aware of the original's appearance or the existence of the brand. The result is a product offering far more varied than the market of authentic things, and a market of highly diverse expectations and knowledge on the part of both buyers and producers. To police such a wide range of products uniformly would require a draconian enforcement and interpretation of intellectual property laws.

Indeed, much of the "fake" consumer goods made and bought in Shenzhen would be more accurately described as variations and adaptations, many of which are highly inventive. For example, at Shenzhen's Huaqiangbei—a very long street of more than ten city blocks lined on both sides with massive multistory malls selling digital, electronic, audiovisual, and high-tech consumer products—one can find products with a far greater range of functionality than exists in the branded market. Some of these products, like touchscreen mobile phones, anticipated major developments in the legitimate but highly belated mobile phone industry in the United States. Others, like dual-SIM-card mobile phones with something like an Apple logo on them, merge new functions with branded designs that remain unavailable outside Shenzhen. This sort of creative adaptation and reverse engineering has been more widely celebrated in China as *shanzhai* culture—a guerrilla spirit for appropriating international brands and products for Chinese demands, uses, and contexts. But in Shenzhen, this kind of guerrilla appropriation has been going on for decades, and what we have called "fakes" have long become so different (and often so much better) than the "originals" that this value-laden distinction needs to be closely historicized for every product to which it is applied.

We should also consider the extent to which this dynamism is the product of highly consumer-centric practices of entrepreneurship. That is, Shenzhen's entrepreneurs have proven to be highly flexible and adaptive at meeting the demands of a wide range of users and consumers, and at personalizing their product without regard for other cultural, traditional, or institutional presumptions about authenticity and history. We can see this in Dafen Village's galleries and workshops, where the experience of shopping for a painting is remarkably different than one would expect in Chelsea, New York. Paintings in Dafen's galleries are stocked in the hundreds and thousands and laid out in stacks that any customer can flip through. They are not organized by the individual names of the original artists or their titles and dates but rather by size, genre, and quality—landscapes, portraits, street scenes, figurative paintings, and so on. Most importantly, Dafen paintings can be made to order, with easily customizable adjustments. A consumer can order oil paintings of any image so long as the consumer supplies an image the painter can work from. They are available in any size or proportion that the buyer might require (the painter just rescales the picture freehand or with a pantograph). And they can ask for details that improve or alter the composition. Signatures, too, are applied according to the client's demands with whatever name (real or pseudonymous) that the client might want. Paintings are priced according to size, quality, and the time it takes to paint them. Abstract paintings are usually cheaper than realist ones. Figurative works cost more than landscapes or flowers. Portraits are priced according to the number of people that appear in them.

This consumer personalization has also led to a consolidation of labor specializations determined by a calibration of price and market demand. By the

2000s, Vincent van Gogh painters were among the lowest paid and considered the lowest skilled because demand was high in the 1990s and most migrants first trained by learning to paint van Goghs (therefore, most Dafen painters are able to paint van Goghs). On the other end of the spectrum, realistic figurative paintings bought to match traditional European furniture were more difficult and time consuming and thus became a higher paid specialty. It would be the middling specialties, such as ocean waves, photorealist portraiture, or "little Impressionism," that would be most highly profitable among Dafen painters— they afforded good prices per piece, were not so time consuming to produce, and were regularly in demand. Consequently, these are also the paintings that are most prevalent in the global market.

These aspects of the Dafen painting trade—its flexibility, small-batch production, customization, and simple subcontracting—are not unique to Dafen but rather represent a continuation of many of the individual and entrepreneurial practices prevalent in Shenzhen's earlier development. Throughout Shenzhen, large-scale development within urban villages was led by the village corporation. In Dafen, this took the form of creative industry.

The Gender of Authenticity

When Dafen Village became a model cultural industry in 2004, Chinese officials from the local to the national level began to focus on promoting "originality" and "creativity" in Dafen. This was in keeping with the cultural and creative industries policy discourse that had taken international policy circles by storm and that was also current in Shenzhen. In particular, officials in the architecture design department of the Shenzhen Urban Planning and Land Use Commission supported design and arts initiatives throughout the city, including the establishment and sponsorship of the Shenzhen–Hong Kong Architecture and Urbanism Biennial. One official, Zhou Hongmei, who had previously been promoted from the government of Longgang District (where Buji and Dafen are both located), also took a special interest in Dafen and oversaw an architecture design competition and successful construction of an art museum in Dafen by the Shenzhen-based design firm Urbanus.

At the street and village level, however, officials also implemented many policies targeted at fostering creativity and originality. The head propaganda chief of the Buji Street Office, Shen Shuren, worked to convince retired artists (many from his own home province of Heilongjiang) who were art academy graduates to move to Dafen. Those who came were given studio space in a government-owned building dubbed the "Red Building," sponsored to take sketching and painting trips that retreked the historical route of the Communist Party's Long March, and accompanied officials and Dafen bosses on international art market research and painting trips (even as far as Dubai). These artists were regularly commissioned to paint works for government spaces and exhibitions, their works were bought privately by these same officials, who simultaneously encouraged a newly established Dafen auction firm to auction their works. Although there were only several dozen "original artists" at Dafen who had the academic credentials to benefit from these policies, these activities gave prominent attention within Dafen circles to the definition, value, and official legitimacy of "originality" (*yuanchuang*). And lest we worry that Shenzhen's village-level propaganda officials lacked access to world-class knowledge, the Buji Propaganda Department would soon retain Ethan Cohen, the influential New York art dealer of Chinese contemporary art, as the "special advisor" of Dafen village.

For the larger population of painters, Dafen propaganda officials initiated public activities that focused on encouraging Dafen painters to become creative artists too. They held events and exhibitions celebrating skill and original art. They handed out free publications on copyright laws and intellectual property rights. They instituted a (later cancelled) system for painters and dealers to register original works so that legal redress might be sought against "copyists." They even considered creating a trademark logo Dafen galleries could display that would authorize their paintings as "original." Along with these public education efforts, original artists were given constant attention through exhibitions at the Dafen museum, officially sponsored promotions, publications, and media coverage.

In addition to implementing policies favoring "originality," officials produced an extensive range of propaganda products promoting notions of creativity and originality, in each case directly branding the usual business of Dafen as "copying." In September 2008, the national CCTV-1 program *Banbiantian (Half the Sky)* aired a six-part documentary on Dafen Village. A well-respected television program that explored women's and social issues, the documentary told the stories of three women working in Dafen—an original artist, a painter aspiring to become one, and a dealer. The program set forth a romantic definition of originality. It proclaimed that original art cannot be made to order, cannot be commercial, and cannot be copied. It is art that comes from within each individual, that is an expression of her own heart and mind, that is therefore authentic and priceless. Financial success does not matter to a true artist, and even when it is achieved, it comes not from greed but from being true to one's innermost self. Portrayed as bohemians, in official discourse, true artists are outsiders to the market and care nothing for the inauthenticity and reproduction that it demands.

Though such a definition of originality sounds just like the romantic myth of the artist, used to tell the stories of artists from Montmartre to Greenwich Village, here it was importantly taken up by a national Chinese television program as a feminized discourse. The "self" that could produce true art was represented in this television documentary exclusively by *women* who settle into urban lives of marriage and take up art as a nonprofessional leisure activity. Thus much of the profile of these three women is spent on describing the "true love" that they have found with their supportive husbands, who each become foils to the women's pursuit of their individual dreams. "True love" becomes the driver of each model woman's success—not profit, greed, or the cold logic of the market.

As an official portrayal of female migrant workers (*dagong mei*), this represents a significant departure from the political rhetoric that dominated official discourse in earlier decades, as Eric Florence and Dong and Cheng detail in this volume. In the 1980s, female migrant workers were encouraged to see themselves as docile country girls ("little sisters") who should reject romance and instead be self-reliant, hardworking, and eventually return to their hometowns.[13] In the 1990s, popular and official narratives instead celebrated their independence and sexual agency as single and mobile young women in the city ("working girls").[14] In the narratives of Dafen in 2008, however, Dafen's female migrant workers were portrayed as women who had found true love, marriage, and a settled and creative life in the city. If Dafen is exhibited as a bohemia of sorts, in official discourse it is nevertheless a "model" bohemia—a place where the political ideals of artistic freedom and rebelliousness from the market is remade into a state-sanctioned creative industry. Dafen becomes, in other words, a model bohemia, or a bohemia without any avant-garde challenge to social and political norms.

Though the individual biographies of many Dafen painters are indeed uplifting, this feminized construction of creative labor belies more pressing problems of migrant labor and gender inequality in Shenzhen,[15] as well as in Dafen. While the percentage of female migrant workers who make up the migrant labor force has been estimated at 70 percent to 90 percent in Shenzhen,[16] in Dafen, painters, artists, and bosses are predominantly male. When official discourse such as the *Half the Sky* television program focuses exclusively on the success of women in Dafen, it is in fact using the tropes of "original art" as convenient ways to delve into social issues of gender while ignoring the realities on the ground. In the documentary, women are shown passionately working to rediscover themselves as creative agents in a new economy, as though the continued inequalities assumed by female migrant workers can be resolved by "creativity" or "love" alone. The transformation of the female migrant worker into an artist-wife, in other words, legitimizes the transformation of the south's factory regime into a creative society while obscuring the actual gendered divisions of labor on which it rests. The far more difficult questions of the actual market (how the wives' painter-husbands support them, how other trade painters work and succeed, or how women's labor is undervalued within the trade) are entirely ignored in favor of this seemingly progressive exploration of gender equality.

Propaganda portrayals notwithstanding, it is important to emphasize that in all areas of the painting trade in Dafen, women work. However, the familial and social networks on which the trade rests are definitively not equal. In the development of Dafen, nearly all men who have worked as painters have had their girlfriends or wives to assist them by painting the backgrounds. Eventually, many of these women learned to paint from their male partners, some taking on exclusive areas of specialties within his larger practice. When a male painter became a boss, establishing a small firm, his wife tended to work alongside him as the firm's boss-lady (*laoban niang*), managing the firm's business, putting out orders, overseeing painters, setting prices, and managing clients. When a man was an original artist, his wife or girlfriend often ran a gallery selling his work. Some learned English to translate for him and to promote him to foreign buyers.

These kinds of codependencies are not gender-mutual. It is unheard of in Dafen for a woman working as a painter to have a husband or boyfriend who would paint her backgrounds or learn from her to paint. Very few women work as independent bosses, but of those I met, their husbands were working in other industries altogether. Finally, a female original artist once explained to me that a working female original artist in Dafen could really only hope to marry a male original artist—if she were to marry at all. When it comes to the working life, it would appear that the women of the painting trade largely fail to appropriate the labor of their husbands in the way that husbands uniformly appropriate the labor of their wives. While there are larger contexts of the gendered division of domestic and professional labor to be considered, what is important to note here is the ways in which even the transgressive picture of artistic life and work in Dafen does not challenge the male dominance in the broader culture of art—a condition as true in the elite Chinese art world as it is in the Western one. Though Dafen propaganda has often focused on women artists in Dafen in order to counter bourgeois or capitalist biases, the gendering of creative labor in official discourse thus reflects a different set of concerns in Shenzhen—namely, the urban and social status of female migrant workers. The imagined space that Dafen Village provides with official discourse is thus a particular one. The women celebrated in this model bohemia are not just creative workers; they are artist-wives, safely domesticated by their pursuit of their feminized creative dreams.

To summarize, in appropriating the creative industry's powerful vision of semiurban artist communities and utilizing it to promote Dafen Village as a model, Shenzhen's officials first had to confront a major problem: the production of Dafen Village was not "creative" by any definition of the socially hierarchized art world—whether in China or in the West. This problem, however, was resolved by returning to the socialist values that champion the creativity of the nonelite—peasants, workers, and women. By this valance, Dafen Village was far from a counterexample of creativity. From the socialist point of view, Dafen Village represented a utopian achievement. Even the rural-born female migrant workers of China could, in Shenzhen, realize their creative dreams and attain a livelihood of freedom, urbanity, and self-expression. Notsurprisingly, the presentation of female artists in Dafen ended up portraying little that was accurate about the trade itself or the conditions of labor there. Nevertheless, it did not prevent the discovery of a successful cultural model out of what was initially its exact opposite. This model was ultimately extended to cities throughout China, as proof of the national ideology of an already realized "China Dream."

Notes

This chapter is based on research conducted between 2006 and 2010. For extended discussion and detailed sources, see the author's *Van Gogh on Demand: China and the Readymade* (University of Chicago Press, 2014). The author would like to thank Mary Ann O'Donnell, Winnie Wong, and Jonathan Bach for their expert editing, advice, and support. The author also acknowledges the Social Science Research Council, Wenner-Gren Foundation, Fulbright-IIE program, Association for Asian Studies, and Harvard University Asia Center for funding the research on which this chapter is based. The Robert Wood Johnson Foundation Health & Society Scholars Program provided financial support for the write-up.

1. "Shenzhen you ge huajia cun," *Yangcheng wanbao*, 1999.

2. Philip Tinari, "Original Copies: The Dafen Oil Painting Village," *Art Forum* (October 2007): 344–51; Christopher Beam, "How to Make It in the Art World: 6. Outsource to China," *New York Magazine*, April 22, 2012.

3. G. S. Shieh, *"Boss" Island: The Subcontracting Network and Micro-Entrepreneurship in Taiwan's Development* (New York: P. Lang, 1992), 82–89.

4. Yao Dingkang, "Dafencun youhua yu Meiguo shichang [Dafen Oil Paintings and the US Market]," United States–China Chamber of Commerce, New York, December 20, 2007, http://www.ccmedu.com/bbs26_56112.html.

5. Zhou Min, "Zhongguo youhua diyi cun ruhe shengji? [How Can China's Model Oil Painting Village Ascend?]" *Yangcheng wanbao*, November 22, 2004.

6. Jia Hezhen, "Dafen De Na Xie Shi'r," *Artron*, http://blog.artron.net/indexold.php?68529/viewspace-240116.

7. César Graña and Marigay Graña, *On Bohemia: The Code of the Self-Exiled* (New Brunswick: Transaction Publishers, 1990).

8. CCTV-1, *Banbiantian* [Half the Sky], "The Story of the Painting Village," 2008.

9. See O'Donnell, chapter 5 in this volume.

10. Extending the discourse of civilization discussed in Bach, chapter 7 in this volume.

11. See O'Donnell, chapter 2, and Florence, chapter 4, in this volume.

12. See Ma and Blackwell, chapter 6 in this volume.

13. Eric Florence, "Migrant Workers in the Pearl River Delta: Discourse and Narratives about Work as Sites of Struggle," *Critical Asian Studies* 39, no. 1 (2007): 121–50.

14. Pun Ngai, "Becoming *Dagongmei* (Working Girls): The Politics of Identity and Difference in Reform China," *The China Journal* 42 (1999): 16–17.

15. See Dong and Cheng, chapter 8, and Mason, chapter 10, in this volume.

16. Pun Ngai, *Made in China: Women Factory Workers in a Global Workplace* (Durham: Duke University Press, 2005), 40.

10

Preparedness and the Shenzhen Model of Public Health

KATHERINE A. MASON

In 2009, the first influenza pandemic in forty years made its way to Shenzhen. After crossing the ocean from California, the H1N1 virus traveled to Europe and Japan before landing on the shores of Hong Kong on May 1, crossing the border to Shenzhen four weeks later. In Shenzhen, city public health officials had been preparing for this moment for six years, ever since the 2003 severe acute respiratory syndrome (SARS) epidemic shed a harsh light on China's disintegrating disease control system.[1] They rapidly isolated the cases, traced their contacts, and herded potentially infectious people and their contacts into a quarantine center in an undisclosed location in the outer districts of the city. They gathered suspect case data from local hospitals and reported cases upward to the provincial and national level using a real-time online surveillance system established in the years following SARS. They continued to track everyone who crossed over the border and reported a plan to stay in Shenzhen. They even called, for a period of several weeks, all those who had no reported symptoms at the time of their crossing—just to be certain that no new symptoms developed.

By August, the outbreak was over. In Shenzhen, as elsewhere, H1N1 had transformed into just another mild, endemic infection from which most people were expected to recover without a problem, and the disease had all but disappeared from the news, both in Shenzhen and elsewhere. Those deaths that did occur in Shenzhen were mostly among the usual victims of any influenza outbreak: the old, the very young, and the infirm. The young adult and middle-aged victims that had suffered disproportionately in the United States and that had lent this otherwise relatively mild flu strain a sense of heightened danger largely escaped unscathed in Shenzhen. The astronomical case counts in migrant and other at-risk communities—widely expected and feared at the time of the initial outbreak in the city—did not really materialize. In fact, a 2012 study found that despite Shenzhen's high population density and mobile population, only around 5 to 6 percent of Shenzhen's population contracted H1N1 in 2009, compared to nearly 20 percent of the US population during a slightly longer time period.[2]

Shenzhen's model of public health—a high-tech, biomedical model oriented around real-time surveillance conducted by highly educated professionals—was developed after the SARS outbreak in consultation with national and global health authorities to rapidly detect novel diseases and prevent them from spreading out of control. By all measures, the first real test of the model had been a great success. Six years of preparedness had worked. Nevertheless, as H1N1 faded from the horizon, the outbreak response team entered a period of ennui. Some described the extensive training they had received since SARS as having been wasted on an unworthy enemy. Comparing their city's response with the more measured responses elsewhere around the world and even elsewhere in China, they suggested that city leaders could have called off the hunt for virus carriers earlier and could have avoided seemingly unnecessary measures like following nonsymptomatic travelers by phone. At the same time, having finally brought their model for public health to fruition—having both developed and implemented a sophisticated disease surveillance and containment plan nearly from scratch with characteristic "Shenzhen speed"—Shenzhen's public health professionals were at a loss. What did they do now that the system had been built and tested? What path forward was there for public health in Shenzhen now that they had succeeded? And would they continue to be a model of excellence if Shenzhen's speed began to slow down?

The public health workers' rapid and effective response to the H1N1 pandemic and their equally quick-to-follow disillusionment with their success highlight how speed had been valued and was being revalued as Shenzhen entered its fourth decade. It also raised the question of what the city could do next to top all the first-place awards and accolades it had garnered during its vibrant youth.[3] Within and against the totalizing model of "Shenzhen speed," both individuals and institutions have grappled to increase efficiency, which is not only measured against global health statistics but also functions as an element of the city's competitive advantage when positioning itself within global production chains.[4] Consequently, an analysis of the value, production, and implicit dangers of "Shenzhen speed" within the context of an epidemic makes salient how so-called economic values have shaped the city's administration of what are (usually) taken to be noneconomic values, such as public health.

Ethnographic analysis of Shenzhen's public health professionals' response to the H1N1 pandemic also recasts the early assertion that Shenzhen was to be used as a laboratory for political experiments that could not be conducted in Beijing, highlighting the technocratic determinism of reform policies and programs with respect to one of its model constituencies—health care professionals. When Deng Xiaoping and his reform allies called the Special Economic Zones (SEZs) "laboratories," they placed reform praxis within the rhetoric of the Four Modernizations campaign to strengthen the nation's agriculture, industry, national defense, and science and technology. The campaign glorified modern science as a means through which the country could be (quickly) saved from the excesses of Maoism and the general standard of living (*xiaokang*) raised. The ideological assertion that Shenzhen was a laboratory implied (as discussed below) the establishment of technocratic administration to solve social problems. In this sense, the Shenzhen response to H1N1 clearly fulfilled an ideological mission that had been voiced by Zhou Enlai as early as 1963 and only taken up by the central government in 1978. Nevertheless, the ideological use of the laboratory metaphor to lionize reform praxis elided one of the fundamental truths of scientific experiments—outcomes are uncertain and not always beneficial. More important, when Beijing leaders and Shenzhen health workers figured Shenzhen as a laboratory, they conflated the practices of laboratory science and its emphasis on repeatable and consistent results with a social and explicitly political concern—getting people to behave in predictable ways.

Mao and the Heyday of Communist Public Health

Few Shenzheners are nostalgic for the Mao years, but at least some of Shenzhen's older and more experienced public health workers harbor nostalgia for the national public health infrastructure associated with those years. Mao made the improvement of public health a major priority of the new Communist regime, declaring that the strength of his new nation depended on the strength of the bodies that made it up.[5] The Maoist model of public health emphasized that health care could also be a means of ameliorating economic class antagonisms that had structured historic access to health care. With a slogan of "prevention first" (*yufang weizhi*), Mao made it clear that public health, with its focus on preventing diseases in whole populations, was more important and more patriotic than clinical medicine, with its focus on treating disease among affected individuals. In addition, Maoist public health approaches attempted to ameliorate class differences that were produced through unequal access to education and training. While public health could be carried out by the people (*renmin*), in service of the people (*wei renmin fuwu*), clinical medicine required the participation of a despised group—the "expert." Mao's claim that mobilization of the masses achieved better health than reliance on professionals was tested during the Cultural Revolution when Red Guards took over urban hospitals and physicians were sent to the countryside. During these years, it became clear that although patriotic fervor was an excellent tool for mobilizing the masses to perform public sanitation and basic preventative work, it failed to perform surgery.

Even as the Mao regime vilified its professional doctors, it glorified its public health pioneers and their aim to raise the general level of health of the country's poorest and least-serviced populations. Some of the most visible manifestations of this were the Patriotic Health Campaigns (*aiguo weisheng yundong*) that swept through both the countryside and the cities in the 1950s and 1960s. The most famous of these included the still ongoing Four Pests (*sihai*) campaign, which sought to eradicate flies, rats, cockroaches, bedbugs, and, in an early version, even sparrows; antiprostitution campaigns, which sought to eradicate venereal diseases; and antischistosomiasis campaigns, which sought to eradicate a common and debilitating rural river parasite.[6] Supporting all these campaigns was a corps of peasant health care workers known as "barefoot doctors" (*chijiao yisheng*). Barefoot doctors were usually ordinary farmers who received several months of training in basic preventive measures like sanitation and vaccination, as well as primary clinical care for minor ailments. Many were based in local Anti-Epidemic Stations (AES), which were small health posts that Mao established at all administrative levels to assist with basic preventive services. Indeed, although statistics from this period should always be interpreted with caution, historians generally have found that Mao's public health campaigns were impressively effective. According to statistics widely published in English-language literature in the 1970s, the infant mortality rate in the 1950s and 1960s fell from two hundred out of one thousand live births to forty-seven out of one thousand births, with life expectancy almost doubling from thirty-five to sixty-five. Many infectious diseases were brought under control for the first time, most famously schistosomiasis and syphilis.

The AES system began rapidly deteriorating with the death of Mao and the birth of the reforms, as public health institutions around the country were defunded, barefoot doctors left their posts or privatized their practices, and Patriotic Health Campaigns wound down (though they never actually disappeared). By the early 1990s, government subsidies accounted for only 10 percent of the total revenues of public health facilities that had once been fully government funded.[7] To make up the lost revenue, AESs supported themselves by charging businesses for mandatory sanitation inspections (required by law in most localities, including Shenzhen) and charging service workers for mandatory health certifications (also required by law). Active prevention work slowed down and public health campaigns all but disappeared. Infectious and parasitic diseases that had plummeted during the Mao years, including both schistosomiasis and syphilis, began to return.

Civilizing Shenzhen, One Mosquito at a Time

The Bao'an County AES, as Shenzhen's health station was known prior to 1979, was a small health post that, like other Mao-era AESs, mostly distributed vaccines and assisted in building latrines. With the founding of the SEZ, it was renamed the Shenzhen AES. For the most part, the fate of the Shenzhen AES mirrored that of AESs in other regions with one crucial exception: at the same time that local governments began deprioritizing public health in most places in favor of investment in economic development, Shenzhen's public health system managed to prove its continued relevance by playing a crucial role in saving the country's most famous economic experiment from collapse. They did this in the early 1980s and mid-1990s, as they would again in 2003 and 2009, by effectively controlling an infectious outbreak through technocratic expertise rather than a popular campaign.

Working without adequate protection or native immunity in the swampy waters of early Shenzhen, construction workers building the new city began to die of malaria almost as quickly as they arrived. In response to the malarial threat—and in stark contrast to the approaches taken during the Mao period—the Shenzhen AES began almost as soon as the SEZ had been established to recruit experienced malaria "experts" from rural antiepidemic stations elsewhere in China to help get Shenzhen's malaria problem under control.[8] These experts, mostly consisting of those with a vocational education (*gaozhuan* or *dazhuan*) in "public sanitation" (*gonggongweisheng*), were poorly educated compared with those dominating public health in Shenzhen today—but at the time, right after the Cultural Revolution, they were the most highly educated people available.

Public health workers arriving from rural areas described the conditions in early Shenzhen as "chaotic," "backward," and a "big mess"—much worse, they said, than the rural counties where they had previously worked.[9] They spoke of working day and night soaking bed nets, spraying the toxic pesticide DDT, and even standing in the most densely affected areas for hours at a time counting mosquitoes. Some were injured by the strong chemicals being used or contracted malaria themselves. The discourse surrounding these dangers was one of sacrifice in the name of development, a theme common to Maoist public health programs. In addition, these conditions were similar to malarial outbreaks during major engineering projects in other underdeveloped regions around the world, where the conquest of malaria has often been equated with the civilizing of a primitive land.[10] Shenzhen was no exception: Shenzhen public health workers who participated in the malaria campaigns of the early 1980s often described their jobs as consisting of making Shenzhen more "civilized" (*wenming*). Consequently, public health workers who had participated in the malaria campaigns found pride in having contributed to the civilizing project of building the SEZ.

After briefly getting the epidemic under control, a second wave of malaria hit in 1995–96. This time, public health workers were able to get it under control and keep it that way, thanks to rising living standards, improved sanitation, and the rapid disappearance of the swampy rural areas in which mosquitoes carrying malaria thrived. By this time, Shenzhen's public health system was a mostly privatized, profit-driven system dependent on sanitation inspection fees for its livelihood following the divestment of the state in public health. Despite all this, the second antimalaria campaign easily succeeded. Malaria largely disappeared from Shenzhen by 2000.

The malaria epidemics that threatened Shenzhen's early development had at least two lasting effects on public health and disease control in China. First, they set a precedent in Shenzhen for the recognition of outbreak control as a crucial companion to economic development during the reform era, even as government funding for public health faded over the next two decades. This was not the case in most other locales, where public health was consistently a very low priority compared with reform-era projects with clearer ties to economic growth. Second, they populated Shenzhen's local public health system with a group of people who were moderately educated and who considered themselves to be experts. Both of these factors would help establish Shenzhen as a national model for a new era of public health governance once funding was restored in the early 2000s.

SARS and Surveillance

The development of Shenzhen's public health system following the malarial outbreaks illustrates the steady evolution of Shenzhen's governance, described throughout this volume, from a regime of discipline to one of surveillance. The first major public health campaign in the new city of Shenzhen aimed to build and discipline a dangerous, uncivilized piece of land through the eradication of an infection. It was another outbreak event that precipitated the transition from discipline to surveillance.

In 2002, just before the SARS virus appeared on the scene in Shenzhen and throughout the Pearl River Delta, the Shenzhen AES was renamed the Shenzhen City Center for Disease Control and Prevention (*Shenzhen shi jibing yufang kongzhi zhongxin*), after the US institution of the same name. At the time of the name change, Shenzhen's AES had diversified into a complex hierarchy of institutions, with AESs established in each of Shenzhen's districts, hospital-managed disease control units established at the "street" (*jiedao*) level, and vaccination stations and health branches of residents committees (*juweihui*) working at the community level. In Shenzhen, as elsewhere, upper-level AESs had little legal or practical authority over lower-level institutions, even within their own supervisory lines, and relied heavily on the building of personal relationships (*guanxi*) in order to gain cooperation to implement public health projects. As elsewhere in China, AES professionals at all levels created and maintained their personal relationships through elaborate banqueting rituals (*yingchou*), smoothing relationships both within Shenzhen's unwieldy public health hierarchy and with other governmental institutions. Thus, although Shenzhen's pre-SARS model had been recognized for its technocratic contributions to economic growth, its day-to-day practices reproduced the messy conditions in other local Chinese public health departments specifically and local governmental agencies more generally.

Unlike the US CDC, which is solely a national institution, the Shenzhen CDC was one of more than two thousand local CDCs in China, established as part of an effort on the part of the central government to reinvigorate the local AESs.[11] It was also one of perhaps a dozen CDCs in Shenzhen alone, which also has CDCs at the district level as well as at the "street" level in the large outer districts of Bao'an and Longgang. Nevertheless, as in other Chinese cities, the significance of this name change was at first unclear at best. The only thing that changed immediately was that the right to levy fines for failed health inspections was transferred to a new institute, the Shenzhen Health Inspection Institute (*weisheng jiandusuo*), to which half of the AES's staff was also transferred. For those left behind, however, life at first continued as usual. Thus, even for those public health workers who understood and admired the US CDC, how that highly funded, research-intensive institution known around the world for its disease control capabilities could be translated into a new model for the half-century-old AESs remained a mystery. That mystery was solved with the arrival of SARS later that year.

When SARS, a deadly new flu-like virus that emerged just outside of Guangzhou in late 2002, exploded onto the global health scene in the spring of 2003, China's crumbling public health system suddenly became a major liability in the country's bid for global respect. Disease containment and reporting failures across the country were laid bare for all the world to see. Moreover, when China's central government admitted several months into the epidemic that it had covered up SARS cases and misreported disease prevalence to the international community, top leaders also committed to a new tactic of aggressive SARS containment measures. In Beijing, which by April 2003 had the most serious outbreak, SARS treatment hospitals were built in a matter of days, entire schools were quarantined, movie theaters were shut down, and the epidemic was quickly stamped out—though luck probably helped in keeping the epidemic from spreading farther in areas that were less well equipped. Nevertheless, travel bans, stigma, and fear took a major, if temporary, toll on the country's bottom line, with China experiencing significant losses in GDP in 2003.[12]

When the call to take aggressive action came down to the local levels in Shenzhen, Shenzhen's public health workers felt they were up to the challenge. Rolling out case identification and rapid mobilization techniques they had developed during the malaria outbreaks, and drawing on their intimate knowledge of the city from having seen it grow up from scratch through the 1980s and 1990s, malaria veterans described their response to SARS as a natural outgrowth of previous battles. They were proud of the outcome. According to official numbers put forth by the Shenzhen Bureau of Health, Shenzhen City recorded zero deaths and only fifty-six cases of SARS in 2002 and 2003—compared to one thousand cases each in Guangzhou and Hong Kong, along with dozens to hundreds of deaths. Furthermore, there were no reported deaths in Shenzhen among health care workers, a group that suffered devastating losses in Hong Kong. Shenzhen's economy took a hit, but not nearly as hard as the one Hong Kong experienced across the border.

The aftermath of SARS brought not only recognition for this accomplishment but also a new mission for the new CDCs: to prevent another SARS-like event from happening. Funding returned to China's public health system at large, with Shenzhen and other Pearl River Delta cities taking in some of the greatest amounts of funding and receiving outsized amounts of national and international attention. Due to its long border with Hong Kong, its extremely high proportion of migrant workers, its highly diverse population, and its positioning in the heart of a region known for its incubation of influenza viruses, Shenzhen was seen as particularly dangerous from the standpoint of infectious disease control. Due to its wealth, it was also considered an ideal place to develop the kind of preparedness infrastructure that the US CDC and WHO insisted was necessary in order to prevent another deadly virus from emerging in the Pearl River Delta.

Dual Use: Remaking the Shenzhen Model of Public Health

After SARS, Shenzhen's local public health model was transformed through engagement with the more technocratic US CDC and WHO models of public health. These foreign health organizations took an active role in promoting the importance of building local emergency response and preparedness infrastructure in order to stop novel viruses "at their source"—the "source," in the case of flu-like viruses, meant southeastern China and parts of Southeast Asia. A defining feature of the international model was the principle of dual use—similar to that of "externalities" in the field of economics—which posits that in preparing for something like an influenza pandemic or Ebola outbreak, benefits well beyond the initial target benefits will accrue. For example, when one buys a machine capable of testing samples for a novel influenza strain, that machine can also be repurposed to test for other, more common ailments. Or, when one builds a sentinel surveillance system, this system can also be used to detect other types of outbreaks.

However, the undercurrent of this discourse of "global health preparedness" was highly skewed toward meeting the interests of American and other Western governments that wanted to make certain that frightening infectious diseases were identified and contained before they could reach Western shores. As such, this model of global health was not necessarily designed to maximally benefit local populations in Shenzhen or elsewhere (see discussion below).

Thanks to funding from the central government and local government in the immediate post-SARS period, the Shenzhen Bureau of Health approved a burst in hiring, and the Shenzhen CDC staff doubled in just a few years. The city government also committed RMB 370 million to the city CDC to build a new campus in Nanshan District that was outfitted with biosafety laboratories, luxurious offices, and even a tennis court. The Shenzhen CDC relocated to the new campus in late 2010. The "experts" with vocational degrees who had dominated leadership positions in the AES/CDC since the malaria campaigns made way for a new set of experts (*zhuanjia*) with master's and PhD degrees from prestigious epidemiology and laboratory biology programs in China, Hong Kong, and Europe, which promoted an international model of public health. Many had trained in a program called the Field Epidemiology Training Program (FETP), run by the US CDC in Beijing, and had spent time abroad on postdocs or visiting scholar positions. They also tended to be very young, with the average age of new hires in the late twenties or early thirties. They considered themselves to be "professionals" (*zhuanye*) compared to the previous generation of experts, many of whom were largely self-taught.

Shenzhen's reputation as a rich, "open" city with plenty of opportunities for migrants from other parts of China made the CDC there an especially attractive place to work for aspiring young people. This internal restructuring thus emphasized technocratic expertise and familiarity with the international model with its emphasis on outbreak preparedness, biomedical research, and surveillance systems.

The new experts set about constructing a sentinel surveillance system to detect outbreaks of novel diseases (such as novel influenza strains) or novel outbreaks of old diseases (such as malaria; hand, foot, and mouth disease; or foodborne pathogens) and to act in response to suspected cases. Shenzhen was one of a few large cities in China to pilot such a system; Shenzhen was also a sentinel city in a national surveillance project. Hospitals, community health centers, schools, and other community- and street-level institutions were chosen as sentinel sites and ordered to report any suspect cases of infectious illness immediately using an online system. Outbreak teams from the districts investigated suspicious clusters, or "epidemic situations" (*yiqing*). Particularly worrying outbreaks—such as potential novel flu outbreaks—were evaluated by specially trained teams from the city level. Case reports were instantly accessible by infectious disease team leaders, who received updates by text message as well as computer.

CDC members at all levels received regular trainings in emergency response, and Shenzhen even started its own homegrown branch of the FETP program. Those specializing in influenza were sent to Hong Kong for training. Beginning in 2006, Shenzhen's public health experts began chasing sporadic human cases of H5N1 avian influenza, a particularly fatal viral strain that international experts feared could cause a catastrophic pandemic reminiscent of the Great Influenza of 1918, which killed tens of millions of people worldwide. Dressed in biohazard suits and disinfectant backpacks and equipped with rapid diagnostic equipment, they identified and "dealt with" (*chuli yiqing*) each case in line with international recommendations.

All this was made possible by a level of funding and local government commitment (*zhengfu zhongshi*) unmatched almost anywhere else in China. Such was the political commitment to this project that local laws made considerably higher demands on the surveillance system than what was recommended or requested by the central or provincial governments. National regulations, for example, defined a *yiqing* of ordinary influenza as twenty clustered cases, Guangdong defined it as fifteen, and Shenzhen defined it as only five cases. Funds for laboratory equipment, personal protective gear, expert salaries, and computer systems flowed unabated throughout the mid- to late 2000s.

H1N1 Pandemic: Testing the Preparedness Model in Shenzhen

After years of training, simulation exercises, infrastructure development, and overall systems building, Shenzhen finally rolled out its technocratic preparedness model with the arrival of the H1N1 influenza virus in the spring of 2009. Dressed in spacesuit-like biohazard suits, public health workers rushed to the designated flu case intake hospital at all hours of the night to receive suspect cases, evaluate the cases for potential infections, and take down lists of contacts before shipping many travelers off to quarantine for seven days. Laboratory technicians worked around the clock to process nasal swabs and identify positive cases. Teams went anywhere suspect cases or the contacts had stayed or visited and sprayed facilities down with backpacks full of powerful disinfectants. Workers were pulled from other jobs at the CDC to man the twenty-four-hour hotline, where Shenzheners could call in with questions about the disease or suspicious case sightings, or to help trace contacts. When the disease began spreading in Hong Kong and the threat to Shenzhen's border became even more pressing, teams began departing for two-week stays at the intake hospital, where they slept in dorms and were on-call twenty-four hours a day for testing and evaluation. And the online surveillance system lit up day and night with reports from sentinel sites of newly suspect cases.

This aggressive virus chasing succeeded in delaying the arrival of H1N1 into Shenzhen for nearly a month after it first appeared across the border in Hong Kong. Once it did arrive, case tracing and quarantine continued through July—long after such rigorous epidemiological tracking had ceased in other parts of China and around the world. By then, the disease had become sufficiently established in Shenzhen and the fatality rate recognized to be sufficiently low that leaders finally acknowledged there was little point in continuing case tracking and controls. Vaccination campaigns to protect against the H1N1 virus were under way soon after, and H1N1 was a nonissue by early 2010, even as it continued to spread in other places. In the end, little in the way of major disruptions to the economy or to daily life in Shenzhen occurred. Although outbreaks in factories and schools did occur, most students and workers recovered without incident.

By all accounts, the preparedness model developed after SARS was a success. Sentinel systems identified suspect cases. Containment teams contained the threat. Laboratory tests produced rapid answers. Highly educated professionals questioned foreigners in their native languages and provided culturally appropriate meals and activities in quarantine, all while (for the most part) avoiding self-contamination. As a result, the spread of the disease inside

Shenzhen did not fully get under way until well into the summer—after schools had let out and shortly before a vaccine was available. This delay likely cut down significantly on case incidence and fatalities. Research teams even produced copious publications in collaboration with foreign and Hong Kong experts after the outbreak was over, making use of the massive amounts of data collected through the system to raise the city's profile and provide career advancement opportunities for those studying and working in the Shenzhen health system.

And yet, beginning in late 2009 and continuing in the ensuing months and years, as the CDC moved into its expansive new space in Nanshan and the caliber of both public health expert and public health science in Shenzhen continued to rise, a sense of malaise set in for some of those most heavily involved in the H1N1 response. Some of this was a simple case of letdown—after six years of preparation and a few months of excitement, the big event that they had all been waiting for was now over. But some public health workers also emerged from the H1N1 successes skeptical about the model itself.

First, the massive expense, organization, training, energy, and infrastructure utilized to respond to what ended up being a relatively mild case of the flu felt both nonreplicable and unsustainable. It was nonreplicable because few other cities in China (other than perhaps Shanghai and Beijing) could possibly fund things like million-dollar assay machinery or free cell phone deliveries to disgruntled patients staying in quarantine centers. Smaller and less desirable locales also had little chance of recruiting scientists with PhDs to run the high-tech machines or conduct epidemiological investigations. And even if they were able to, there was much less urgency about pandemic preparedness in places that did not have major border crossings or sit alongside major international financial centers with heavy transnational air traffic. The latest Shenzhen model of public health was unsustainable because even Shenzhen was unable to maintain the momentum of an emergency response longer than a couple of months—due partly to burnout and partly to the fact that such a response required virtually shutting down the center's other functions, including the tracking of more common and, in some cases, more deadly infectious diseases such as hand, foot, and mouth disease. Outbreaks of this latter disease during the summer of 2009 went unrecorded at the district levels in large part because personnel were too busy trying to account for H1N1.

Second, even in Shenzhen, many of the new professionals recruited to the city to build public health systems found it difficult to discern what outbreak preparedness and emergency response were really contributing to the public's health. Those who continued to work on surveillance and emergency response activities long after H1N1 faded into the background found the sense of urgency difficult to maintain in the face of what they found to be little evidence that what they were doing was necessarily making a big difference in the health of the population. In effect, these professionals repudiated the "dual use" model that Western countries relied upon to justify the investment of large amounts of resources in preparedness for catastrophic disease events that might never occur. While microlevel examples of dual use did indeed occur in Shenzhen, the larger claim of dual use—that in pouring money into preparedness for a specific disease, one was essentially pouring money into public health more generally—did not really pan out. Expensive machines might be able to test for other viruses, but they did little to address diabetes, occupational safety, or nutrition. Surveillance systems were good at detecting cases of a new disease during the high drama of an acute outbreak but proved to be poor at detecting cases of more common diseases day after day—when motivation, time, and resources at the lower levels to support reporting compliance was low. And outsiders with PhDs were great for designing research projects and managing quarantine procedures, but they often lacked the *guanxi* connections or local knowledge that were crucial for collaborating with district-, street-, and community-level institutions. The increasingly academic atmosphere at city-level and even some district- and street-level public health institutions in Shenzhen also frustrated some who had come to the field in search of a stable government job or a career in public service.

Global Circulation of Public Health Models

In the face of this post-H1N1 malaise, the Shenzhen public health system began to experiment with a wider range of international models even as it settled into a new campus that was built largely for surveillance and laboratory work. Antismoking messages began to proliferate, with smoking banned from the CDC campus in recognition of a similar ban at the US CDC. Drinking was increasingly recognized as an important public health program, and drinking at banquets was radically curtailed after a series of high-profile drunk driving accidents among local bureaucrats in 2010. Emphasis on researching and preventing chronic diseases like diabetes, stroke, and obesity grew. Moral concerns about social justice issues like migrant health inequalities took on a new countenance, with Shenzhen's new public health experts taking active roles in investigating and implementing new migrant health care and disease prevention schemes. Collaborations with foreign partners and professional trips abroad to research or study provided the opportunity to learn from other public health models in foreign countries while drawing some lessons for Shenzhen.

Public health models also did not flow in only one direction. Whether consciously or not, during the Ebola outbreak of 2014, the United States briefly adopted a version of disease response that Shenzhen modeled so well during the 2009 H1N1 outbreak, including border quarantines, mass fever checks, and incubation period tracking.[13] The states of New York and New Jersey even took steps to mandate the quarantining of anyone returning from Ebola-affected nations even if they had no symptoms—a step that went further than anything Shenzhen attempted during SARS or H1N1.

The strong local resistance to draconian quarantine policies in the United States—and subsequent backtracking on the part of American politicians—highlights the difficulty of replicability on a global scale. As both Chinese and US officials readily acknowledge, measures workable in an authoritarian state often do not translate well to other political contexts. Chinese public health officials had taken a model originally developed and promoted by American public health authorities and had, thanks to a lack of democratic process, improved it in ways that their American counterparts were not capable of doing. Shenzhen in turn took that Chinese model and built it into the best, most well-resourced, most effective version of itself. The outcome of all this perhaps reveals less about Shenzhen and more about the fantasy of the model itself as a fix-all for disease control around the world. Bigger machines, faster computers, and more PhDs might do less for the public's health than the everyday drudgery of diet and exercise.

Notes

The author would like to thank Mary Ann O'Donnell, Winnie Wong, and Jonathan Bach for their expert editing, advice, and support. The author also acknowledges the Social Science Research Council, Wenner-Gren Foundation, Fulbright-IIE program, Association for Asian Studies, and Harvard University Asia Center for funding the research on which this chapter is based. The Robert Wood Johnson Foundation Health & Society Scholars Program provided financial support for the write-up.

1. See A. Kleinman and J. Watson, eds. *SARS in China: Prelude to Pandemic?* (Stanford, CA: Stanford University Press, 2006); Katherine A. Mason, *Infectious Change: Reinventing Chinese Public Health after an Epidemic* (Stanford, CA: Stanford University Press, 2016).

2. X. Xie, S. Q. Y. Lu, J. Q. Cheng, X. W. Cheng, Z. H. Xu, J. Mou, S. J. Mei, et al., "Estimate of 2009 H1N1 Influenza Cases in Shenzhen: The Biggest Migratory City in China," *Epidemiology and Infections* 140 (2012): 788–97; S. S. Shrestha, D. L. Swerdlow, R. H. Borse, V. S. Prabhu, L. Finelli, C. Y. Atkins, K. Owusu-Edusei, et al., "Estimating the Burden of 2009 Pandemic Influenza A (H1N1) in the United States (April 2009–April 2010)," *Clinical Infectious Diseases* 52, S1 (2011): S75–S82.

3. See Florence, chapter 4 in this volume.

4. See Bach, chapter 7, and Wong, chapter 9, in this volume.

5. S. R. Schram, *The Political Thought of Mao Tse-Tung* (New York: Praeger Publishers, 1969). See also C. C. Chen, *Medicine in Rural China: A Personal Account* (Berkeley: University of California Press, 1989); J. S. Horn, *Away with All the Pests: An English Surgeon in People's China* (New York: Paul Hamlyn, 1969).

6. See R. Rogaski, *Hygienic Modernity: Meanings of Health and Disease in Treaty-Port China* (Berkeley: University of California Press, 2004); M. Cohen, G. Ping, K. Fox, and G. Henderson, "Sexually Transmitted Diseases in the People's Republic of China in Y2K: Back to the Future," *Sexually Transmitted Diseases* (March 2000): 143–45.

7. William C. Hsiao and Linying Hu, "The State of Medical Professionalism in China: Past, Present, and Future," in *Prospects for the Professions in China*, ed. W. P. Alford, K. Winston, and W. C. Kirby (New York: Routledge, 2011), 111–28.

8. For more on the implications of the malarial roots of Shenzhen's public health system, see the author's "Mobile Migrants, Mobile Germs: Migration, Contagion and Boundary-Building in Shenzhen, China after SARS," *Medical Anthropology* 31, no. 2 (2012): 113–31.

9. On the reversal of urban-rural migration of professionals, see Mary Ann O'Donnell's conclusion in this volume.

10. See R. Packard, *The Making of a Tropical Disease* (Baltimore: Johns Hopkins University Press, 2007).

11. See L. Lee, "The Current State of Public Health in China," *Annual Review of Public Health* 25 (2004): 327–39; J. Peng et al., "Public Health in China: The Shanghai CDC Perspective," *American Journal of Public Health* 93, no. 12 (2003): 1991–93; Lu Y. and L. M. Li, "Woguo Jikong He Jiandu Tixi Zhineng Yu Xiandai Gonggong Weisheng Tixi Zhineng Neihan de Bijiao [Comparison of Our Country's CDC and Health Inspection System Functions and Current Public Health System Functions and Meanings]," *China Journal of Public Health Management* 22, no. 5 (2006): 365–67.

12. J. W. Lee and W. J. McKibbin, "Estimating the Global Economic Costs of SARS," in *Learning from SARS: Preparing for the Next Disease Outbreak: Workshop Summary*, ed. S. Knobler, A. Mahmoud, and S. Lemon (Washington, DC: National Academies Press, 2004).

13. E. Uretsky and A. Roess, "Can the Lessons of SARS Help Stop Ebola?" *National Interest*, October 30, 2014.

11

Simulating Global Mobility at Shenzhen "International" Airport

MAX HIRSH

Flush with cash and hubris, Shenzhen's urban ambitions had, by the early twenty-first century, gone global. With its status as a model for socioeconomic experimentation firmly entrenched within China, the city aimed to become, like Shanghai and Hong Kong, a "world-class" center for the international exchange of goods, services, and ideas. In the fields of architecture and urban design, Shenzhen's upwardly mobile aspirations were articulated in the city's active courtship of foreign architects, who were recruited to construct iconic public spaces and invest the city with sophisticated infrastructure systems. Articulating a deeply rooted sense of status anxiety, these show projects were a defining visual element of special border zones (SEZs) across Asia—a form of aesthetic globalization *avant la lettre* designed into an environment where, until very recently, nonlocal faces were met with goofy grins and inquisitive stares. And in the specific context of the Pearl River Delta, these projects represented an attempt to transcend Shenzhen's regional reputation as the scrappy upstart. Having neither the administrative clout of Guangzhou—the capital city of Guangdong and the center of power in Southern China for several millennia—nor the cosmopolitan flair and global financial weight of Hong Kong, Shenzhen sought to deploy architectural bling in order to improve its regional stature and international visibility.

The competition for the expansion of Shenzhen's Bao'an International Airport (SZIA) fit squarely within this paradigm. In 2007, the Shenzhen airport authority invited submissions from six foreign architects. The competition's brief called for the construction of a 400,000 m2 terminal that would double the capacity of the current airfield and would be capable of handling both domestic and international flights. Many of the competing architects, such as Norman Foster, Kisho Kurokawa, and Meinhard von Gerkan, had previous experience with large-scale aviation projects.[1] Yet the winning entry came from the office of the Italian designers Massimiliano and Doriana Fuksas, making SZIA Fuksas's first airport project, as well as their first foray into China.[2] For the jury members, composed of representatives from the airport authority and the planning bureau, technical expertise in airport planning was ultimately less important than architectural bravado—a spectacular design that would identify Shenzhen as a cutting-edge, cosmopolitan space of global connectivity.

Fuksas' striking terminal facade—a patterned glass and steel canopy studded with hexagonal windows that deploy the parametric design technology en vogue in architecture culture—unquestioningly signaled Shenzhen's recently acquired wealth and its desire to increase its cultural capital (see figure 21).[3] Yet the functional goals underlying the airport's expansion—that is, to better connect Shenzhen with the rest of the world—have been hampered by the politics of national security and intercity rivalries within the Pearl River Delta. China's military severely limits the operation of foreign airlines within Chinese airspace, and its civil aviation authority channels most international flights through three designated airports in Beijing, Shanghai, and Guangzhou.[4] At the same time, an agreement between the airport authorities in Hong Kong and Shenzhen has led to a rough division of labor whereby Hong Kong International Airport (HKIA) serves as the twin cities' international gateway—with frequent departures to more than seventy countries—while SZIA functions as a hub for cheap domestic flights within China. Offering cut-rate fares and nonstop service to dozens of Chinese provincial capitals and secondary cities, SZIA attracts both Hongkongers flying to the Mainland, as well as Mainlanders transiting to Hong Kong.[5] Despite the publicity surrounding the Fuksas terminal, inaugurated with much fanfare in 2013, the revamped airport thus offered surprisingly few international flights—an indication of the disconnect between the special border city's cosmopolitan ambitions and the requisite policy changes needed to turn Shenzhen into a global hub.

The recent influx of foreign visitors to Shenzhen, along with the much larger flows of passengers traveling between Hong Kong and the Mainland, has led to a rapid increase in the demand for international flights in Shenzhen and for China-bound ones in Hong Kong. Addressing the gap in supply and demand, the two airport authorities have sanctioned the development of "cross-boundary" airport bus terminals that allow passengers in Hong Kong to check in for flights departing from SZIA and enable travelers in Shenzhen to do the same for flights leaving from HKIA. In both cities, passengers check in for their flights at cross-border terminals located inside shopping malls and transportation hubs, as well as at hotels such as the Metropark in my neighborhood. They then board a bus to the border and, after clearing customs and immigration, transfer to a second coach that takes them to the airport. The companies that operate these routes have developed a flexible and loosely organized network of city-to-airport bus and minivan services that piggyback on existing, nonairport infrastructures in order to shuttle thousands of air travelers across the Hong Kong–Shenzhen border every day. In the process, the infrastructure of aerial mobility has been extended far beyond the confines of the two airports and inserted into the everyday urban fabric of Shenzhen.

This chapter traces the movement of air passengers as they travel from the airport check-in terminals in Shenzhen, cross the border into Hong Kong, and transfer to international flights at HKIA. In so doing, it posits the cross-boundary bus network as a useful lens for investigating the tension between Shenzhen's global aspirations on the one hand and the constraints entailed by its subordination to regional economic interests and national security concerns on the other. Moreover, the chapter contends that that tension has critically shaped Shenzhen's urban development. Investigating changes in the city's transport and border infrastructure since the turn of the twenty-first century, it argues that the cross-boundary check-in system has produced a variety of spatial and typological experiments that have fundamentally reordered the way in which Shenzhen is connected to the outside world.

Building Shenzhen, Building SZIA

Inaugurated in 1980, the Shenzhen Special Economic Zone was designed as both a laboratory and a model city: a laboratory for experimentation with market-driven modes of management and production and a model city for displaying the benefits of Deng Xiaoping's Reform and Opening policy, which aimed to transform the stagnant Chinese economy through an influx of foreign capital, expertise, and technical equipment. Much of that policy was predicated on engaging overseas Chinese investors in Hong Kong, Singapore, and Taiwan. Guangdong, with its proximity to Hong Kong and with family links to Cantonese-speaking diaspora communities scattered throughout East and Southeast Asia, played a substantial role in Deng's broader plans to jump-start the Chinese economy. As the historian Ezra Vogel notes, "If there was a single magic potion for a Chinese economic takeoff, it was Hong Kong. Roughly two-thirds of the direct investment in China between 1979 and 1995 came through . . . Hong Kong . . . Deng's experiment to open the 'great southern gate' between Guangdong and Hong Kong [became] China's most important channel through which flowed investment, technology, management skills, and ideas about the outside world."[6]

Planned as a linear city straddling the border to Hong Kong, Shenzhen was strategically sited at the foot of that "southern gate." Its intellectual genesis lay in the Chinese government's quiet study of SEZs elsewhere in Asia, which provided a model through which Deng and party officials in Guangdong sought to interact with foreign and overseas Chinese investors.[7] Together, they

> proposed that the entire province be allowed to implement a special policy that would give Guangdong the flexibility to adopt measures to attract foreign capital, technology, and management practices necessary to produce goods for export. China would supply the land, transport facilities, electricity, and labor needed by the factories, as well as the hotels, restaurants, housing, and other facilities needed by foreigners . . . The special policy for Guangdong and the unique leeway given to the special economic zones made these areas into incubators for developing people who would be able to function well in modern factories, stores, and offices in cosmopolitan settings.[8]

With its future inexorably tied to Hong Kong, Shenzhen's urban plan developed along a series of transport axes that ran perpendicular to the Hong Kong border. Elsewhere in this volume, Weiwen Huang, formerly one of the city's chief planners, periodizes the history of Shenzhen's spatial layout according to the establishment of the so-called Shen-Kong transport corridors—outfitted with imposing border control checkpoints—that were designed to move goods and people efficiently between the two cities.[9] At the same time,[10] a second boundary, likewise equipped with highway checkpoints, separated the SEZ from the rest of Guangdong Province, thus marking Shenzhen both socially and spatially as a zone of exception. In the 1980s, Shenzhen's first cross-border transport axis developed in the city's Luohu District at the intersection of the border and the main north-south railway that connected Hong Kong to Guangzhou. The thrust of urban development moved progressively westward as new highways, commuter rail lines, and long-distance bus terminals opened in Futian District in the 1990s. That trajectory continued into the twenty-first century, as the focus of urban development coalesced around road and rail corridors in Nanshan District that connected the airport in Hong Kong to its counterpart in Shenzhen (see figure 22).

The development of Shenzhen's Bao'an International Airport mirrored both the linear evolution of Shenzhen's urban fabric as well as the roll-out of Deng's economic reform policies. Until 1978, all Chinese airports were directly controlled by the People's Liberation Army. As part of the Reform and Opening policy, China moved toward a market-driven approach to air travel, formalized by the separation of the Civil Aviation Administration of China (CAAC) from the military in 1980.[11] The CAAC subsequently transferred the responsibility for airport operation to local governments, broke up the national airline into several smaller entities, and encouraged the development of private regional carriers to compete with state-run ones.[12] Shenzhen was one of the most aggressive early adopters of the liberalized aviation regime. In the 1980s, the closest airport to Shenzhen was in Guangzhou, a four-hour drive away along bumpy country roads. Believing that aerial connectivity was a precondition for accelerating the city's economic growth, the municipality began constructing its own airport in 1989 and established its own carrier, Shenzhen Airlines, three years later.[13] At that time, responsibility for building major urban infrastructure projects in Shenzhen was delegated to the state-run architectural design institutes and urban planning bureaus of China's thirty-odd provinces and provincial-level municipalities, each of which was obligated by the central government to contribute to Shenzhen's urban development.[14] Boxy and functional, and designed by low-profile employees of the Northeast Institute of Architectural Design based in China's Manchurian rust belt, the original terminal served a very small number of civil servants and employees of state-owned enterprises commuting between Shenzhen and their home provinces.[15] When the airport opened in 1992, urban planners in neighboring Hong Kong expressed skepticism about the city's potential to support a major air hub, citing the underdeveloped state of Shenzhen's economy and the low income levels of its inhabitants.[16] Yet by 2000, Bao'an had become the fourth-largest airport in Mainland China, handling more than six million travelers a year. That figure quadrupled in the following decade. By 2013, more than thirty million passengers flew in and out of Shenzhen, placing it on a par with Barcelona, Newark, and Sao Paolo.[17]

In 1996, China's civil aviation authority reclassified SZIA as an international airport, a formal prerequisite for handling flights from abroad, which mandates the installation of customs and immigration facilities. Yet despite its upgraded status, national aviation policy has consistently omitted the city from the elite triumvirate—Beijing, Shanghai, Guangzhou—that constitutes China's officially sanctioned gateways for international aviation.[18] The operation of international flights is further complicated by Bao'an's proximity to China's southern frontier, as the military does not allow commercial airplanes to fly across the border below five thousand meters.[19] As a consequence of these constraints, 96 percent of the flights that flew in and out of SZIA in 2014 were domestic. Of the eighty-two destinations served by SZIA, only eleven were located outside Mainland China, and the airport did not offer a single intercontinental flight.[20]

Prior to the opening of the Fuksas terminal in 2013, Shenzhen's inferior status within China's aviation hierarchy was rendered all too apparent by the airport's division into three separate buildings. The smallest was the International, Hong Kong, Macau, and Taiwan Terminal—the politically correct, if somewhat convoluted, term used for nondomestic facilities at Mainland Chinese airports. The building felt like a mock-up, or simulation, of an actual airport. Prominent wall clocks displayed the current time in Paris, Berlin, Cairo, Moscow, New Delhi, Tokyo, Sydney, Los Angeles, and New York; yet the tiny terminal was more Potemkin than real, accommodating only a handful of flights to Southeast Asian cities at much higher prices than comparable routes from Hong Kong. Deafeningly silent, the "international" terminal never operated at full capacity.

By contrast, a walk through the arrival halls of the two adjacent domestic terminals was tantamount to an aural and olfactorial obstacle course. Security guards strutted around a mass of meeters and greeters, taxi touts, and thousands of passengers arriving from Chinese cities both large (Beijing, Shanghai, Chengdu) and small (Hailaer, Yibin, Zhangjiajie). As passengers left the baggage claim area and moved toward the exit, they encountered a row of booths, similar to the ones occupied by the travel agents upstairs, that advertise cross-border bus and minivan services to HKIA. Administered by no-nonsense saleswomen in form-unflattering orange and magenta uniforms, the booths were operated by three competitors called Eternal East Cross-Border Coach Management, Trans-Island Chinalink, and Sinoway HK-China Express. Founded in the early 1990s by Hong Kong entrepreneurs to facilitate the growth in travel to and from Guangdong, these companies used their expertise in cross-boundary transport management as a springboard for developing transit connections to and from the region's airports.[21] These routes play a crucial role in the Pearl River Delta's aviation infrastructure. In 2013, five million people flying out of HKIA—about every tenth passenger—arrived at the airport on a cross-border bus originating in Guangdong Province.[22] The staff working at these booths directed passengers bound for HKIA toward the Shenzhen–Hong Kong Airports Link Passenger Lounge. Adorned with the twin logos of the Hong Kong and Shenzhen airport authorities, the entrance to the waiting lounge was staffed by teenage greeters wearing identical neon orange polo shirts. Inside, two clerks sat at a desk advertising "FREE one stop check-in service" for travelers departing from Hong Kong's Chek Lap Kok airport. In an adjoining room, small groups of passengers lolled about in oversized maroon armchairs, waiting to be transferred by bus to HKIA.

The Airport in the Urban Village

Easily overlooked by the casual passerby, the Shenzhen–Hong Kong Airports Link Passenger Lounge is the product of complex negotiations between airport authorities on both sides of the border. Since 2008, HKIA and SZIA have embarked on a path of careful cooperation, promoting themselves as complementary rather than in competition with one another and inviting travelers in Shenzhen who are flying out of Hong Kong to check in at SZIA.[23] Yet this strategy, dreamed up by airport planners, defies Shenzhen's underlying spatial logic. Shenzhen's airport is located on the city's northwestern haunch, a forty-five-minute drive from the skyscrapers of Luohu and Futian. Unless you happen to live near SZIA, or have just landed there from elsewhere in China, the airport's peripheral location makes it an inconvenient place to hop on a bus to HKIA. Recognizing this, in 2009 the cross-boundary bus companies began to develop a network of "in-town" check-in terminals that corresponded to the city's extensive, multipolar layout. Spread across the three central districts that border Hong Kong, these terminals were inserted into an unusual variety of urban spaces. In Futian District, the check-in terminal abuts the main entrance to the city's central bus station. Ten kilometers to the east, Shenzhen Grand Theatre, the city's largest performing arts center, houses an in-town check-in terminal between its box office and a shop selling watercolor prints and vases.[24] Finally, in Nanshan District, the Hong Kong International Airport Check-In Service is sandwiched between two of Shenzhen's quintessential typologies: a shopping mall and an urban village.[25]

Alternatively translated as "villages in the city," urban villages are the result of peculiar planning policies that guided Shenzhen's spatial development.[26] In 1980, more than three hundred villages existed on the territory that was designated as the future SEZ. Huang notes that "these spaces existed outside the city's master plan in the cracks between designated [development areas]. Local villagers began to develop this space independently, outside of the plan, without municipal administration such as planning, design and building approval, quality control, property registration, and any other regulatory procedure."[27]

Characterized by six- to eight-story buildings with extremely high plot ratios, warrens of narrow alleyways, and some of the cheapest rents in Shenzhen, urban villages attracted "low-income families, migrant workers, low-cost business, and entertainment service industries."[28] In so doing, the villages "supplemented what the urban plans had clearly overlooked: sufficient housing for low-income workers and recent arrivals."[29] Although they account for less than 10 percent of Shenzhen's total land area, in 2007 more than half of Shenzhen's inhabitants lived in urban villages, producing population densities exceeding seventy thousand people per square kilometer.[30] Many of these villages have become the object of land speculation due to their prime location within Shenzhen's inner districts. Some have been razed and entirely replaced by office towers and high-end gated communities, while others have been "progressively enclaved" by a ring of shopping malls and condominiums built on their periphery.

The latter condition accurately describes the immediate surroundings of the cross-border airport check-in terminal in Nanshan. Housed along a service road inside a multistory parking garage, the terminal is flanked on one side by an upscale shopping mall and condominium complex called the Kingkey Banner New Lifestyle Center and the other side by Baishizhou, a sprawling urban village that provides an affordable home to thousands of migrant workers and young college graduates who power the city's services and manufacturing industries (see figure 23).[31] Like other urban villages, Baishizhou totters precariously on the edge of Shenzhen's relentless social and spatial cleavages, its densely packed tenements encircled by office towers and condo developments that cater to the special border zone's affluent middle and upper classes. Messy, lively, and overcrowded, Baishizhou's streets are lined with small shops devoted to plastic kitchenware, cleaning supplies, and plumbing fixtures. Street vendors sell *jiaozi* and fish balls on wooden skewers in front of busy sheet metal workshops. At night, the road facing Kingkey Banner Plaza transforms into an outdoor eating and entertainment area filled with seafood restaurants, pool halls, and noodle stalls.

Baishizhou's southern frontier is demarcated by a manicured plaza and promenade dotted with palm trees and skyscraper apartment buildings. At its center, the Kingkey Banner New Lifestyle Center was part of a crop of upscale shopping centers that were built by real estate developers all over Shenzhen in the late 2000s, often on former village land. The Kingkey Banner Center offers the typical accoutrements of the Chinese middle class: cleaned-up versions of dim sum parlors, exhibitions of luxury cars, and domestic clothing chains posing as foreign brands. Situated in between the mall and the tenements, the airport terminal consists of two small rooms: a reception area with a row of check-in counters where passengers can print out a boarding pass and buy a ticket for the bus and minivan transfers to the airport and a small waiting lounge where they can check their flight status on giant flat-screens that display upcoming departures at HKIA. Every fifteen minutes, a small group of female employees, armed with clipboards, megaphones, and walkie-talkies, shepherds passengers down a flight of stairs to the parking garage, where idling buses and minivans wait to take them across the border.

The Airport at the Border

The border between Hong Kong and what later became Shenzhen was formally established in 1898, when Qing officials and their British counterparts established the Shenzhen River as the boundary between colonial Hong Kong and the Chinese Mainland. For much of its history, the border was porous and provided plenty of opportunities for the smuggling of both people and contraband. It continued to be permeable even after the establishment of the People's Republic of China in 1949, when hundreds of thousands of Chinese fled to Hong Kong. Stricter enforcement was not implemented until a mass exodus of Guangdong residents, escaping the famines induced by the Great Leap Forward in the early 1960s, obligated British officials to stem the flow of refugees.[32]

In 1997, the United Kingdom formally ceded its control over Hong Kong to the People's Republic of China, in what is commonly referred to as "the handover."[33] In the decades following the handover, the urban economies of Hong Kong and Shenzhen grew increasingly interdependent, as the governments of both cities initiated a series of regulatory changes and infrastructure projects designed to accelerate and simplify the movement of goods and people across the border. In an idealized future, the two cities will fuse together into one gigantic megalopolis, their physical integration emblematizing the harmonious rapprochement between Hong Kong and the Mainland.

Nevertheless, first-time visitors to the region are often surprised to find that the border between Hong Kong and Shenzhen is demarcated by a heavily patrolled no-man's-land lined with two parallel sets of fences and a string of guard towers (see figure 24). On the Hong Kong side, a rural "frontier closed area" that can only be accessed by special permission serves as an additional buffer zone between the two cities.[34] Hugging the southern edge of downtown Shenzhen, the border is undoubtedly the most vivid spatial manifestation of the One Country, Two Systems policy that structures Hong Kong's relationship with China. Under that policy, Hong Kong's status as a Special Administrative Region (SAR) guarantees it a high degree of autonomy in key areas of governance, including the ability to implement immigration and trade policies that are distinct from those on the Mainland. In the first five years after the handover, the fear of being overwhelmed by an influx of poor mainlanders led Hong Kong to severely restrict access to the city for PRC passport holders who could only enter the SAR as part of a registered tour group or if they could demonstrate that they were coming to conduct legitimate business transactions. Moreover, until 2002 Hong Kong enforced a quota system that limited the total number of Mainland visitors to two thousand per day. That changed in 2003 when—reacting to a sharp decline in tourism following the outbreak of the SARS epidemic—Hong Kong introduced an experimental individual visit scheme (IVS) that allowed Mainlanders who held an urban *hukou* from four midsized cities in Guangdong to enter Hong Kong for up to seven days.[35] After a successful trial period, IVS was granted to *hukou* holders from all cities in Guangdong, including Shenzhen, and beginning with Beijing and Shanghai in 2004, IVS was progressively extended to dozens of cities throughout China.[36]

The overarching motivation behind these policy changes was the growing purchasing power of PRC citizens. In effect, once a city or province achieved a respectable level of per capita GDP, Hong Kong's immigration authorities allowed its *hukou* holders to participate in IVS. And in 2008, Hong Kong began issuing one-year, multiple-entry visas to Shenzhen *hukou* holders as well as seven-day transit visas to any PRC citizen who could produce evidence of a plane ticket for a flight departing from HKIA. In essence, these measures represented vague proxies for wealth and status on the Mainland, with regulators seeking to profit from the growing affluence and free-wheeling spending habits of Chinese consumers while at the same time excluding poorer Mainlanders who were likely to place a strain on Hong Kong's welfare systems. These shifts in Hong Kong's migration regime radically increased the number of Mainland visitors, whose shopping and leisure predilections represented a crucial source of tourism dollars as well as a daily reminder of the inexorable social and cultural divide that separates Hong Kong from the Mainland. However, they did not much help the majority of Shenzhen's inhabitants, many of whom, as migrants, held a rural *hukou* and were thus effectively barred from entering Hong Kong. Rather, they exacerbated the inequalities embedded in Shenzhen's demographic composition, with less than one-third of the population who had a Shenzhen *hukou* enjoying not only better access to municipal services but also freer movement abroad. In response to complaints about the restrictiveness of its policies toward Shenzhen residents, in 2010 Hong Kong's immigration authority expanded IVS to employees of state-owned enterprises working in Shenzhen and, two years later, extended the scheme even more dramatically by granting entry permits to anyone who had been legally registered as a Shenzhen resident for at least one year.[37]

The relaxation of entry requirements for Mainlanders over the past decade required a substantial physical overhaul of Hong Kong's border infrastructure, as existing boundary control facilities became overwhelmed by the increase in cross-border flows that these policy changes entailed. Whereas at the turn of the twenty-first century only two thousand Mainlanders per day were permitted to enter Hong Kong via its land border with Shenzhen, a decade later more than half a million people were crossing between the two cities every day.[38] Hong Kong's transport and immigration authorities tried to reduce the pressure on Shenzhen's oldest checkpoints—Luohu and Huanggang—by constructing two new border crossings with streamlined immigration processes: Lok Ma Chau station, a terminal connected to Hong Kong's suburban KCR rail line that opened in 2003, and Shenzhen Bay Border Control Point, a facility that anchors the Shenzhen side of the Western Corridor Link, a cross-border suspension bridge that opened in 2007.

These changes in the border infrastructure had significant implications for the operation of the cross-boundary airport buses. While most traffic to HKIA and SZIA had been channeled through the decaying border crossing at Huanggang, from 2007 onward the airport bus companies gradually shifted their operations to Shenzhen Bay.[39] The new checkpoint had two distinct advantages. Located on the far western end of Shenzhen on an oblong parcel of reclaimed land at the southeastern tip of Shekou peninsula, Shenzhen Bay is closer to both SZIA and HKIA than any of the inner-city checkpoints. It was also the first "co-located" immigration and customs control point—a new type of border crossing, commissioned by Hong Kong's government-owned Architectural Services Department and designed by local engineering consultants, whose purpose was to expedite movement across the border by improving on the basic design of the twin cities' boundary control facilities.[40] Most border crossings between Hong Kong and Shenzhen consist of two parallel checkpoints, located several hundred meters from one another, where Mainland Chinese and Hong Kong immigration officers carry out two separate sets of customs, immigration, and quarantine (CIQ) inspections. By contrast, Shenzhen Bay combines, or "co-locates," Hong Kong and PRC formalities under one roof, a bureaucratic feat made possible through an elaborate process of jurisdictional and infrastructural gerrymandering. Though geographically in Shenzhen, the "co-located" facility is legally a part of Hong Kong and is jointly administered by immigration officers from both sides of the border.[41] On the Hong Kong–Shenzhen Western Corridor, a five-kilometer suspension bridge that connects Shenzhen Bay to Hong Kong, PRC law applies to the structural elements that anchor it to the seabed while the bridge's deck is a territorial extension of the Hong Kong SAR. Spurred by the growth in cross-border air passengers, Shenzhen Bay's daily traffic grew by 20 to 30 percent a year; in 2014, nearly 100,000 people passed through the "co-located" facility every day.[42]

Due to the special role that Shenzhen Bay assumed in facilitating the movement of passengers between HKIA and SZIA, the border crossing effectively became a functional extension of the two airports. Located roughly at the halfway point between them, the checkpoint is flanked on both sides by massive bus depots and smaller airport check-in lounges, built into the ground floor of the boundary control building, for those passengers who have not yet received a boarding pass. These companies maintain separate fleets on both sides of the frontier in order to avoid the logistical complexities involved in bringing a bus across the border. In Shenzhen, as everywhere on the Mainland, vehicles drive on the right; in Hong Kong, reflecting the city's colonial heritage, they do so on the left. Moreover, authorities in Hong Kong and Guangdong Province maintain separate vehicle licensing systems. A car registered in Hong Kong can thus only cross the border if the owner applies for a special cross-border license, denoted by a special black plate that is affixed to the front and back of the vehicle. Buses arriving from the in-town check-in terminals deposit passengers at the entrance to the five-story checkpoint, where they queue up at two sets of customs and immigration inspections. Depending on the time of day, the process takes between ten minutes and an hour. They then emerge on the other side of the checkpoint and board a second bus that takes them to HKIA.

For twice the price of a bus ticket, travelers can choose to ride in a seven-seat minivan, which has the distinct advantage of offering "on-vehicle CIQ clearance at Shenzhen Bay Port."[43] In effect, the differing price structure represents another proxy for passengers' wealth and status, as those who can pay for the more expensive van tickets are afforded both additional comfort as well as a less rigorous inspection. Rather than proceed to the border control building, the van drives through an imposing tollbooth-like structure where the driver passes a thick wad of passports to a Mainland Chinese immigration officer sitting inside a small cubicle. Using a special set of mirrors, the officer checks the faces of each of the van's occupants against their passport photo, occasionally asking the driver to pull up or reverse in order to get a clearer view. The officer then returns the passports to the driver and waves him through to his counterpart from the Hong Kong Immigration Department, who repeats the procedure at an identical booth several meters away.

The final stop for the Hong Kong–bound airport buses is Terminal 2 at HKIA. "T2" was designed by the corporate architecture firm Skidmore, Owings, and Merrill in partnership with the local consultant OTC. The terminal is unusual insofar as it features most of the typical design elements of an airport terminal—check-in desks, shopping and entertainment facilities, lounges, immigration and security controls—yet it has no departure gates and is not connected to the airfield. Inaugurated in 2008, the terminal's primary purpose was to accommodate the needs of so-called intermodal air passengers—people transiting by bus, ferry, or minivan to and from Mainland China. In the early 2000s, these travelers constituted only a small fraction of all passengers flying in and out of Hong Kong and could be processed through a few service counters located in the arrivals hall of Terminal 1. A decade later, however, the proportion of Mainland passengers had grown to more than a quarter of all travelers at HKIA.[44] Unable to accommodate that growth within the existing terminal, Hong Kong's airport authority shifted them into a new building designed explicitly for their needs. Terminal 2 is anchored by the Mainland Coach Terminus, a spacious expanse of cross-border bus transfer desks, a "Hong Kong Shenzhen Airports Link Passenger Lounge," and a bus depot with thirty-four departure bays.[45] Illuminated signs advertise property developments in Shenzhen, while television screens indicate the departure times of buses bound for destinations across Guangdong Province. Passengers arriving from Shenzhen walk through a tunnel to the main terminal at HKIA. Having already received a boarding pass in Shenzhen, they continue directly to security and immigration and then on to their international flights.

Conclusion

Western observers of Chinese urbanism tend to focus on large-scale infrastructure projects like the Fuksas terminal at SZIA. Their attention is infused with an element of "megastructure porn," a visual attraction to the sheer scale of what is being built in China, which readily ignores its structural flaws (such as the rapid senescence of shoddy building materials) and programmatic deficiencies (e.g., an international airport with few international flights). That fascination is likewise tinged with a deep nostalgia for a bygone era when Western governments, and the voting publics who elected them, demonstrated the political will to invest in major public works projects of their own.

From the perspective of the Chinese client, usually a municipality or a state-owned institution like an airport authority, megaprojects like the Fuksas terminal at SZIA are crucial for not only upgrading a city's basic infrastructure but also increasing their city's name recognition and cultural capital. Moreover, the timely completion of these "face" projects within the cycle of the five-year plan—which remains the basic temporal frame for China's economy—are essential for securing future promotions for the urban leaders involved. As central as these infrastructure projects may be to the design ambitions of the urban planner or to the career aspirations of the politician, it is questionable whether they are really at the forefront of urban innovation—or whether, like many things in Shenzhen, they merely represent a short cut to achieving global recognition through visually striking, yet functionally inadequate, architectural simulacra. In effect, the expansion of SZIA represented a highly formal approach to globalization devoid of the requisite content—an imitation of an international airport terminal that aestheticized, rather than actualized, Shenzhen's ambitions for cross-border connectivity. The parametric design of the Fuksas terminal, with its unabashed privileging of form over function, lent itself well to this pursuit, as it captured the arriviste's desire to show off newly acquired wealth as well as Chinese leaders' obsession with bombastic displays of technological novelty.

Meanwhile, the bulk of Shenzhen's international air traffic is channeled through a network of exceptionally banal check-in terminals and bus depots whose ultimate goal is nevertheless quite remarkable—namely, to compensate for the global connectivity that Shenzhen's megaprojects cannot provide. These facilities are not the subject of international design competitions; rather, they are engineered by local technical service providers and are articulated in an explicitly understated aesthetic register that aims to deflect attention from the larger, transnational objectives informing their development. Tucked away into the corners and interstitial spaces of much larger structures, the check-in terminals are characterized by very little signage and almost no advertising. While they host hundreds of passengers per day, they are built to process travelers as quickly as possible, checking them in and loading them onto Hong Kong–bound buses in less than fifteen minutes. Their barebones layout tersely acknowledges the indispensability of the cross-boundary bus system yet discourages a broader reflection upon the strange set of political and economic contradictions that led to its genesis.

Yet precisely such a reflection is necessary in order to understand the underlying processes that are guiding Shenzhen's urban development at the beginning of the twenty-first century. In effect, the cross-border check-in system spatializes the inherent conflicts between the structures of a planned national economy, predicated on rigid geographic hierarchies, and the exigencies of a multipolar global capitalism that fetishizes unfettered point-to-point mobility across national frontiers. For the time being, Shenzhen's contact to the outside world is being channeled via HKIA. This mediated form of global connectivity is designed in equal measure to assuage Hong Kong's existential anxiety about its declining role as a broker between China and the West as well as to placate the PRC's military and security apparatus, which is deeply reluctant to countenance any increase in cross-border activity over Guangdong's airspace. Thus as Shenzhen seeks to extend its influence beyond the Mainland, it finds its aspirations circumscribed by the dictates of the military and by its subordinate role to established centers of international trade like Hong Kong and Guangzhou, who are loathe to be bypassed by the upstart metropolis in their midst. Shenzhen exercises an extraordinary amount of autonomy within the confines of the special zone, but where its ambitions intersect with more powerful regional interests and national security concerns, it quickly becomes apparent that Shenzhen lacks the regulatory framework, and the political clout, to become a global city.

Economic elites in the Pearl River Delta have promoted investments in cross-boundary infrastructure as a means of achieving "win-win situations"—a diplomatic shorthand for the pragmatic compromises that are needed in order to circumvent the region's sociopolitical cleavages and intercity rivalries. Quite unintentionally, this strategy has become a source of inspiration for urban design, as the attempt to interpolate between Shenzhen's worldly ambitions and subaltern constraints has yielded a profusion of spatial and typological innovations. The cross-border check-in system removes international passengers from the remote confines of the airport and inserts them into a variety of inner-city locations, investing these distinctly local spaces with a global significance that explodes their typological definition. As the cross-boundary coaches channel passengers between the check-in terminals and the border, the flow of international air traffic along Shenzhen's streets and alleyways thus becomes a facet of everyday life in the city. In effect, the in-town check-in terminal represents the foil of the Fuksas project at Bao'an—aesthetically understated and deeply entrenched in the city's urban fabric but also remarkably effective at providing Shenzhen with an infrastructure of international air travel.

As efficient as the cross-boundary coach system may be, it is undeniably time consuming. Travelers are advised to arrive at the check-in terminal in Shenzhen three to four hours before their flight is scheduled to depart from Hong Kong. They then board a bus to the border, haul themselves and their luggage through two sets of customs and immigration controls, board another bus to HKIA, and proceed yet again through security and immigration. If the lines at the border are short and the buses leave promptly, the seventy-five-kilometer trip takes a little more than an hour. If there are hiccups along the way, the trip can be much longer. For a metropolis of fifteen million inhabitants, with rising levels of income and an attendant eagerness to travel abroad, this is clearly not an ideal solution. Borne of the three traits that pervade urban development in the Pearl River Delta—pragmatism, improvisation, and entrepreneurialism—the cross-border check-in system points to the unavoidable conflict between those who would like to increase Shenzhen's global connectivity and those who would like to arrest it.

In *The Anxieties of Mobility: Migration and Tourism in the Indonesian Borderlands*, the anthropologist Johan Lindquist studies the movement of migrant workers and working-class tourists on Batam, a Special Economic Zone located off the coast of Singapore in Indonesia's Riau Archipelago. Focused on an island "located at the periphery of a global city," Lindquist's book seeks to "consider the social organization of human mobility from the vantage point of the spatial and temporal border between the 'developing' and the 'developed' worlds."[46] In so doing, he creates "an alternative topography of Southeast Asia ... that allows us to understand Batam not strictly as a place that is 'offshore' in relation to Singapore ... but rather as a node in a system of human mobility that is territorially and culturally unbounded."[47] Lindquist's book serves as a useful model for probing the distinctive agency embedded in special border cities such as Shenzhen. While they originate as an economic appendage to established centers of wealth and power, over time these special zones take on a dynamic of their own and seek to assert autonomy from their more dominant twin city by engendering multidirectional mobility flows across the border. This phenomenon can be observed not just in Shenzhen but in special border cities throughout Asia.[48]

Yet Shenzhen is unique among Asia's special zones insofar as it has already surpassed its conjoined twin in many respects. By the first decade of the twenty-first century, its population was more than twice that of Hong Kong's; and with a median age of just thirty-three years, its residents were, on average, a decade younger than those across the border and more likely to be gainfully employed.[49] From an architect's perspective, the continued availability of both open space and open-minded clients made the prospect of building and designing in Shenzhen an appealing one. And while Hong Kong's residents were, on the whole, much wealthier, more educated, and cosmopolitan, it remained a city whose education system and stratified social hierarchies offered relatively few opportunities for upward mobility. For even the most grossly underpaid migrant worker, moving to Shenzhen afforded a dramatic step up in terms of economic status and prestige. And for a growing number of middle-class Hong Kongers, relocating to Shenzhen presented itself as a viable and even attractive alternative to Hong Kong's extremely confined living conditions.

Thus Shenzhen finds itself at an odd moment of urban adolescence: still subordinated to wealthier regional centers of power like Hong Kong and Guangzhou, yet faced with the prospect of greater autonomy in the not-too-distant future. How Shenzhen renegotiates its relationship with Hong Kong is instructive not only in terms of its role at the forefront of Mainland China's reconciliation with Hong Kong but also insofar as it charts an evolutionary path for less developed special border cities elsewhere to follow. Until now, these zones' self-definition, as well as their perception from outside, was largely determined by practical exigencies: this city exists as a cheap place to produce t-shirts or assemble computer chips. Residents, too, saw their sojourn in the zone as a means to an end, connected to aspirations for social and material improvements in hometowns located far from their place of employment. Over time, however, these zones evolve, and its predominantly migrant population acquire a modicum of civic pride, along with more bourgeois expectations about life quality, that demands a certain normalization of the "special" border city. In so doing, however, they discover a peculiar contradiction in the mobility regimes that govern special zones. On the one hand, these cities owe their very existence to the privileging of movement—of goods, of people, of foreign capital—and their spatial layout revolves invariably around infrastructures of mobility (train stations, highways, air- and seaports) that connect them to the wider world. At the same time, residents of special zones find their cross-border mobility circumscribed by the vested interests of their interreferenced twin—established entrepôts such as Hong Kong, who expend much energy in order to ensure that they will not be bypassed by their scrappy frontier-town double across the border. Unfettered mobility in border cities is likewise viewed negatively by the very national governments who created them, as they are unwilling to grant further autonomy to these special zones for fear that they will challenge the highly centralized power hierarchies, concentrated in capital cities like Beijing, that are endemic to the Asian nation-state.

The conflict between mobility and immobility can be seen most palpably in Shenzhen's continuous and simultaneous upgrading of its border control facilities and its cross-border transportation systems. That conflict is likewise echoed in the spatial tactics of Shenzhen's inhabitants, producing a nonchalant attitude toward the region's complex and at times opaque regimes of cross-border movement. That blasé approach reflects what the historian Willem van Schendel has called the "everyday transnationality" common to people living in border cities who "continually explore and challenge the territorial pretensions" of neighboring states.[50] In the case of Hong Kong and Shenzhen, the pragmatic attitude regarding where and when one flies out of which airport—if you're heading to Taipei or Bangkok, choose HKIA; for Kunming or Zhengzhou, better to pick Shenzhen—reflects a broader attitude of comparison and contrast between systems, cities, and currencies that is interwoven into quotidian routines in special border cities across Asia.

The writer Katha Pollitt has remarked upon the "strange internal logic" of a dream: the irrational chains of cause and effect that seem wholly plausible until the moment of awakening, when you realize the spurious nature of the causalities that your brain concocted while you were asleep.[51] Shenzhen, too, has its own peculiar internal logic, as its organizing principles depend on creative rationalizations in order to make dissonant social, economic, and judicial regimes cohere with one another. In the overarching absence of historical precedents, those who dwell in Shenzhen—a city that cannot be accurately described as either a cosmopolitan metropolis or a provincial hinterland, existing instead in taxonomic limbo—do not perceive anything unusual about a $1.4 billion dollar international airport terminal that has almost no international flights, or about checking in for a trip to New York in a bus terminal located inside an urban village.[52] Rather, the contrast between highly representative, yet ineffectual, spaces of global connectivity and hastily erected transport nodes that ingeniously evade regulatory restrictions on cross-border mobility testifies to the compensatory design strategies devised by a municipality that is trying to transcend its hinterland role and interact directly with the outside world, but cannot do so just yet.

Notes

1. "Shenzhen International Airport Terminal 3," *Dezeen*, April 12, 2008; Andrew Yang, "Fuksas Wins Shenzhen Airport Competition," *Architectural Record*, June 10, 2008.

2. Fang Tian (project manager, Fuksas), interview with the author at Shenzhen Bao'an International Airport, March 7, 2012.

3. Fuksas developed the terminal facade in collaboration with the German engineering firm Knippers Helbig, which specializes in parametric design. See Florian Scheible and Milos Dimcic, "Parametric Engineering: Everything is Possible" (conference paper delivered at annual symposium of International Association for Shell and Spatial Structures, September 2011), http://www.programmingarchitecture.com/publications/ScheibleDimcic_IASS_2011.pdf.

4. See Yang Xiuyun and Yu Hong, "Deregulatory Reform of China's Airports: Attracting Non-State Investors," *East Asian Policy* 2, no. 2 (April 2010): 62–72. See also Zhang Anming and Chen Hongmin, "Evolution of China's Air Transport Development and Policy towards International Liberalization," *Transportation Journal* (Spring 2003): 31–49.

5. This arrangement was formalized in October 2008, when HKIA and SZIA signed an agreement that "enables passengers at HKIA or SZIA to check in and obtain boarding passes for connecting flights at either airport." "Key Dates and Events," *Hong Kong International Airport*, http://www.hongkongairport.com/eng/media/key-dates-events.html. See also Jianfa Shen, "Assessing Inter-City Relations between Hong Kong and Shenzhen: The Case of Airport Competition or Cooperation," *Progress in Planning* 73 (2010): 55–73. On average, flights to Mainland Chinese destinations that depart from Shenzhen are 30 to 40 percent cheaper than those that leave from Hong Kong. Moreover, SZIA offers much more frequent departures to major hubs like Beijing, Shanghai, and Chengdu, as well as nonstop service to dozens of provincial capitals and secondary cities.

6. Ezra F. Vogel, *Deng Xiaoping and the Transformation of China* (Cambridge, MA: Harvard University Press, 2011), 403, 404, 406.

7. Vogel, *Deng Xiaoping*, 399.

8. Vogel, *Deng Xiaoping*, 398, 411.

9. Weiwen Huang (director, Shenzhen Center for Design), interview with the author at Shenzhen Center for Design, March 7, 2012. Huang also notes that "these sites of integration with Hong Kong transportation and logistic networks stretched between Mainland China and

Hong Kong, informing the development of each stage of Shenzhen's development as well as the conditions through which land use changed. To date, however, there has been insufficient research from this perspective on the development of the city." See Huang, chapter 3 in this volume.

10. Ma and Blackwell discuss this in chapter 6 in this volume.

11. Shih-Lung Shaw et al., "China's Airline Consolidation and Its Effects on Domestic Airline Networks and Competition," *Journal of Transport Geography* 17 (2009): 293–305.

12. Civil Aviation Administration of China, "Quan guoyou duzi dao chanquan duoyuanhua: Minhang xiandai qiye zhidu zhubu jianli wanshan [From Full State Ownership to Pluralization of Property Rights: The Modern Civil Aviation Administrative Regulations are Progressively Perfected]" (press release, December 16, 2008), http://www.caac.gov.cn/D1/30years/200812/t20081216_20823.html.

13. See Shaw et al., "China's Airline Consolidation."

14. Weiwen Huang, interview.

15. Weiwen Huang, interview.

16. Writing in 1991, the authors of the master plan for Hong Kong's new airport argued that the construction of Shenzhen's airport would have little effect on demand at HKIA and would serve chiefly as an "overflow facility" for Guangzhou's Baiyun airport. Moreover, they projected that traffic at Bao'an would amount to no more than 3.4 million annual passengers by the year 2000. Greiner International and Maunsell Consultants Asia, *New Airport Master Plan: Final Report; Planning* (Hong Kong: Provisional Airport Authority, 1991), 3–2.

17. Center for Asia Pacific Aviation, "Shenzhen Reports 6% Traffic Growth in 2011," January 12, 2012, http://www.centreforaviation.com/news/shenzhen-airport-reports-6-traffic-growth-in-2011-136349. Author's own calculations.

18. This position was reaffirmed most recently by the twelfth five-year plan adopted in 2011. See Xin Dingding. "Airspace Management Reform Urged," *China Daily*, October 30, 2012.

19. According to the One Country Two Systems Research Institute, "Study of Hong Kong's Aviation Industry: Current Challenges and Future Strategies" (Central Policy Unit of the Government of the Hong Kong Special Administrative Region, September 2010), 62:

> The airspace of the PRD region is very complicated. Three different air traffic management authorities oversee flight movements in the area, using different operation procedures and standards. All are subordinate to the Chinese military in terms of rights of access to the region's air space . . . In addition, an 'invisible wall' exists between Hong Kong and Macau on the one hand, and the Mainland on the other, the result of three separate aviation information zones. When aircraft leave HKIA airspace, they are required to reach a specific altitude before they are allowed to climb 'over the wall' into Chinese Mainland airspace. This results in extra flight time and fuel consumption, as well as logistical complications for air traffic controllers.

20. Author's own calculations based on data provided by Shenzhen Airport Group. See also Keith Wallis, "HK Airport Outpaces Shenzhen," *South China Morning Post*, January 26, 2012. An article in the British *Independent*, written on the occasion of T3's inauguration, noted that "Shenzhen airport's website boasts a wide range of international destinations, including Dubai, Cologne, and Sydney. But a trawl of sources, including airline websites and the aviation data specialist OAG, failed to find any services to or from these airports. Existing links seem to be purely regional, to cities such as Bangkok, Kuala Lumpur, and Singapore." Simon Calder, "Shenzhen Airport Terminal Three: Vast, Shiny, New—and Empty?" *The Independent*, November 28, 2013.

SIMULATING MOBILITY AT SHENZHEN AIRPORT   247

21. Trans-Island Chinalink Ltd., "Cross-Border Transportation Service of Trans-Island Chinalink," accessed November 1, 2012, http://www.trans-island.com.hk/eng/china.html; Eternal East Cross-Border Coach Management Ltd., accessed November 1, 2012, http://www.eebus.com/combg.asp.

22. Airport Authority Hong Kong, *Together We Make the Future: Annual Report 2011/2012*, 43. See also Huang Fang, "Bao'an guoji jichang yingsong luke 70 wan [Bao'an International Airport Transports 700,000 Travelers]," *Baoan Daily*, October 8, 2013.

23. Similarly, passengers in Hong Kong can check in for flights leaving from Shenzhen at HKIA. Airport Authority Hong Kong, "HKIA, SZIA Jointly Launch Hong Kong-Shenzhen Airport Links" (press release, October 8, 2008).

24. Site visit to Futian Coach Terminus, Shenzhen, March 8, 2012; Site visit to Shenzhen Grand Theatre, Shenzhen, November 29, 2012; Site visits to Kingkey Banner Center, Shenzhen, December 1, 2009, February 22, 2010, and March 8, 2012.

25. Eternal East Cross-Border Coach Management Ltd., *EEBus Service 2009 Guide Book* (pamphlet, November 2009), 4.

26. For more on Shenzhen's urban villages, see Wang Ya Ping, Wang Yanglin, and Wu Jiansheng, "Urbanization and Informal Development in China: Urban Villages in Shenzhen," *International Journal of Urban and Regional Research* 33, no. 4 (December 2009): 957–73.

27. See Huang, chapter 3 in this volume.

28. See Huang, chapter 3 in this volume.

29. See Huang, chapter 3 in this volume.

30. See Huang, chapter 3 in this volume.

31. As Mary Ann O'Donnell notes, the area known as Baishizhou emerged out of a collective farm and five adjoining villages. See Mary Ann O'Donnell, "Of Shahe and OCT," *Shenzhen Noted*, January 31, 2010, http://maryannodonnell.wordpress.com/2010/01/31/tangtou-baishizhou.

32. See Shun-Fai Tsang, "Border Control in Colonial Hong Kong, 1958–1962" (undergraduate thesis, University of Hong Kong, 2010).

33. For an in-depth account of the handover, see John Carroll, *A Concise History of Hong Kong* (Lanham, MD: Rowman & Littlefield, 2007).

34. For more on the history of the Hong Kong–Shenzhen border, and on Hong Kong's immigration regulations, see Y. M. Yeung and Shen Jianfa, eds., *The Pan-Pearl River Delta: An Emerging Regional Economy in a Globalizing China* (Hong Kong: Chinese University of Hong Kong Press, 2008).

35. Commerce and Economic Development Bureau of the Government of the Hong Kong Special Administrative Region, accessed May 19, 2013, http://www.tourism.gov.hk/sc_chi/visitors/visitors_ind.html. As noted in the previous chapter, China's *hukou*, or household registration system, defines how often Mainland citizens can travel outside China according to their place of birth.

36. A complete list is available at http://www.tourism.gov.hk/english/visitors/visitors_ind.html.

37. "Shenzhen Extends Individual Visit Scheme," China Internet Information Center, accessed May 12, 2013, http://www.china.org.cn/travel/2011-01/05/content_21674188.htm.

38. Government of the Hong Kong Special Administrative Region, "Annex to Legislative Council Question 16: Number of Passengers Using Various Boundary Control Points from January to September 2011" (press release, September 11, 2012).

39. Site visits to Shenzhen Bay Port Border Control Point, Shenzhen, March 5, 2010, April 3, 2010, and January 26, 2011.

40. The Shenzhen Bay border control facility was designed and constructed by JRP, a professional engineering consultancy based in Hong Kong. See J. Roger Preston Ltd., "Hong Kong Shenzhen Western Corridor, Co-location of Boundary Crossing Facilities," accessed November 15, 2012, http://www.jrp-group.com.

41. Government of the Hong Kong Special Administrative Region, "Shenzhen Bay Port Hong Kong Port Area Ordinance" (ordinance no. 4 of 2007, April 26, 2007). For more on the design of Shenzhen Bay, see Max Hirsh and Jonathan D. Solomon, "Does Your Mall Have an Airport?," *Log* 19 (Summer 2010): 99–106.

42. Government of the Hong Kong Special Administrative Region, "Legislative Council Question 16: Utilisation of Boundary Control Points and the SkyPier" (press release, September 11, 2012).

43. Trans-Island Limousine Ltd., "HKIA Guangdong Deluxe Cross Border Limousine" (brochure, n.d., retrieved November 2009).

44. According to Victor Fung, the former CEO of HKIA, the airport authority built Terminal 2 in response to an increase in the number of passengers traveling to and from Mainland China: "Terminal 2 . . . isn't really meant as an airport terminal, but rather as a transport hub for land and sea passengers who are coming to the airport." Victor Fung (chief executive officer, Li & Fung), interview with the author at Alexandra House, Hong Kong, February 11, 2010.

45. Site visits to Mainland Coach Terminus, Hong Kong International Airport Terminal 2, June 29, 2009, February 22, 2010, December 10, 2011, and November 28, 2012.

46. Johan Lindquist, *The Anxieties of Mobility: Migration and Tourism in the Indonesian Borderlands* (Honolulu: University of Hawaii Press, 2008), 17.

47. Lindquist, *Anxieties of Mobility*, 8, 12.

48. At the southern tip of Malaysia, for example, the border city of Johor Bahru has labored to shed its image as a supplier of cheap groceries and illicit activities for day-tripping Singaporeans, investing instead in middle-class housing developments and institutions of higher education in an attempt to attract a more cosmopolitan population of foreign students, young families, and retirees. Zhuhai, the special border city adjacent to Macau, has repositioned itself as a bucolic retirement destination, touting itself as the "Chinese Florida." Similar to the case of Shenzhen, which Mainland Chinese tourists fly through in order to get to Hong Kong, Batam is developing a new identity as a point of transit for Indonesia's emerging middle class, who fly to the island's Hang Nadim International Airport and then continue by ferry to spend their holidays in Singapore. Dendi Gustinandar (commercial director, Hang Nadim International Airport, Batam, Indonesia), interview with the author at Hang Nadim International Airport, March 11, 2013.

49. Government of the Hong Kong Special Administrative Region, Census and Statistics Department, *Hong Kong: The Facts*, June 2014.

50. Willem van Schendel, "Spaces of Engagement: How Borderlands, Illegal Flows, and Territorial States Interlock" in Itty Abraham and Willem van Schendel, eds., *Illicit Flows and Criminal Things: States, Borders, and the Other Side of Globalization* (Bloomington: Indiana University Press, 2005), 62. Van Schendel argues that this attitude is a common facet of daily life, and a necessary survival strategy, throughout Asia's border cities:

> Inhabitants of borderlands share . . . an uneasiness about dominant conceptions of spatial reality. Their lived experience makes it impossible for them to accept as given, and unproblematic, the contemporary organization of the world as defined by state elites . . . They cannot restrict their imagination to the territory of a single state, and they see those who do so as imprisoned in a delusion . . . Around the world, inhabitants of

border regions have developed practices and worldviews that take account of the state but never as an undisputed, overarching entity. [They] scale their world in ways that do not coincide with state borders [and that] spill over the spatial limits set by the state's territory. (Abraham and van Schendel, 54, 56)

51. Katha Pollitt, "Learning to Drive," *New Yorker*, July 22, 2002, 36.

52. Vlad Savov, "Shenzhen's New Airport Terminal Puts the Wonder Back into Flight," *Verge*, November 28, 2013, http://www.theverge.com/2013/11/28/5154484/baoan-international-airport-terminal-3-studio-fuksis-design.

CONCLUSION

Learning from Shenzhen

MARY ANN O'DONNELL, WINNIE WONG,
AND JONATHAN BACH

Shenzhen is as improbable as it has been successful. Its origin in the logics of policy experimentation and model making continue to define the city, even as the model subject shifts from political leaders to migrant workers to technocrats and artists, and even as its model spaces have been continually remade. Shenzhen is therefore itself a shifting subject, refusing easy incorporation into metanarratives of economic development, urban planning, or neoliberalism. The authors in this volume have collectively uncovered processes and moments throughout the history of Shenzhen that contest linear developments from plan to policy to execution. Whichever grand narrative one chooses, we find that the crucial moments happened much earlier or later than would be expected, or did not happen in a way that accorded with either plan or theory, or that held a very different significance from a highly local perspective than as it might for an outside observer with hindsight. The empirical investigations in the preceding chapters present a history of Shenzhen as important not only because the city is synonymous with cutting-edge changes in China but also because Shenzhen seriously complicates how those changes can be theorized for China specifically and for contemporary urban transformations generally. In the transformation of rural Bao'an into metropolitan Shenzhen, for example, urban planning has succeeded to the extent that it allows for flexible responses and situated appropriations—zones, new districts, and urban villages all exemplify the economic effects of the political decision to promote and experiment with and within exceptional spaces.

CONCLUSION

Territorial Exceptionalism

The many stories that this volume has told about the urban, about China, and about structural changes at the level of the world system coalesce around China's effective, if not always predictable, use of territorial exceptionalism as a strategy for managing the socioeconomic changes that come with the market. The story of Shenzhen's external and internal borders related here by Ma and Blackwell, Shenzhen's clumsy attempts to configure its global space for international travel explored by Hirsh, and the fluid borders between urban and rural that appear in O'Donnell's and Bach's analyses of urban villages are all manifestations of Shenzhen's active restructuring of the relationship between territory, politics, and economy by way of spaces that either explicitly license or tacitly condone special privileges therein.

Each successive iteration of the modern world system introduces a new logic of territory as the locus of political and economic experimentation and the new spatial forms that condense and consolidate these experiments.[1] In our era, the trade zone reorganizes both global and national logics of territory and identity by allowing for exceptions to national law within bounded territories. More than thirty years after its establishment, the Special Economic Zone (SEZ) as innovated in Shenzhen represents an evolving strategy of territorial exceptionalism now seen around the world where not only trade laws but the form of the city itself are up for grabs.[2]

The exceptional activities that define free-trade or foreign-trade zones allow for goods to be landed, handled, manufactured, processed, and reexported with minimal intervention by custom authorities. Within Shenzhen, there are a variety of free trade zones. However, what made Shenzhen the world's iconic SEZ is its history as the place in the immediate post-Mao era where the Chinese planned economy—a centralized system that integrated economics and politics through rural collectives and urban work units—could be experimentally reformulated as a market economy, even though no one knew what that might mean at the time. By being the locus for these experiments, Shenzhen became essential to post-Mao structural change and all that followed, including new understandings of society and the goals of political life. In other words, in contrast to early free trade zones elsewhere, the Shenzhen Special Economic Zone became the prime example of how zones can afford the possibility for far-reaching social transformations by being as much a spatial solution to a political problem as it is an economic innovation.

The chapters in this book illustrate how a key feature of Shenzhen's territorial exceptionalism, in both its national and global dimensions, has been the coevoluntionary development of planned and unplanned spaces, contingent subjects, and the metamorphosis of socialist era ideals and values. In the officially sanctioned exceptional spaces of Shenzhen, the Chinese government could selectively experiment with dismantling the planned economy and reconstituting it as an urban capitalist economy under party control, while generating competitive economic and symbolic advantages in the global economy. At the same time, the adjacent urban villages functioned as unofficially tolerated exceptions to the exception and provided the unacknowledged preconditions for experimentation as O'Donnell, Bach, Mason, and Wong have shown in different ways through their chapters.

The urban villages appear as both the problem and the solution to rapid urbanization and have acted consistently as a catalyst for much of Shenzhen's experimentation, intertwining their "informal" urbanism with "formal" planned spaces in the making of Shenzhen's polycentric urban geography. Ironically, these original settlements in the counties that became Shenzhen also became the city's negative spaces, often left literally as blank spaces in master plans. This anomalous situation invited experimentation, turning the villages into structural holes that allowed the city's experiments to take on new dimensions. For example, many factory workers in the 1980s Shekou District lived in village handshake rentals in neighboring Shuiwan and Wanxia New Villages, while the young architects and scientists of Overseas Chinese Town (OCT) and the Science and Technology Park continue to live in the neighboring village of Baishizhou. One of the most commercially vibrant districts in Shenzhen, Luohu is a tight checkerboard of planned and unplanned spaces that abut and spill into each other, simultaneously accommodating state-owned industries, international manufacturing, piecework manufacturing, and cross-border shopping sprees for both shanzhai and mutated innovations. At the same time, these nested territorial exceptionalisms also took the old Maoist division of Chinese society into urban and rural territories and reconstituted it as a new spatial division of planned and unplanned experimental spaces within the city.

The chapters in this book reflect how each urban village's story is also about the process of disembedding China's economy from the Maoist era via the process of turning land into a commodity. Karl Polanyi famously detailed the fraught, difficult, and even deadly negotiations between the English state and its subjects when the state sought to enclose common land as private property, to establish wage labor, and to turn money into a financial product. Land, for Polanyi, is one of the three key commodities alongside labor and capital that he calls "fictitious" because they are not produced for sale in the market. Turning these *into* commodities requires negotiation between the state and the actors for whom these played different social roles prior to capitalism. In China, the

CONCLUSION 253

ongoing alienation of peasants from the land requires their disenfranchisement from collectivist status in the state apparatus. In this process, the market appears as increasingly autonomous in the guise of a priori laws of market rationality, though upon closer examination, any claim to autonomy is, as Polanyi argues, only a new type of reembedding, or as we see in Shenzhen a recombining, of the social and the economic. The creative process of recombining is perhaps an even more interesting story than the disruptive process of disembedding, for each form of recombination affords different opportunities to construct society and its environs, as shown in O'Donnell and Bach's respective discussions about the ongoing tension between city expropriation and local appropriation of villages and their histories.[3]

Exceptional Migrants

Shenzhen has spearheaded the changing dynamics of the urban-rural relationship in China, due both to its early role in Reform and Opening and its particular approach to urban space. The structure of its urban villages shaped Shenzhen's ability to accommodate and integrate migrants who flocked to the city from its earliest days. Indeed, if this book has a central protagonist, it is the figure of the migrant worker as an ambiguous subjectivity, one that can simultaneously accommodate subjugation and individual transformation. This well-known figure of contemporary China took its form in Shenzhen first rather than in established cities such as Beijing and Shanghai, because the whole city was an experiment. Urban villages in older cities first evolved as tightly packed extensions of hometown societies that excluded rural migrants from the forms of urban citizenship and belonging that accrue to the native city dwellers.[4] Unlike native Beijing or Shanghai residents, however, native Shenzheners—formerly Bao'an peasants—were the most likely to be excluded from a city identity because they exist in an uneasy space between the city residents, themselves white-collar migrants, and the migrant workers, who in Shenzhen have ironically been recast as the urban with respect to the local rural.

As O'Donnell explained in chapter 5, and Mason in chapter 10, Shenzhen's exceptionalism opened a space for migrants to come in large numbers from other cities to provide professional labor. Migration appears as the precondition of belonging to the city rather than something that prevents belonging (as captured in the hackneyed phrase "come and you are a Shenzhener"). Indeed, while urban villages exist throughout China, in Shenzhen the residents of these spaces are not only young, single, rural migrants and factory workers but also college graduates and families comprising three generations.

Moreover, the city's decision to renovate centrally located urban villages has strengthened Shenzheners' awareness of the importance of urban villages for migration to Shenzhen as well as their centrality to identification with the city. This awareness builds on the attempts from the mid-1990s, as detailed in Florence's chapter, to construct a narrative of rural migrant workers that fits within the dominant imaginary for Shenzhen.

The result is Shenzhen as a space where both workers and villages come together as complimentary subjects of model making—as model workers and model villages. Bach's ethnography of entrepreneurial villagers, Dong and Cheng's portrayals of migrant sex workers, Mason's study of public health professionals, and Wong's analysis of gender and artist-workers in Dafen village all show how the urban village appears as space for the forging and contesting of urban identities. Each urban village story highlights not only a municipal work-around for problems that the city cannot, or will not, address but also a national level experiment in determining who is the subject of the urban, as expressed through ongoing reforms about how to allow "desirable" migrants to achieve legal urban residency. The question of who is the subject of (and subject to) the urban, is the flipside of when and how officially "agricultural" land can become urbanized. In this context, the chapters collected here have shown how municipal, developer, villager, and migrant identities and interests diverge despite moments of accommodation, leaving urban identity a subject in formation rather than converging into one model type of Shenzhen person.

No Chinese city has arguably benefited from internal migration to the same extent as Shenzhen, resulting in the city's heightened attention to migrants' subjectivity and their legality, as made evident through the cases of the model worker described by Florence and the stigmatized female migrant sex workers described by Dong and Cheng. This provides a model for more established cities that historically lacked the same flexibility to accommodate and integrate different levels of migrants, lumping them all together in a more general and debased class of "rural migrants" but who now are also looking for ways to manage the unprecedented levels of rural to urban and urban to urban migration across China.

Modeling the Future?

The constant reconfiguration of the Shenzhen subject that we see in Florence, Mason, Dong and Cheng, and Wong's chapters thus must be understood through its coevolution with the physical configuration of the city, as explored in chapters by O'Donnell, Huang, Ma and Blackwell, Bach, and Hirsch. In tracing the relation between subject formation and city making, we see a shift from experiment to model that is predicated on an accompanying rhetorical move that justifies the model ex post facto, as if the city's contingent origins had never existed. In recent years, the party apparatus has backed away from early descriptions of "reform and opening" (*gaige kaifang*) as the kind of open-ended experimentation captured in part I. Instead, the party increasingly claims that the measurable successes of Shenzhen have been the result of technocratic planning, with the ironies visible in Mason, Wong, and Hirsh's chapters. The contingency of the Special Economic Zone's early years becomes elided in self-congratulatory rhetoric of planned urbanization. In turn, these ex post facto plans become the basis for new forms of common background knowledge, against which new plans can be proposed.[5]

What is the future of Shenzhen's storied history of model making as the city and country increasingly deploy it to justify and inform urban planning and urbanization? For instrumental reasons alone, there will be pressure for continued model making in Shenzhen because it is a strategy for political mobility, despite the fact that experimentation remains the condition for viability and success. Receiving the designation of a "model" allows for an advance of rank within the municipal apparatus and at the national level, as with the recent rise of Shenzhen's exemplary Qianhai District (which includes Shekou) to the national designation of self-governing trade zones (*zi mao qu*). Indeed, many Shenzhen businessmen and officials speak of seeking ways to create "national value" (*guojia jiazhi*) that will bootstrap sections of the Shenzhen economy into higher levels of national integration via ever-exceptional spaces. This serves as a reminder that Shenzhen itself is part of a national system in which other cities are also competing for the benefits brought by the strategic use of exceptions throughout the global economy. These spatialized exceptions are both the motor of economic growth and part of a naturalized system of inequality precisely because in China, economic inequality is usually interpreted as a result of political decisions.

In thinking about the future, and as way of closing, we draw attention to three levels at which Shenzhen continues to function as a strategically ambiguous space for experimentation and ex post facto model making. First, at the level of city formation, the story of urbanized villages continue to embody the potential to create exceptional spaces in response to specific situations. Although the space of the villages is becoming increasingly formalized, with smaller and smaller spaces for flexible response, news of their demise has proven premature many times before. Shenzhen's urbanized villages will continue to demonstrate a responsiveness to Shenzhen's future development, as they have done in the many cases explored in this volume. Villages, villagers, migrant residents, and city interlocutors have shown a flexibility that has consistently exceeded the best laid plans of Shenzhen functionaries and even the city's well-intentioned intelligentsia. Villages form a type of architectural vernacular that continues to influence the spatial structure of the city and an institutional legacy as joint stockholding companies that enables locals to support and promote community identities and incorporate Hong Kong and overseas relatives.

Today, many urban villages in the outer districts are just starting to undergo redevelopment, while in the inner districts, renovation and upgrading continues apace. This means that the great transformation made tangible in this volume continues, as villager's collective land continues to be turned into capital through compensation and into a commodity through real estate development. The tension between Shenzhen's rural roots and its urban identity thus also continues. Yet, as this volume's emphasis on the sociospatial dimensions of exception has highlighted, this tension is less about differences between country and city folk (real though they may be) than about the management of flows across and within the borders that shape Shenzhen's urban subjectivity. Shenzhen's villages are again at the vanguard of a shift from the enclosure of workers in inflexible residency classifications to the modulation of workers through differential forms of access to urban rights, where rights refers to both virtues and prerogatives. While the old improvisations of the urban villages as sites for quasi-licit migrant housing are decreasingly central to Shenzhen's urban form, the new challenges of modulating urban citizenship will spark no less improvisation in the years to come.

Second, at the level of national formation, Shenzhen's legacy illuminates how central the logic of territorial exceptionalism remains for testing and introducing countrywide reforms. In Shenzhen's earliest days, finding ways to allow once forbidden foreign direct investment (FDI) in China was the prime goal of the SEZ, and it met with great success. Today, more than thirty years later, a new 15 km^2 exceptional space called the "Qianhai Shenzhen–Hong Kong Modern Service Industries Cooperation Zone" is being formed in Shenzhen. The goal of this new space is to experiment with allowing offshore banks to make loans in China's domestic market for the first time since the founding of the People's Republic.[6] This decision remains politically delicate because of the importance of central government control over credit to try to manage an often overheated economy and because one of the apparent goals of this economic displacement is the political marginalization of Hong Kong.[7]

This leads to the third level of global formation—Shenzhen's function within the One Country, Two Systems policy makes salient the selective redeployment of borders, both domestic and international, for political restructuring. In the political and economic context of repatriating money, people, and territory to the home country, we see a strategic ambiguity that is also an extension of Shenzhen's earlier experiments—the border with Hong Kong. Just as the Second Line, as shown in Ma and Blackwell's chapter, was used to generate hierarchically ranked spaces of exception, China is strategically using its borders with Hong Kong, Macao, and Taiwan to create blurred spaces that leverage China's scale. (Already, for example, Shenzhen has surpassed Hong Kong to become the third biggest port in the world after Shanghai and Singapore.[8]) Thus, as borders between the PRC and culturally Chinese territories become increasingly porous, the SEZ serves as an instrument for new formations of zonal logics in the context of geopolitical realignments, extending both within and beyond China's ostensive borders. This includes the idea of the Special Administrative Region (SAR) within China, the proposed construction of the China-Pakistan Economic Corridor, which begins in Kashi (Kashgar), Xinjiang, and culminates at Gwadar, Pakistan, where a Special Economic Zone will be built over the next twenty years, and China's development of Special Economic Zones in Africa as part of the larger Chinese government policy of "going out" (*zou chuqu*).[9]

The contributors to this volume have illuminated how these processes result from a history of Shenzhen's accommodation and anticipation of social transformation at municipal, national, and international levels. We have privileged thick description and ethnographic accounts of Shenzhen to denaturalize received wisdoms about Shenzhen as a planned city or neoliberal dystopia. Shenzhen *is* planned, but it would be a mistake to conflate the plan with the city. Prevailing neoliberal trends can be seen in Shenzhen, but the city is the product of situated political logics, and Western neoliberal assumptions about the relation between economics and politics do not easily translate to the Chinese context. Ultimately, what we see in Shenzhen is the continued evolution of a strategic ambiguity that works because nearly everyone can project their fantasies onto the city—and act upon them. This works precisely because of a double movement that rests on the public secret of the economy as a space outside of politics—everyone knows that it is not, but everyone pretends that it is. As the diverse cases in this book have amply described, this ambiguity is constantly shifting in the specifics of urban practices that are structured ostensibly as merely economic but are deeply political in their effects. The historic trajectory of the city from a zone of experimentation to a model city is in that sense broadly indicative of the Chinese party-state's attempt to encircle all economic successes in technocratic, nationalistic, and socialist rhetorics to solidify its legitimacy. But in fact, on the ground, each new configuration of that ambiguous space between politics and economics—be it about property rights, citizenship, or border controls—affords different outcomes that are not wholly predictable in advance. If this book's exploration of the slippery, messy, and often counterintuitive process by which Shenzhen was produced in the late twentieth and early twenty-first century is any indication, we will still be able to learn from Shenzhen as both a model and an antimodel for many years to come precisely because as a city, it continues to afford spaces of unanticipated exceptionalism.

Notes

1. In *The Long Twentieth Century*, Giovanni Arrighi explains how the shifting centers of global capitalism correspond to different strategies of territorial control, as states seek to displace and manage capitalist crises that could otherwise threaten a given regime. Italians needed to control their city-states, the Dutch their relatively small colonies, and the British their large colonies, and Americans had to invent new forms of control over their mainland and areas of direct interest. See Giovanni Arrighi, *The Long Twentieth Century* (New York: Verso, 2010).

2. See Bach's discussion in chapter 1 and also the discussion of zones in Keller Easterling, *Extrastatecraft: The Power of Infrastructure Space* (London: Verso, 2014).

3. Karl Polanyi, *The Great Transformation* (Boston: Beacon Press, 1944). To these, one could add many more cases that we lacked space to explore fully. For example, in the village of Xixiang and the outer districts, the dis/reembedding occurred when early factories used imported materials from overseas clients (*san-lai-yi-bu*) to supplant farming, leading eventually to the Pearl River Delta's Township and Village Enterprises (TVEs), which in turn formed the basis for Shenzhen's early economy. As the TVE model became disembedded from their reliance on collective village land privileges, we can trace their transformation into today's stock-holding "village" corporations that reembed village structures of power and profit-sharing into what are putatively "urban" state-run firms, such as the Dafen Oil Painting Village. At the same time, in Huanggang, Shuiwei, and Shazui, reembedding happened less through foreign direct investment in export industries and more through the sexualization of service industries for Hong Kong men until (after the 2006 reforms) these villages had to restructure their businesses toward more legalized consumption. In Shatoujiao villages and Baishizhou, the process of turning land into a commodity resulted in dispossession when the negotiating power of the villagers vis-à-vis the city was weak, and instead of co-optation, the result was confrontation.

4. See Li Zhang, *Strangers in the City: Reconfigurations of Space, Power, and Social Networks within China's Floating Population* (Stanford: Stanford University Press, 2001), for the classic study of migrants in Beijing. Recently, urban villages have been attracting increasing and more varied attention from scholars. See, among others, Fulong Wu, Fangzhu Zhang, and Chris Webster, eds., *Rural Migrants in Urban China: Enclaves and Transient Urbanism* (New York: Routledge, 2014); Leonardo Citterio and Joseph di Pasquale, *Lost in Globalization: The Paradigm of Chinese Urban Housing* (Milan: Jamko Edizione, 2015); Stefan Al, *Villages in the City: A Guide to South China's Informal Settlements* (Honolulu: University of Hawai'i Press, 2014); and Bruno de Meulder, Kelly Shannon, and Yanliu Lin, *Village in the City* (Zurich: Park Books, 2014).

CONCLUSION

5. On this point in a very different context, see Lucy Suchman, *Plans and Situated Action* (Cambridge: Cambridge University Press, 1987).

6. Since 1980, FDI in China grew from effectively zero to more than $128 billion by 2015 (excluding Hong Kong's FDI), overtaking the United States as an investment destination. But China's national currency is not fully exchangeable, and billions of dollars' worth of its national currency lives overseas. More important, finding ways to repatriate Chinese money is important for China's ambition to make the RMB a global currency. While Hong Kong has been a destination for such funds, it is closed out of the domestic lending economy. It is in this context that the Qianhai zone assumes national importance. See "China Overtakes US for Foreign Direct Investment," *BBC News*, January 29, 2015, http://www.bbc.com/news/business-31052566.

7. See "Qianhai Taking RMB to the Next Level," HSBC News, January 31, 2013, http://www.hsbc.com/news-and-insight/2013/qianhai-taking-rmb-to-the-next-level.

8. "Shenzhen Knocks HK out of Top Three," *The Standard*, January 17, 2014, http://www.thestandard.com.hk.

9. Thanks to Nausheen Anwar for drawing our attention to this connection. See, for example, "Corridor of Power," *Herald*, June 2, 2015, http://herald.dawn.com/news/1153156.

A Shenzen Glossary

People

Bosses (*laoban* 老板) or executives (*zong* 总) are nonworkers and capitalists owning any size of an enterprise. They are entrepreneurs who work for themselves, while red bosses (*hong laoban* 红老板) are those in charge of state-owned industries. Entrepreneurs (*getihu* 个体户) are small-time cottage-industry entrepreneurs—simple subcontractors much discussed in early SEZ development but less commonly used in recent years.

Co-villagers (*Laoxiang* 老乡) are people from the same hometown.

Foreigners (*Laowai* 老外) are people from outside China, usually Euro-Americans.

Illegal residents or "the three withouts people" (*sanwu renyuan* 三无人员) are people residing in Shenzhen lacking one of three statuses: a legal residence, a job, or a fixed abode. In contemporary slang, "a three withouts person" refers to a man without a high-paying job, a car, and a house—the three items necessary for marriage.

Intellectuals or literati (*zhishi fenzi* 知识分子 or *wen ren* 文人) are people who live off knowledge and cultural work, whether within or outside the official apparatus.

Leaders (*Lingdao* 领导) are bosses, superiors, and supervisors. Whether they are in the private or state sectors, *lingdao* are the people subordinates have to publicly agree with (or else lose their jobs).

Locals (*bendi ren* 本地人) are people who were born in pre-Shenzhen Bao'an County or born to people who were born in pre-Shenzhen Bao'an County. They may also be referred to as villagers (*cunmin* 村民), especially when talking about a specific township or village enterprise.

Officials (*guanliao* 官僚), or bureaucrats, are people who are paid a salary by the municipal and/or state apparatus.

Ordinary folks or "the Old One Hundred Surnames" (*Lao baixing* 老百姓) are people who are not part of the municipal and/or state apparatus, used in contrast with the State (*Guojia* 国家) as an agent.

Peasant workers (*nongmin gong* 农民工), or the shortened migrant workers (*mingong* 民工), are rural-to-urban economic migrants—in essence, dislocated peasants whose primary livelihood is wage labor.

Shenzhen inhabitants are people who reside in Shenzhen. There is no direct Mandarin Chinese equivalent. Shenzhen's municipal administration refers to them as the administrative population (*guanli renkou* 管理人口), which is constructed out of mandatory registration with local police stations.

Shenzheners (*Shenzhen ren* 深圳人) are people residing in Shenzhen who hold a Shenzhen household registration (*hukou* 户口) and self-identify with the city. The children of Shenzheners are called Shen-2s (*Shen Er* 深二), a shortened form of the expression second-generation Shenzheners (*Shenzhen er dai* 深圳二代). Unlike their parents who still have hometown identities and usually speak their hometown language, Shen-2s do not split their allegiance between Shenzhen and elsewhere.

Temporary residents (*changzhu renyuan* 常住人员) are members of society legally residing in Shenzhen on a temporary residence permit (introduced in 1985) on a long-term basis.

Wage laborers (masculine, *Dagong zai* 打工仔, and feminine, *Dagong mei* 打工妹) are colloquial terms borrowed from Hong Kong Cantonese and refer to anyone who has a wage job. The term comes from the expression to work for wages (*dagong* 打工) and can refer to any form of alienated labor. The least secure of these laborers take odd jobs (*da linghuo'er* 打零活儿).

White collars and clerks (*bailing* 白领 and *xingzheng yuan* 行政员) are office workers who have monthly salaries and some form of welfare benefits. However, they are not officials. They go to the office (*shangban* 上班) instead of laboring (*dagong* 打工).

Places

Hometown (*laojia* 老家) is a migrant's ancestral home and source of geographic identity, the status of household registration notwithstanding. People from the same hometowns refer to each other as compatriots (*lao xiang* 老乡).

Inner districts, or "within the checkpoints" (*guannei* 关内), refer to the territory of the original Special Zone.

The interior (*neidi* 内地) refers to Shenzhen's hinterlands. As a more general sociogeographic referent, it refers to someplace less "open" than Shenzhen. Importantly, the meaning of the interior has been constantly changing, and thus at different moments in the SEZ's history, the interior has meant "everywhere else" until 1984 and then "not the coastal cities" from 1984 forward. In recent years, the term has also been used to refer to only relatively impoverished rural areas.

New villages (*xincun* 新村) are postreform village settlements that were built and often planned by specific village joint ventures and were sometimes planned to include renovated ancestral halls, temples, and public plazas. Although the expressions urban village (see below) and new village refer to the same neighborhoods, the terms convey specific positions; outsiders use the expression urban village and villagers refer to their villages as the village, the old village, or the new village depending on whether they are speaking of the village in general, the original settlement, or their commercial and rental properties.

Officialdom (*guanfang* 官方) usually refers to the position of the government in a particular negotiation in opposition to an ordinary person. Officialdom is contrasted with society (see below).

Outer Districts, or "outside the checkpoints" (*guanwai* 关外), refer to the territory of the former New Bao'an County.

The Second Line (*er xian* 二线) was the border that divided the Special Zone from what became New Bao'an County and the rest of China or the interior (*neidi* 内地). Beginning in 2003, the Second Line became increasingly porous, and by 2004, it no longer existed as a functioning part of SEZ administration. Shenzhen decommissioned the Second Line in 2010 and in 2015 began demolishing checkpoints.

Shenzhen Municipality (*Shenzhen Shi* 深圳市) refers to Shenzhen as a component of the state apparatus and its concomitant bureaucratic hierarchies. This system (see below) was under construction from the elevation of Bao'an County to Shenzhen Municipality in 1979 through the official decommissioning of the Second Line and the administrative integration of the Special Zone and the outer districts.

Society (*shehui* 社会 or *minjian* 民间) is the everyday world with rural and traditional connotations and is inhabited by ordinary folks without power. Society is the opposite of officialdom (see above).

Special Zone (*Tequ* 特区) was the abridgement of the expression Special Economic Zone (*Jingji Tequ* 经济特区) and was used in official speeches, academic papers, and ordinary conversation to refer to Shenzhen (especially during the 1980s and 1990s), such as in Shenzhen's official newspaper, the *Special Zone Daily*. It was the exception to the system (see below).

System (*tizhi* 体制) refers to the Chinese state apparatus and its concomitant bureaucratic hierarchies. During the 1980s and 1990s in Shenzhen, "system" referred to the organization of the Maoist state and planned economy. With the establishment of the Special Zone, the system became the explicit object of Reform and Opening policies and experiments.

Urban villages (*chengzhongcun* 城中村) are informal settlements that grew on and around historical village settlements, and in many cases, the term refers to either working-class or low-income neighborhoods. Historically, the expression village (*cun* 村) has referred to the land held in common by villagers. However, since the annexation of village lands into the Shenzhen municipal apparatus, the expression "urban village" has referred to the clusters of commercial and rental properties held singularly or collectively by villagers and managed by joint venture (*gufen gongsi* 股份公司). Despite the importance of urban villages to the success of Shenzhen as a city, in 2004 Shenzhen became technically the first Chinese city without villages when the municipality finished annexing village lands into the state apparatus.

Shenzhen's Geopolitical Structure
(in Ranked Order)

Center (*zhongyang* 中央) refers to the national level government.

Province (*sheng* 省) is the location of regional governance. For Shenzhen, this is Guangdong.

Municipality (*shi* 市) is the highest level of territorial administration. In terms of its administrative ranking, Shenzhen is a subprovincial city.

District (*qu* 区) is a middle administration level that has a certain amount of independence in the development of economic policies and concomitant urban planning.

Street Office (*jiedao banshichu* 街道办事处) is the precinct office that implements policy on the ground.

Community or Neighborhood (*shequ* 社区 or *juweihui* 居委会) is the lowest level of administration in Shenzhen and is responsible for registering inhabitants and door-to-door implementation of policies.

Ideas

China Dream (*Zhongguo Meng* 中国梦) is an initiative put forth by Xi Jinping. The China Dream brings together allegiance to the Party, neo-Confucian morality, and national ambition as an organizing ethos for the nation-state.

Demolition and Relocation (*chaiqian* 拆迁) is the colloquial expression to describe mass evictions that precede the demolition of an urban village. In official literature, the process is glossed as urban redevelopment (*chengshi gengxin* 城市更新), emphasizing the replacement of informal settlements with high-end developments.

Propaganda (*xuanchuan* 宣传) includes propaganda, public relations, public education, marketing, and promotion. In Mandarin, it has a positive connotation.

Reform and Opening (*gaige kaifang* 改革开放) is the name of the post–Mao era policy meaning "reforming the system" and "opening to the outside world."

Shanzhai (山寨) literally means a "mountain stronghold," but in the twenty-first century came to designate commercial goods made in the spirit of righteous mountain bandits or guerillas. Shanzhai can range from counterfeited, pirated, illegitimate, unauthorized, and/or fake goods, yet many shanzhai products are considered cute, daring, ingenious, patriotic, or parodical forms of cultural appropriation.

Shenzhen Speed (*Shenzhen sudu* 深圳速度) in the 1980s referred to the rate of construction of Shenzhen's skyscrapers: three stories per day. Later on, it came to refer to the visceral experience of how urbanization transformed rural Bao'an County (pop. 300,000) into the megacity Shenzhen (pop. 20 million) in one generation.

Contributors

Jonathan Bach is an associate professor of international affairs and chair of the Global Studies Interdisciplinary Program at The New School in New York City. His work explores the intersection of culture and politics, with a focus on how microlevel practices reformulate received notions of sovereignty, territory, and identity. He has written on postsocialism in Germany and China and on political and cultural theory. His work on Shenzhen includes "Shenzhen: Constructing the City, Reconstructing Subjects," *Open Democracy*, March 14, 2013; "Modernity and the Urban Imagination in Economic Zones," *Theory, Culture & Society* 28, no. 5 (2011); "Shenzhen: City of Suspended Possibility," *International Journal of Urban and Regional Research* 35, no. 2 (2011); and "Peasants into Citizens: Urban Villages in the Shenzhen Special Economic Zone," *Cultural Anthropology* 25, no. 3 (2010).

Adrian Blackwell is an artist and urbanist whose work focuses on the relation between urban spaces and political/economic forces. His current research examines the polarities of neoliberal urbanization, using Shenzhen as a case study. This work has been published in *Urban China*, *Shaping the City*, *Networked Cultures*, and *Architecture and Ideas* and is forthcoming in *Volume*. Blackwell is an assistant professor in the School of Architecture at the University of Waterloo, a visiting professor at Harvard University, and a founder and editor of the journal *Scapegoat: Architecture/Landscape/Political Economy*.

Yu Cheng is an associate professor at the School of Sociology and Anthropology and deputy director of the Center for Migrant Health Policy of Sun Yat-sen University in Guangzhou, China. His English-language works on health needs and risks among female migrant workers and factory workers in Shenzhen have been published in *Occupational Environmental Medicine* (2012) and in *Migrants and Health in Urban China*, ed. B. Gransow and D. Zhou (Berliner China-Hefte, 2010).

Willa Dong is a doctoral student at the Gillings School of Global Public Health, University of North Carolina at Chapel Hill. She obtained her MSPH from the Johns Hopkins Bloomberg School of Public Health in 2012 and conducted her fieldwork in Shenzhen from March 2012 to February 2013. She then worked as a research fellow at the Chinese Centers for Disease Control from 2013 to 2015. Other relevant publications include "Risk Factors for Congenital Syphilis and Adverse Pregnancy Outcomes in Offspring of Women with Syphilis in Shenzhen, China: A Prospective Nested Case-Control Study," *Sexually Transmitted Diseases*

41, no. 1 (2014), and "High Prevalence of HIV and Syphilis and Associated Factors among Low-Fee Female Sex Workers in Mainland China: A Cross-Sectional Study," *BMC Infectious Diseases* 14, no. 225 (2014).

Eric Florence is the director of the French Center for Research on Contemporary China in Hong Kong and an assistant professor at the University of Liege. His publications include "Migrant Workers in the Pearl River Delta: Between Discursive Inclusion and Exclusion," in *Migration between States and Markets*, ed. H. Entzinger et al. (Ashgate, 2004); "Migrant Workers in the Pearl River Delta: Discourse and Narratives about Work as Sites of Struggle," in *Media, Identity and Struggle in Twenty-first Century China*, ed. V. Fong and R. Murphy (Routledge, 2009); and *Towards a New Development Paradigm in Twenty-First Century China: Economy, Society and Politics*, edited with Pierre Defraigne (Routledge, 2012).

Max Hirsh is a research assistant professor at the Institute for the Humanities and Social Sciences at the University of Hong Kong. He is the author of *Airport Urbanism: Infrastructure and Mobility in Asia* (University of Minnesota Press, 2016). Hirsh's writing on Shenzhen and the Pearl River Delta has appeared in the *Journal of the Society of Architectural Historians*, *History and Technology*, *Places*, and *Log: Journal for Architecture*.

Weiwen Huang was formerly the vice-chief urban planner of the Urban Planning and Land Use Resource Commission of the Shenzhen Municipal Government. Currently he is the director of Shenzhen Center for Public Art and the Shenzhen Center for Design. From 2009 to 2010, he was the Loeb fellow at the Graduate School of Design at Harvard University. Since 2005, he has served as the main organizer of the Shenzhen–Hong Kong Urbanism\Architecture Bi-city Biennale (UABB).

Emma Xin Ma holds a masters in architecture from the University of Waterloo, where her graduate thesis, "The Second Line," examined Shenzhen's social and urban morphology in relation to its internal boundary. She is currently practicing as an intern architect in Toronto.

Katherine A. Mason is a Robert Wood Johnson Foundation Health and Society scholar at Columbia University and an assistant professor of anthropology at Brown University. Mason received her PhD in social anthropology from Harvard University in 2011, with a concentration in medical anthropology. Between 2008 and 2010, Mason conducted ethnographic fieldwork on the professionalization of public health in China following the 2003 SARS epidemic. She is the author of *Infectious Change: Reinventing Chinese Public Health After an Epidemic* (Stanford University Press, 2016). She has published numerous articles on Shenzhen and the surrounding areas, including work on the H1N1 pandemic response, Shenzhen's floating population, and female banqueting. Her work has been published in *Medical Anthropology*, *The China Journal*, *The Journal of Infectious Diseases*, *Anthropology and Medicine*, *AIDS and Behavior*, and *Behemoth*.

Mary Ann O'Donnell is an anthropologist by training who has conducted ethnographic research in Shenzhen since 1995. Her interests and interventions include Handshake 302, a collectively run artspace in Baishizhou; "Shenzhen Noted," a blog that theorizes postsocialist urbanization in Shenzhen; and ongoing collaboration with Fat Bird Theatre, Shenzhen. Her research has been published in *TDR: The Drama Review*, *Positions: East Asian Cultures Critique*, and *Cultural Studies*' special issue on Hong Kong. O'Donnell is an editor at *Architectural Worlds* (Shenzhen University) and the director of the CZC Special Forces, a citizen group that aims to bring Shenzhen's urbanized villages into public discussions about urban planning and renewal projects.

Winnie Wong is an assistant professor of rhetoric and art history at the University of California, Berkeley. She is a historian of modern and contemporary art and visual culture, with a special interest in fakes, forgeries, frauds, copies, counterfeits, and other nonart challenges to authorship and originality. Her study of contemporary art and Shenzhen's Dafen Village, *Van Gogh on Demand: China and the Readymade* (Chicago: University of Chicago Press, 2014), was awarded the Joseph Levenson Book Prize by the Association of Asian Studies in 2015. Her other work in the region includes a forthcoming article on the Shenzhen genomics firm BGI, "Speculative Authorship and the City of Fakes," and a 2015 research studio on art villages in the Pearl River Delta (http://artvillage.city) sponsored by the Mellon Foundation's Global Urban Humanities Initiative at UC Berkeley. She is currently writing a history of export painting in eighteenth- and nineteenth-century Guangzhou.

Acknowledgments

This volume developed out of two conferences: "Shenzhen+China, Utopias+ Dystopias," held at the Massachusetts Institute of Technology in 2011, and "Learning from Shenzhen," held at the Shenzhen Land Use Resources and Planning Commission in 2011 as part of the Shenzhen Urbanism Biennale. We thank MIT's School of Architecture and Planning and, in particular, Mark Jarzombek and Adele Santos for their support of the conference at MIT. For their support of the conference in Shenzhen, we thank Curator Terence Riley, Director Weiwen Huang, and the staff of the Shenzhen Biennale. In addition to the contributors in the current volume, we are most grateful to James Corner, Larissa Heinrich, Tunney Lee, Samuel Liang, and Alex Lui for their papers at these conferences, which enriched our understanding of Shenzhen's significance in national and transnational contexts. Additionally, at the New School, Ashok Gurung and the India China Institute provided generous support. At the University of California, Berkeley, the Center for Chinese Studies supported the manuscript preparation, and Liang Hao assisted in supplementary research. In Shenzhen, the Shenzhen Center for Design, Shenzhen University School of Architecture, the Sino-British Street Museum, and the China Merchants Shekou Archive generously researched archival images and granted permission for their use. In assembling this volume, we thank the two anonymous reviewers for their feedback and Susan Bielstein for guiding the project. Most important, the editors are grateful to the many inhabitants of Shenzhen who helped us understand the nuances of being "special" and the consequences of experimenting, including Cai Chusheng, Cheng Jun, Pauline Cheung, Lynn Fu, Guo Ruoqing, Hua Rongzhen, Weiwen Huang, Gigi Liang, Liang Xiaoli, Liu He, Liu Zhen Lei Sheng, Rao Xiaojun, Tan Ziqing, Teng Fei, Wu Dan, Yi Zhifang, Yin Xunzhi, Yu Haibo, Zhang Kaiqin, and Zhao Xiaoyong. Extra special thanks, as always, to Yang Qian.

Index

Figures are italicized at the end of relevant entries.

Agamben, Giorgio, 5
airports
    check-in terminals, in-town network of, 234, 239, 242
    and Chinese Mainland airspace, 229, 232, 246n19
    Civil Aviation Administration of China (CAAC), 232
    cooperation between SZIA and HKIA, 229–30, 234, 245n5, 247n23
    Individual Visit Scheme, 237
    Studio Fusksas design of Shenzhen Bao'an International Airport terminal, 228–29, 240–41, 242, 245n3
    See also bus system; Hong Kong International Airport; Shenzhen Bao'an International Airport
Anagnost, Ann, 101n21, 157–58, 164n8
aphorisms. *See* slogans
architecture
    awards
        Abercrombie Urban Planning Award honorable mention, 66
        Shenzhen designated a UNESCO "City of Design," 160
    construction of the Second line, 127–29
    "handshake" buildings (also known as "kissing" or "can't get a coffin in" buildings), 115–16, 145–46, 150, 156, 157, 252
    and identity
        civilization by design, 160–61
        village identity of Shenzhen's villages, 12–13
    and international visibility, 228–29
    Las Vegas as a model of, 2, 16

MixC Mall ("Mix City"), 159, 169n69
    rural-urban imbrication of Shenzhen, 6, 8–10, 80–84, 140, 143, 150–52, 157–58, 161, 163
    Shenzhen International Trade Building, 58, 60, 65
    Studio Fusksas design of the Shenzhen Bao'an International Airport terminal, 228–29, 240–41, 242, 245n3
    See also real estate development; urban planning
Arrighi, Giovanni, 258n1

Baishizhou 白石洲
    area of, 72, 119, 247n31, *figure 12*
    dispossession resulting from the commodification of land in, 258n3
    plans for rebuilding of, 121
    vibrancy of, 119–20, 235
Bamboo Curtain, 60, 68, 114
Bao'an County 宝安县
    Anti-Epidemic Station of, 217–18
    British sovereignty over, 125–26
    during the Great Leap Forward, xi, 125
    elevation to Shenzhen Municipality, 65, 67–69, 108–9, 111, 118
    flight of residents to Hong Kong, 68, 78, 123n13, 256
    locals
        defined, 261
        and Hong Kong, 78, 123n13, 256
        and the total population of Baishizhou, 119
    maps of between 1953–78, *figure 10*
    mythologization of, 32
    restructuring as New Bao'an County, 69–70, *figure 11*

Bao'an County 宝安县 (cont.)
    autonomous spaces created by villagers in, 132
    emergence of TVEs in, 77, 116–17
    1992 map of, 117
    as a window for China's first post-Mao experiments, 68–69
    Xin'an County renamed as in 1914, 67, 113
    See also Buji Street Office; Dongmen Walking Street; First Line
Barthes, Roland, 138–39, 140–41, 144, 150, 157, 159, 162
Beijing
    pro-reform leaders of, xii, 54
    urban planning
        China Academy of Urban Planning and Design, 74
        circle delineation, 71–72, figure 4
    See also 1989 democracy movement
border crossings
    and the agency embedded in special border cities, 242–43
    arbitraged difference between rural and urban Shenzhen, 4–7, 13, 124–25, 130–31, 134–35, 172, 257
    Hong Kong's Individual Visit Scheme, 237
    in the Maoist era, viii, xiii
    permits to cross into Shenzhen SEZ, 5–6
    permits to cross Shenzhen-Hong Kong border, 75
    policy changes
        implications for the cross-boundary infrastructure, 238–39
        and the purchasing power of PRC citizens, 237–39
    See also airports; Bamboo Curtain; bus system; checkpoints; cross-boundary infrastructure; First Line; Kowloon-Canton (Guangzhou) Railway; Second Line
Brecht, Bertolt, *Handbook for City Dwellers*, 32
Buji Street Office 布吉街道办事处 (formerly Buji Township 布吉镇)
    checkpoint, 133
    redevelopment of, 134
    and the elevation of Shenzhen to a municipality, 69
    Propaganda Department, promotion of Dafen by officials of, 193, 199, 207–8
    Township and Village Enterprises in, 77
    unplanned population growth in, 83, 132
    See also Dafen Village; Longgang District
bus system
    impact of border-crossing policy changes on, 237–39
    indispensability to Shenzhen's cross-border connectivity, 241
    in-town network of check-in terminals developed by, 234, 239, 242
    spaces of exception generated by differing price structure of, 239
    time-consuming travel on, 242
    See also cross-boundary infrastructure

Caiwuwei Village 蔡屋围村, 81, 113, 115, 116, 121, 154, 159, figure 16
Cantonese language and culture
    diaspora communities, 230–31
    and different names of checkpoints, 136n10
    influence of, xiii
    and the term *dagong*, 91, 262
central business district (CBD), 155–56
    statue of Deng Xiaoping in, 158
checkpoints
    flower arrangement villages developed around, 131–32
    locations of, 133, 136nn20–21
    See also airports; Buji Street Office: checkpoint; cross-boundary infrastructure; Huanggang Village; Luohu Checkpoint; Meilin Village; Shatoujiao; Shenzhen Bay
China Dream 中国梦 (*Zhongguo Meng*)
    Dafen Village as an emblem of Shenzhen's version of, 200–202, 211
    defined, 264
    "harmonious society" as a precursor to, 201
    See also creative industries; Dafen Village
China Merchants Hong Kong, Yuan Geng as standing vice chairman of, 40, 42, 47–48
China Merchants Shekou
    incorporation as a limited holding company, 48–49
    Ministry of Transportation oversight of, 47
    oversight of Shekou Industrial Park, 69
    Yuan Geng as director of, 40
    See also Shekou Industrial Park
circle delineation
    as Deng Xiaoping's ad hoc planning strategy, 71–72, figure 4
    enclosed enclaves created by development companies, 72–73, 82
    greenbelts, 72, 74–75
    "Story of Spring" allegory, 71
    and the urban structure of the Shenzhen SEZ, 71–75
Clark, Constance, 33, 90–91
Cohen, Robin, 89
Collier, Stephen J., 7
commodification of labor
    dagong defined as, 91, 101n28
    and peasant disenfranchisement from collectivist status, 252–53

INDEX

and the precarious condition of migrant workers, 99
valorized in the "World of Dagong" page in the *Shenzhen Special Zone Daily*, 91–93, 95–96
and the values of the Shenzhen Spirit, 97
copyright
and Dafen propaganda officials, 208
Western views of Dafen painters, 193–94, 203–5
creative industries
and the China Dream, 14, 211
and discursive appropriation, 14, 202–9
and the gendering of creative labor, 194, 205, 209–11
and the shift to design as a major economic sector, 160
See also Dafen Village
crime
crackdowns on sex work, 184, 186
organized crime associated with some villages, 151
public safety, 151–52
villages associated with, 148
cross-boundary infrastructure
development of, 230–31, 233
impact of border infrastructure changes on, 238–40
spatial and typological innovation of, 15, 229–30, 234–35, 241
See also airports; border crossings; bus system; checkpoints; Kowloon-Canton (Guangzhou) Railway
Cultural Revolution
attitudes toward sexuality during, 173
economic paralysis, 1, 23, 24, 32
glorification of Mao Zedong, 51
harshness of, 40, 42, 46
Mao's mobilization of the masses during, 215–16
slogan "Learn from Dazhai," 2, 16
smuggling across the Hong Kong border during, 147

Dafen Village 大芬村
Dafen propaganda officials, 207–8
and discursive appropriation, 14, 202–9
feminized creative dreams of female migrant workers, 197–200, 208–9, 211
participation in the global market, 195–96, 201, 202, 206–7
recognition as a model cultural industry, 207
transformation into stock-holding "village" corporation, 258n3
TVE model compared with, 258n3
Western views of Dafen painters, 193–94, 203–5
dagong 打工
dagong mei 打工妹, 209, 262

dagong zai 打工仔, 262
defined as a term, 91, 101n28
"World of Dagong" page in the *Shenzhen Special Zone Daily*, 91–93, 95–96
See also commodification of labor
Deng Xiaoping 邓小平
circle delineation of Shenzhen SEZ, 71–72
metaphors of
"feeling rocks to cross the river," 41
Special Economic Zones as "laboratories," 215
1984 Southern Tour, 46, 53, 60, 84n3, 115
1992 Southern Tour, xii, 3, 58, 73, 77, 82, 86, 88, 174
See also Reform and Opening
development
compensatory design strategies devised in Shenzhen, 240–45
global reach of Shenzhen developers, 34–36
greenbelts, 72, 74–75
self-organizing development of villages, 75, 80–82, 161
of Shenzhen Bao'an International Airport, 231–32, 246n16
See also circle delineation; cross-boundary infrastructure; enclave development; export processing zones; global mobility; property laws; real estate development; rural urbanization; Shenzhen Special Economic Zone master plans; Shenzhen speed; urban villages and housing concerns
Dongguan Municipality 东莞市
growth linked to Shenzhen's rise, xii, 77–78
TVEs in, 77
Dongmen Walking Street 东门步行街
Baishizhou Pedestrian Street mimicking of, 120
MixC Mall ("Mix City") compared with, 159, 169n69
traditional commercial area of, 44, *figure 10*
dreams
feminized creative dreams of Dafen Village female migrant workers, 197–200, 208–9, 211
as an ideological force, 94–95, 103n45, 200, 211
of the modernist planned city, 31–32, 36
projections of fantasies onto Shenzhen, 12, 15, 36, 138, 257–58
rags-to-riches story of villager success, 142–43
See also China Dream

economy
the contract responsibility system, 43–44, 62n10
economic enclaves
gated enclaves of material wealthy, 66
and the history of trade, 25

economy (cont.)
  economic paralysis of China's economy during the Cultural Revolution, 1, 23, 24, 32
  the *hukou* system as a tool for national economic planning, 70
  PRC socialist planned economy contrasted with Cold War economic development of Hong Kong, 68
  Shenzhen's GDP growth from 1980 through 1995, 65
  socialist planned economy, 93–94
    dismantling of, 67
    as a space outside of politics, 257–58
  See also commodification of labor; free trade; global capitalism; global economy; Reform and Opening; Special Economic Zones; Township and Village Enterprises
enclave development
  "the countryside surrounds the city" inverted to "the city surrounds the countryside," 8, 108, 113, 121, 146
  and the development of OCT and Baishizhou, 72
  "village-in-the-city" enclaves, 80–81, 143–44, 147
exceptionalism
  and activities that define free-trade or foreign-trade zones, 251
  created by the price structure of buses, 239
  global economic participation facilitated by the strategic use of, 27–31, 251, 252, 255–56
  Individual Visit Scheme, 237
  Shenzhen Miracle as, 77–78
  territorial exceptionalism as a strategy of development, 126–27, 251–52, 256–57
  See also exceptional spaces; *hukou*; Special Economic Zones
exceptional spaces
  classification and management of populations in zonal spaces, 28–29
  gray zones of communities, xiii, 6, 12–13, 16, 125–26, 161–62
  Kowloon Walled City as, 125–26, 127, 132
  separation of Hong Kong and Shenzhen airport workers, 7
  spatial and temporal thresholds of the SEZs, 29, 250–52
  transformation via zonal logic into the rule, 11, 27–30, *figure 2*
  villages as, 12–13, 144, 150–52, 252, 256–57
    Luofang Village as, 126, 127
  See also border crossings; Second Line; urban villages and housing concerns; urban villages and space
export processing zones
  and Dafen Village goods, 195
  global forms of capital accumulation facilitated by, 27–31
  Shannon, Ireland, model, 26
  in South Korea and Taiwan, 24, 26, 32–33, 36
  Special Economic Zones compared with, 25, 36–37

female sex workers (FSWs). *See* sex industry
First Line (Sino-British Border)
  number of villages on either side of, 139–40, 164n9
  as a political boundary of the One Country, Two Systems policy, 124–27
flexibility
  flexible production methods
    of Dafen's trade, 196
    rise of, 27
    and Shenzhen's labor repressive system, 89
  as a strategy, 135, 250, 252, 254–57
Foxconn, 4, 31
  labor exploitation at, 17n4
  location beyond the Second Line in Bantian, 118, *figure 5*
  slogans on the walls of, 103n45
free trade
  exceptional activities that define free-trade or foreign-trade zones, 251
  and market economy of Xin'an County, 67
  and Shenzhen SEZ, 76, 125, 251–52
Friedmann, John, 83
Fullilove, Thompson, 177, 184
Futian District 福田区
  and Shenzhen SEZ's early development, 69, 73, 79
  villages of, 82
  as breeding grounds for crime, 148

Gangxia Village 岗厦村, 81, 150, 151
gender
  and divisions of labor, 194, 205, 209–11
  feminized discourse of the artist, 208–9
  and economic inequality, 180, 185
  emphasis on production of a male heir, 173
  second wives, 6, 174
  single-gender dormitories, 174
  wage laborers, 209, 262
  See also sex industry
global capitalism
  female migrant workers as the object of triple oppression of, 103n45
  investment of overseas Chinese in Shenzhen, xii, 25, 29, 33, 117, 120, 147, 230–31
  post-Mao China appropriation of capitalist logic, 1, 2–3, 60, 75–77, 93–94, 98

and Shenzhen's mediated global connectivity, 241
shifting centers of, 66, 258n1
global economy
and conditions of arbitrage at border sites, 6–7, 28, 147
contribution of the villages to, 148–50, 252
Dafen painters' participation in the global market, 195–96, 201, 202, 206–7
Hong Kong's development during the Cold War, 68
imposition of the Second Line as an self-imposed apparatus of segregation from the global market, 126
ongoing reconfiguration of, 16, 23
rise of the export state, 25–26
Shenzhen's participation in
and the links to overseas Chinese, xii, 25, 29, 33, 117, 120, 147, 230–31
rapid transformation as factor, 23, 27, 36, 66, 83
and the reemergence of the Pearl River Delta as a node in, 29, 66–67, 78, 258n3
and the strategic use of exceptions, 27–31, 251, 253, 255–56
and Township and Village Enterprises, 9
See also global capitalism
global mobility
and the agency embedded in special border cities, 242–43
compensatory design strategies devised in Shenzhen, 4–5, 240–45
Hong Kong's mediation of Shenzhen's global connectivity, 241–44, 246n16
See also airports; border crossings; cross-boundary infrastructure
Graeber, David, 54

harmonious society 和谐社会 (hexie shehui), 157, 201
Harms, Erik, 169n67
Hong Kong
agricultural gates, 129, figure 15
as an aspirational model for Shenzhen, 7, 33, 68
and Bao'an County locals, 68, 78, 123n13, 256
border crossings with Shenzhen, effect of the SARS crisis on, 237–38
border infrastructure
Individual Visit Scheme, 237
Kowloon-Canton Railway, 44, 113, 238
and the operation of cross-boundary airport buses, 238
and the circle delineation of the new territories, 71–72, figure 4
and political marginalization, 256
and Shenzhen's rise, 256–57
entropy flows and energy dissipation between Shenzhen and, 78–80, 245–46n9
forced evacuation of Kowloon Walled City, 125–26, 127, 132
interactions with Shenzheners, xiii–xiv, 7, 127
and the H1N1 Pandemic, 223
1985 Sino-British Joint Declaration, 125
1997 handover to the People's Republic of China, 33, 125, 236
PRC socialist planned economy contrasted with Cold War economic development of, 68
and the SARS crisis of 2003, decline in tourism resulting from, 236–37
as a Special Administrative Region, 125–26, 236
See also Lok Ma Chau
Hong Kong International Airport (HKIA)
cross-border airport busses feeding into, 229–30, 233, 234, 235, 238–40
as an international gateway, 229–30, 244
and Shenzhen
mediation of global connectivity of, 241
separation of workers from Shenzhen from their counterparts at, 7
Huang, Weiwan 黄伟文, 245–46n9
Huanggang Village 皇岗村
checkpoint at, 127, 238
development of, 150
land port of, 79–80
sex industry at, 127, 258n3
Huawei 华为, 4, 31, 118
Hubei Village 湖贝村, 116
Hu Jintao 胡锦涛
harmonious society, 157, 201
Shenzhen described as heroic and miraculous, 32
hukou 户口 (household registration system)
"blue hukou" (or Shenzhen "green card"), 93, 102n35
and Hong Kong's Individual Visit Scheme, 237
inclusive exclusion of
constraints on the mobility of migrant workers, 5–6, 70, 108–9
as a marginalizing spatial apparatus, 5–6, 89, 172
registration of migrant workers as temporary residents, 93, 100n14, 166n26, 166n32
Shenzhen as witness to reforms of, 5–6, 70, 73–74, 111
Shenzhen's urban form impacted by, 117–18
and women
female sex workers, 173, 177–78
systematic exclusion of rural women from entrepreneurial classes leading to, 185

Hu Yaobang 胡耀邦
	death of, 57
	loyalty to Deng, 41
	as Luo Zhengqi's political patron, 52
	Shenzhen heroes supported by, 52, 57

identity
	and multidirectional mobility flows across borders, 243, 248n48
	rural and urban class distinctions, 143–44, 157, 181, 253–54
	transformation of China's identity by the creation of Special Economic Zones, 23, 87
	village-based cultural identities, 10, 108, 110
	See also Shenzhen inhabitants: Shenzheners; Shenzhen's new identity; Shenzhen Spirit
industrial parks
	Bagualing Industrial Park 八卦岭工业区, 72
	Shangbu Industrial Park 上埗工业区, 72
	See also Shekou Industrial Park
inequality
	created by the price structure of buses, 239
	downplayed during the Shekou Storm, 56, 58
	exploitation of migrant workers, 2–3, 9, 17n4, 35, 96–99
	and gender, 180, 185
	resulting from rapid urbanization, 70
	and rural urbanization, 120–21, 143–44, 147
	socioeconomic inequality, of Shenzhen's migrants, 13
	See also enclave development; exceptionalism; exceptional spaces; hukou; migrant workers: social justice for
infrastructure
	Shenzhen Municipality citywide transportation network, 73
	and vertical accumulation, 9
	See also airports; bus system; checkpoints; cross-boundary infrastructure; railways; taxi system
intellectual property law
	Agreement on Trade-Related Aspects of Intellectual Property Rights (TRIPS), 202
	and Dafen propaganda officials, 208
	legitimacy of, 201, 202, 205–6
	and Leonardo's Mona Lisa, 193, 203–5

Jiang Zemin 江泽民, 39, 59, 86, 87

King, Anthony, and Abidin Kusno, 163n2, 169n67
Koga, Yukiko, 169n67
Koolhaas, Rem, 23
Kowloon-Canton (Guangzhou) Railway (KCR)
	and economic development, 112–14, 117, 118
	located near Caiwuwei, 113
	station at Luohu, 44, 113, 231, 238
Kowloon Walled City, 125–26, 127, 132

labor
	gendering of creative labor, 194, 205, 209–11
	neoliberal model of work, 56, 98
	reform of the *danwei* 单位 system, 47–48
	small-time bosses (*getihu* 个体户 or *laoban* 老板, entrepreneurs), 195, 261
	See also commodification of labor; migrant workers
land use rights
	Chinese Constitution amended in 1982 regarding, 8, 109
	conflicts between Shenzhen Municipality and its "urban villages" over, 118, 131–32, 140, 144
	as a core element of the economic zone, 30
	hereditary rural land rights turned into expiration-based urban use rights, 155–56
	and "village-in-the-city" enclaves, 143, 147
	and Hong Kong–style land management regimes, 33
	Land Law of 1986, 110
	transferred from collectives to the municipality, 9, 118, 153
	and unsanctioned arrangements along the Second Line, 131–32
	See also property laws; urban villages and housing concerns
Lee, Ching Kwan, 8, 89
legality
	demolition of illegal buildings, 132, 153–55, 156, 157, 163, 264
	gray zones of communities, xiii, 6, 13, 16, 45–46, 125–26, 161–62
	illegal residents ("the three withouts people," 三无人 *sanwu renyuan*), 99n6, 101n21, 103n59, 132, 261
	See also copyright; hukou; intellectual property laws; land use rights; property laws
Liang Xiang 梁湘
	"ant theory" metaphor, 45, 62–63n15
	biographical details, 41–42, 61n2
	expulsion from the party, 42, 58
	loyalty to Deng Xiaoping, 41
	political patronage of, 41
	as a Shenzhen leader hero, 40, 45–46, 59, 62n14
Li Hao 李灏, 59
Lindquist, Johan, 242–43
Lindsay, Greg, 35
Li Youwei 厉有为, 59, 88
Li Zhang, 8, 140, 164–65n15, 169–70n76
Lok Ma Chau 落马洲 (Hong Kong), 83, 127, 238

INDEX

Longgang District 龙岗区, 74, 134, 141, 175, 199, 207
Luofang Village 罗芳村, 126, 127
Luohu 罗湖 (Lo Wu) Checkpoint
　Luohu Bridge, viii, 114, 126
　as a passenger checkpoint, viii, 126, 231
Luohu District 罗湖区
　counterfeit goods available in, 127, 205, 252
　jurisdiction of illegal developments in, 132
　and the Kowloon-Canton Railway, viii, 44, 77, 79
　production areas in early Shenzhen located in, 44, 77
　sex industry in, 172, 174, 178
　urban renewal, 117
　See also Yumin Village
Luo Jingxing, 44–45
Luo Zhengqi 罗征启
　biographical details, 42–43
　death of his brother Luo Zhengfu, 42–43
　expulsion from the party, 58
　Hu Yaobang as his political patron, 52
　letter to Deng during the democracy protests, 57–58
　new model of intellectual citizen advocated by, 57
　as a Shenzhen leader hero, 40–41, 42
　Shenzhen Spirit articulated by, 52–53
　Shenzhen University's construction led by, 43, 52–54, 102n32

Mao Zedong 毛泽东
　death of, 39
　public health campaigns of, 215–17
　See also Cultural Revolution
marginalizing spatial apparatus. *See* border crossings; enclave development; exceptional spaces; First Line; *hukou*; Second Line
Meilin Village 梅林村, 73, 83, 132, 133, 151
migrant workers 外来工 (*wailaigong*)
　contribution to the economic growth of the Shenzhen SEZ, 98
　crime and vice associated with, 148
　exploitation of, 2–3, 9, 17n4, 35, 96–99
　concealment of, 89, 100n15
　"labor-squeezing strategy of development," 88–89
　illegal residents, 132, 261
　　compared with migrants elsewhere, 97
　　plots of land for small-scale agriculture, 145
　peasant workers, 89, 97, 261
　registration as temporary residents, 93, 100n14, 166n26, 166n32
　social isolation of, 177–78
　social justice for, 14, 15, 82
　　Dafen's migrant-worker painters, 198–200
　　health inequalities, 225
　transformation into Shenzheners and "desirable" migrants, 97–99, 253–54
　"How to Be a Shenzhen Person" campaign, 89–91
　and neoliberal conceptions of the market, 88
　valorization of, 11–12, 86–88, 90–94, 96–97, 172
　See also dagong; *hukou*; peasants; sex industry
Ministry of Transportation
　and the Chinese system of *danwei*, 57
　oversight of China Merchants Bureau, 69
　oversight of Shekou, 47
　Yuan Geng as head of Overseas Affairs of, 42
MixC Mall ("Mix City"), 159, 169n69
model making
　as a classic feature of socialist governance, 1
　Las Vegas as a model of architectural design, 2, 16
　model communes under Mao Zedong
　　Daqing 大庆, 1, 16, 25
　　Dazhai 大寨, 1, 2, 16
　by Shenzhen heroes, 11, 40, 43
　Shenzhen's storied history of, vii–viii, xiii–xiv, 23, 35, 66, 250, 255–58
　Shenzhen as a Potemkin village, 1, 233
　Shenzhen University model of intellectual citizens, xiii, 52–54
　See also Dafen Village; public health models; Shenzhen heroes; Shenzhen inhabitants: Shenzheners; Shenzhen speed; Special Economic Zones

Nantou 南头
　checkpoint at, 133
　Chinese maritime access to the South China Sea via, 112
　as the county seat of Xin'an County, 112, *figure 3*
　and the establishment of Shenzhen SEZ, 69
　industrial park at the tip of Nantou Peninsula, 44
　map of concentric occupations of the Nantou Peninsula, *figure 9*
　salt fields harvested for centuries in, 112
New Bao'an County. *See under* Bao'an County
1989 democracy movement
　support by the three Shenzhen heroes, 40, 55, 57–58
　Tiananmen crackdown, 55, 63n27

officialdom 官方 (*guanfang*), 14, 161, 262
Ong, Aihwa, 23, 28
overseas Chinese
　investment in Shenzhen, xii, 25, 29, 33, 117, 120, 147, 230–31
　Overseas Chinese Town (OCT), 72, 119–20, 124, 252

Palan, Ronen, 28
peasants 农民 (*nongmin*)
    and the contract responsibility system, 43–44, 62n10
    disenfranchisement from collectivist status in the state apparatus, 253
    English romantic identification with heroic individualism distinguished from, 166n25
    hukou constraints on the mobility of, 5–6, 70, 89, 108–9
    and the image of Shenzhen as a city of limitless opportunity, 2, 142–43
    Kangxi Emperor's forceable movement of, 112
    peasant workers 农民工 (*nongmin gong*), 89, 97, 261
        transformation into urban migrants, 97–98, 114, 122n9, 143–44, 253
    and sent-down urbanites, 93
    in Shenzhen in 1973, viii–x
    and the slogan "the countryside surrounds the city," 8, 107–8, 113, 121, 146
    transformation into Shenzheners, 13, 88, 89–91, 138–70, 253–54
    as village shareholders, 145–46, 147, 156–57
    See also *dagong*; migrant workers
Peck, Jamie, and Jun Zhang, 33
Polanyi, Karl, 252–53
Pollitt, Katha, 244
population
    classification and management in zonal spaces of, 28–29
    density of Shenzhen's population, 166n33
    hukou population of Shenzhen Municipality, 74, 100n12, 122n8
    police allocation in the city based on, 152
    Shenzhen's rapid growth, vii, 65, 118, 172
    See also migrant workers; Shenzhen inhabitants
property laws
    ambiguous property rights, 118, 163
    resistance to developers, 153–54
        nail house in Chongqing, 153
    See also intellectual property law; land use rights
public health models
    Anti-Epidemic Stations, 216, 219
    Mao's public health campaigns, 215–17
    and sex work, 172, 176–77
    Shenzhen model of public health
        and the Bao'an County Anti-Epidemic Station, 217–18
        and the discourse of global health preparedness, 14, 221–22, 225–26
        and the SARS crisis of 2003, 14–15, 213–15, 218–22, 224
        Shenzhen CDC, 219, 221–22
Pun Ngai, 8, 98, 103n45

Qian Chaoying 钱超英
    "Dear Deng" letter, 50–51
    1989 crackdown process compared with treatment of, 55
Qianhai Cooperative Zone 前海深港现代服务业合作区
    China Merchants holdings in, 63–64n32
    "Qianhai Shenzhen–Hong Kong Modern Service Industries Cooperation Zone" formation, 256
    Qianhai Thoroughfare, 113
    Shenzhen–Hong Kong airport site planned for, 80
    status as a self-governing trade zone 自贸区 (*zi mao qu*), 255
    Xi Jinping's visit to, 60
railways
    connection to the highway system, 79–80, 113–14
    TVEs built in anticipation of, 9, 77, 117
    urban development around rail corridors, 231
    See also Kowloon-Canton (Guangzhou) Railway
real estate development
    circle delineation
        enclosed enclaves created by development companies, 72–73, 82
        as a policy to attract participation of, 71
    commodification of land, 252–53, 256
    Dafen Village as a target of, 197, 199
    inclusion in SEZs, 25
    Kingkey, 154, 235
    Liu Tianjiu's investments in, 45, 62n12
    Overseas Chinese Town (OCT), 72, 119–20, 124, 252
    renovation of urban villages, 82, 111, 132, 154–55, 156, 157, 163, 264
    Shenzhen International Trade Building, 58, 60, 65, 156
    Shenzhen Special Economic Zone Real Estate Company, 45
    Tianmian Village's transformation into a model "civilized village," 155–57
    "vertical accumulation" process exploited by developers, 9
    See also enclave development
Reform and Opening 改革开放 (*gaige kaifang*)
    and the contribution of leader-heroes, 54
    defined, 264
    Deng Xiaoping's furthering of, xii, 40, 61n1, 68, 174
    introduction of, vii
    and the 1978 Plenary Session, 58
    open-ended experimentation of, vii, 65, 230
    downplayed by the party in recent years, 256
    and the system, 108, 263

INDEX

279

practical and measurable results of, 54
and the Shenzhen Miracle, 77
and the Shenzhen Spirit, 87
shifts in lifestyles and values resulting from
    and the migration of rural workers, 93–94
    and the sex industry in Shenzhen, 173–74
utopian ideals of, 198
    effect on young people, 55
    and the formulation of Shenzhen Spirit, 87
Rofel, Lisa, 101n27
Rowlands, Michael, 162
Roy, Ananya, 36
rural urbanization
    and global inequality, 120–21, 143–44, 147
    as an ideological practice, 121
    rural-urban imbrication of Shenzhen, 6, 8–10, 80–84, 140, 143, 150–52, 157–58, 161, 163
    See also urbanization

SARS crisis of 2003
    China's crumbling health system exposed by, 219–20
    and the Ebola outbreak of 2014, 226
    Hong Kong tourism affected by, 236–37
    and Shenzhen's preparedness for the H1N1 pandemic, 213–15, 223–24
    and the transformation of Shenzhen's local public health model, 14–15, 218–22
Scott, James C., 100n15
Second Line 二线 (Er xian, internal boundary)
    construction and delineation of, 109, 127–29, 134–35
    and the imposition of a self-imposed apparatus of segregation from the global market economy, 126, 131
    decommissioning of, 133–34
    defined, 124, 262–63, figure 13
    flower arrangement villages 插花村 (chahua cun), 131–32, 133
    spaces of exception generated by, 13, 124–25, 130–31, 134–35, 172, 257
    See also checkpoints
sex industry
    female sex workers (FSWs)
        health issues, 176–77, 188n4
        from Hong Kong, 79
        and hukou, 173, 177–78
        Pan Suiming's seven-tier classification of, 189n42
        and Shenzhen's new identity, 171–72
        specific forms of exclusion experienced by, 185–86
    HIV prevention, 176
    interviews conducted in Shenzhen, 174–75, 178–84, 187n9, 187n10
    red light districts in Shenzhen villages, 175
    and the reembedding of foreign demand, 79, 186, 253
    stigma experienced by, 179
SEZs. See Special Economic Zones
Shangsha Village 上沙村
    commercial development in, 82, 150
    red light district of, 175
shanzhai 山寨, 4, 17n9, 205–6
Shatoujiao 沙头角 (Sha Tau Kok)
    checkpoint, 127
    commercial development in, 79
    dispossession resulting from the commodification of land in, 258n3
    and the establishment of Shenzhen Municipality, 69
    Sino-British Street located in, 75, 127
Shawei Village 沙围村, 150–51
Shazui Village 沙嘴
    female sex workers in, 175
    reembedding of foreign demand in, 253
Shekou Bulletin News 蛇口通讯包
    critique of Yuan Geng's reforms published by, 49–50
    as experiment in loosening restrictions on journalism, 50
    on the "Shekou Storm," 56
Shekou Industrial Park 蛇口工业区
    acceptance of intellectuals sent home to reflect on their mistakes, 58
    circle delineation of, 71–72
    establishment of, xi–xii, 69, 71–72
    incorporation into Shenzhen Municipality, 58, 59, 63–64n32
    as one of the primary early areas of development in Shenzhen, 50, 79
    as part of the Qianhai District, 255
"Shekou Storm" 蛇口风波 (Shekou Fengbo) 56–57, 58, 102n32
Shenzhen Bao'an International Airport (SZIA)
    development of, 231–32, 246n16
    as a hub for cheap domestic flights, 229, 232, 244, 245n5
    reclassification as an international airport, 232–33, 240–41
    Studio Fusksas's design of, 228–29, 240–41, 242, 245n3
Shenzhen Bay
    checkpoint, 80, 127
    design of the border control facility at, 238–49, 248n40
    flight of Chinese from Bao'an County during Mao Zedong's Great Leap Forward, xi, 125, figure 14
    sex industry near to, 175
Shenzhen heroes
    "heroism" of, 61n3

Shenzhen heroes (*cont.*)
  introduced, 39–40
  1989 democracy movement supported by, xii, 40, 55, 57–58
  party secretaries contrasted with, 59
  progressive laissez-faire governance of, 40, 58–59
  re-remembered Shenzhen model of, 60–61
  See also Liang Xiang; Luo Zhengqi; Yuan Geng
Shenzhen inhabitants
  described by Vogel in 1973, viii–x
  illegal residents, 132, 152, 261
  Shenzheners 深圳人 (*Shenzhen ren*)
    city identity excluded from, 253–54
    defined as a term, 262
    "How to Be a Shenzhen Person" campaign, 90–91
    locals 本地人 (*bendi ren*) distinguished from, 261
    Luohu as an object of nostalgia for second-generation Shenzheners for Luohu, 116
    Maoist model of public health as an object of nostalgia for, 215
    and Shenzhen–Hong Kong interactions, xiii–xiv, 7, 127, 223
    transformation of rural migrants into, 13, 88, 89–91
  See also peasants
Shenzhen Miracle 深圳传奇 (*Shenzhen Chuanqi*), 2–3, 32, 38n20, 86, 98, 148
  temporal predicates of, 77–78
Shenzhen Municipality 深圳市 (*Shenzhen Shi*)
  Bao'an County's elevation to the status of, 65, 67–69, 108–9, 111, 118
  citywide transportation network, 73
  as a component of the state apparatus, 263
  demographics, 74, 81, 100n12, 166n33
  division into New Bao'an County and Shenzhen SEZ, 69–70
  five-tiered bureaucracy of, 107
  incorporation of Shekou Industrial Zone, 58, 59, 63–64n32
  promotion as an agent of economic reform, 59
  See also villages
Shenzhen SEZ Administrative Line. *See* Second Line
Shenzhen's new identity
  and the sex industry, 171–72
  and the valorization of migrant workers, 11–12, 86–88, 90–94, 96–97, 172
  See also Shenzhen Spirit
Shenzhen Special Economic Zone (Shenzhen SEZ) 深圳经济特区 (*Shenzhen Jingji Tequ*)
  as China's southern gate, 231
  creation of, vii, 23–25, 83

and Deng Xiaoping's economic reform policies, 39, 230–32
electronics manufacturers located in, 118
ending of, 59
high-tech factory campuses in, 4, 31, 118
  and circle delineation, 71–72, 74–75, 80–83
  original, 29, 74–75, *figure 5*
  2010 master plan, 66
  See also Foxconn; Tencent
illegal structures in, 139, 143, 151–56
SEZ as an administrative category for Shenzhen, 165n21
Shenzhen Special Economic Zone Real Estate Company, 45
  See also Bao'an County; First Line; Second Line
Shenzhen Special Economic Zone master plans
  cluster model of, 74–75
  Comprehensive Urban Plan 1982, 8, 74, 110
  Comprehensive Urban Plan 1986, 34, 66, 74, 117
  Comprehensive Urban Plan 1996, 117, 122n3, *figure 12*
  Comprehensive Urban Plan 2010–2020, 66, *figure 7*
  five year plan 1996–2000, 35
  1996–2005 master plan, 155–56
  transportation axes perpendicular to the Hong Kong border, 231
  2000 master plan, 34
  2030 urban development strategy, 33–34, *figure 1*
  villages not accounted for in, 234
  See also circle delineation; Shenzhen Special Economic Zone spatial politics; urban villages and housing concerns; urban villages and space
Shenzhen Special Economic Zone spatial politics
  impact of *hukou* on, 117–18
  and the possibility of far-reaching social transformation, 252–57
  and the regulation of movement, 4–5, 7–8
  See also circle delineation; exceptional spaces; First Line; *hukou*; rural urbanization; Second Line; Shenzhen Special Economic Zone master plans
Shenzhen speed 深圳速度 (*Shenzhen sudu*)
  and the achievements of China's post-Mao economic reinvention, vii, 2
  as a challenge to Shenzhen's model of public health, 172, 214
  coined as a phrase, 58, 65, 164n7
  conflation of governance with the simultaneous construction of a government itself, 44
  contribution of the villages to, 139, 148, 154
  mechanisms and logic enabling the success of, 65–67, 70

INDEX

Shenzhen University as a model of, 53, 60
unprecedented pace of urbanization, vii, 65, 118
Shenzhen Spirit 深圳精神 (Shenzhen jingshen)
    Luo Zhengqi's articulation of, 52–53
    and migrant workers
        debasement of, 98–99, 99n6
        valorization of, 86–88, 91–93, 96–97, 172
    "Shekou Spirit" (Shekou jingshen), distinguished from, 46–47
Shenzhen University
    building of, 43, 52–55
    Institute for the Study of Special Zones, 84n3
    Luo Zhengqi's presidency of, 42–43, 52–54, 58
    model of intellectual citizens, xiii, 52–54
    as a model of Shenzhen speed, 53, 60
    prodemocracy activity, 40
        Qian Chaoying, "Dear Deng" letter, 50–51, 55
Shenzhen Youth Herald 深圳青年报 (Shenzhen Qingnian Bao)
    closing of during the 1987 crackdown, 52
    criticism of Shenzhen University construction, 54
    publication of "Dear Deng" letter of Qian Chaoying, 50–51
Shieh, G. S., 194–95
Shuiwei Village 水围村, sex industry, 171, 175, 258n3
Simmel, George, 4
Singapore
    as an aspirational model, 24, 34, 35
    and Batam, 242–43, 248n48
    export-led growth of, 26
    as a low-wage country, 31
    overseas Chinese in, 33, 230
    and the 3+C model, 76
Sino-British Border. See First Line
Sino-British Street 中英街 (Zhong-Ying Street), 75, 79
Siu, Helen F., 8, 140, 141
slogans
    emphasizing pragmatic application of grit and entrepreneurial spirit
        "Construction is a poem, written by poets, who wrote them with steel and cement," 138
        "Dare to Become the World's First," 31
        on the walls of Foxconn, 103n45
        "practical work brings prosperity," 32
        "time is money, efficiency is life," 32, 49, 158, 169n63
        "To get rich is glorious," 158
        "to pay with one's blood and sweat," 93
    "Feeling rocks to cross the river," 41
    "Learn from Dazhai," 2, 16
    politicized aphorisms describing uncertain pragmatism, 44

"prevention first" slogan of Mao's public health campaigns, 216
transformation of
    "the countryside surrounds the city" inverted to "the city surrounds the countryside," 8, 107–8, 113, 121, 146
    "Learn from Dazhai" to "Learning from Las Vegas," 2
    "time is money, efficiency is life," 49
Solinger, Dorothy J., 143
South Korea
    export processing zones
        and Dafen Village goods, 195
        at Masan, 24
    and national development, 26
    New Songdo City, 36
    Shenzhen's emulation of, 32–33
    Special Economic Zones compared with, 25, 36–37
Soviet Union
    implementation of Lenin's New Economic Policy, 131
    Mikhail Gorbachev's visit to Beijing, 57
    private property introduced into former collective farms, 166n30
Soviet cities
    old Soviet-style building design, viii
    planning of, 7, 32
Special Economic Zones (SEZs)
    adaption beyond the East Asian development state, 34–35, 66
    as an administrative category for Shenzhen, 39, 165n21
    Batam, 242–43, 248n48
    metaphorical views of the function of
        and Barthes's reference to the city as a text, 138–39, 140–41, 142, 144, 150, 157, 162
        as like a "bull clearing the wilderness" (tuo huang niu), 87
        as like "laboratories," 65, 77, 162–63, 215, 230
        as like a series of locks in a shipping canal, 79–80
        "opening up" (kaituo) a window, 65, 68, 77, 127, 215
    spatial and temporal thresholds of, 29
    Special Zone 特区 (Tequ) as a term for, 263
    3+C model, 76–77
    zonification of China, 23, 30, figure 2
    See also Shenzhen Special Economic Zone
Stark, David, 3
Street Offices 街道办事处 (jiedao banshichu)
    Buji Street Office, 193, 199, 207–8
    defined as a term, 263
    Shekou Industrial Zone as the Street Office precinct of Nanshan District, 58

Sum, Ngai-Ling, 27
system 体制 (*tizhi*), 108, 263

Taiwan
   export processing zones
      and Dafen Village goods, 195
      at Kaohsiung, 24
      and national development, 26
      SEZs compared with, 25, 36
      Shenzhen's emulation of, 32–33
   subcontracting production practices, 194–95
taxi system, 130, 134, 150
television
   feminized discourse of the artist discussed by *Half the Sky* 半边天 (*Banbiantian*), 208–9
   "How to Be a Shenzhen Person" debate launched on, 90
   programs about Dafen, 199, 201
Tencent, 31, 60
Teyssot, Georges, 4
Tianmian Village 田面村 (*Tianmian Cun*)
   transformation into a model "civilized village," 150, 155–57
   villagers transformed into wealthy landlords in, 145–47
Toulmin, Stephen, 31
Township and Village Enterprises (TVEs)
   as both a symptom and success of the Reform and Opening policy, 123n15
   in Dongguan, 77
   emergence of, 77
   and the localization of global production chains, 9
   in Shenzhen, 77, 116–17
   transformation into stock-holding "village" corporations, 258n3
   and Wanfeng, 167n35

UNESCO "City of Design"
   Shenzhen's designation as, 160
   Sphinx office of, 160–61
urbanization
   reappearance of commercial sex in Shenzhen associated with, 171–72
   Shenzhen's unprecedented pace of, vii, 65, 118, 172
   See also rural urbanization; Shenzhen speed
urban planning
   China Academy of Urban Planning and Design, 74
   contingency and flexibility as a strategy, 135, 250, 252, 254–57
   division of Shenzhen Municipality into New Bao'an County and the SEZ, 69–70
   informal urbanization in the outer districts, 116–18

   and the Maoist segregation of rural and urban societies, 70
   role of China Merchants Shekou, 48
   Shenzhen government documentation of, 84n3
   of Soviet cities, 7, 32
   See also circle delineation; enclave development; infrastructure; rural urbanization; Shenzhen Special Economic Zone master plans; urbanization
Urban Planning and Land Use Commission, 207
urban villages 城中村 ("villages in the city," *chengzhongcun*)
   conflicts with Shenzhen Municipality over land use rights, 118, 140, 144
   flower arrangement villages, 131–32, 133
   defined as term, 139–40, 263
   as enclaves within Shenzhen, 80–81, 143–44, 147
   illegal buildings in, 139, 143, 152–56
   and the inequalities of rapid urbanization, 70
   as a phenomenon across China, 164–65n15
   self-organizing development of, 80–82, 161
   See also urban villages and housing concerns; urban villages and space; villages
urban villages and housing concerns
   gentrification, 66, 82–83, 162
   and low-income workers, 81–82
   quasi-licit forms of, 151–52, 163, 234, 256
   plots of land allotted to villagers, 145, 147, 152–53, 168n49
   Article 8 of the Chinese Constitution, 109
   See also land use rights; urban villages
urban villages and space
   location on either side of the First Line, 139–40, 164n9, 165–66n23
   and the rural-urban imbrication of Shenzhen, 8, 12–13, 80–84, 140, 143, 150–52, 157–58, 161, 163
   and the slogan "the countryside surrounds the city," 8, 108, 113, 121, 146
   spatial differentiation of villages, 110, 150–52, *figure 6*
   See also land use rights
use rights. *See* land use rights

Van Schendel, Willem, 244, 248–49n50
Venturi, Robert, Denise Scott Brown, and Steven Izenour, 16
villages 村 (*cun*)
   competition between, 167n43
   as outside the city's master plan, 74, 234
   Shenzhen as China's first "city without villages," 9–10, 107, 140, 141–42
   See also Baishizhou; Caiwuwei Village; Dafen Village; Futian District; Gangxia Village; Huanggang Village; Hubei Village; Luofang Village; Meilin Village; Shangsha

Village; Shatoujiao; Shawei Village; Shazui Village; Shuiwei Village; Tianmian Village; Township and Village Enterprises; urban villages; urban villages and housing concerns; urban villages and space; Wanfeng Village; Xiasha Village; Xinzhou Village; Xixiang; Yumin Village

Vogel, Ezra F.
　on direct investment in China via Hong Kong, 230–31
　on foot traffic between Shenzhen and Hong Kong, viii, 126

Wanfeng Village 万丰村
　joint-stock company formed by, 146–47
　TVE contribution to growth of, 167n35
Wang, Jin, 30, *figure 2*
Webster, Chris, and colleagues, 148, 161

Xiasha Village 下沙村 (*Xiasha Cun*)
　commercial development in, 82, 167n43
　red light district of, 175
Xi Jinping 习近平
　China Dream initiative
　　defined, 264
　　"harmonious society" as a precursor to, 201
　father Xi Zhongjun's support of Liang Xiang, 42
　Shenzhen inspection tour (2012), 59–60
Xin'an County 新安县 (*Xin'an Xian*, also Sim On)
　Nantou as the county seat of, 112, *figure 3*
　renamed as Bao'an County, 67, 113
　Shenzhen Market as the political and economic center of, 67, 113
Xinzhou Village 新洲村 (*Xinzhou Cun*), 151
Xixiang 西乡 (*Xixiang*), 69, 116, 258n3

youth and youthfulness
　the "Shekou Storm," 56–57, 58, 102n32
　Shenzhen as a space for realizing aspirations of social mobility of, 86, 93–94, 98, 102n42
　success promised through humility and hard work by, 95–97
　See also 1989 democracy movement; *Shenzhen Youth Herald*
Yuan Geng 袁庚
　biographical details, 42
　as director of China Merchants Shekou, xi, 40, 42, 47–48
　Shekou Industrial Park development directed by, xi–xii, 46–47, 48, 71
　as a Shenzhen leader hero, xi–xii, 40
　slogan "time is money, efficiency is life," 32, 49, 158, 169n63
　as standing vice chairman of China Merchants Hong Kong, 40, 42
Yu Cheng, 174
Yumin Village 渔民村 (*Yumin Cun*)
　Dan households in, 115
　mistaken for Shenzhen's original settlement during Deng Xiaoping's 1984 Southern Tour, 115
　rebuilding of, 115–16

Zeng Xianbin, 56
Zhang, Jun, 33, 34
Zhao Ziyang 赵紫阳
　arrest and dismissal of, 57
　loyalty to Deng Xiaoping, 41
　son Zhao Erjun, 42, 58
Zhou Weimin, 49, 58